SUSTAINABLE TOURISM ON A FINITE PLANET

Sustainable Tourism on a Finite Planet challenges readers to consider the new skills, tools, and investments required to protect irreplaceable global resources from the impacts of escalating tourism demand in the next 50 years. This volume documents how technology is driving a travel revolution and propelling the growing global middle class to take leisure trips at unprecedented rates. Travel and tourism supply chains and business models for hotels, tour operators, cruise lines, airlines, and airports are described with key environmental management techniques for each sector.

This book recommends that decision makers assess the current and future value of natural, social, and cultural capital to guide investment in destinations and protect vital resources. Case studies illustrate why budgets to protect local destinations are consistently underestimated and offer guidance on new metrics. Innovative approaches are proposed to support the transition to green infrastructure, protect incomparable landscapes, and engage local people in the monitoring of vital indicators to protect local resources.

Sustainable Tourism on a Finite Planet provides students, professionals, and policy makers with far-reaching recommendations for new educational programs, professional expertise, financing, and legal frameworks to lower tourism's rapidly escalating carbon impacts and protect the health and well-being of local populations, ecosystems, cultures, and monuments worldwide.

Megan Epler Wood founded and led The International Ecotourism Society (TIES) from 1990 to 2002. She is the Director of the International Sustainable Tourism Initiative at the Center for Health and the Global Environment, Harvard T.H. Chan School of Public Health, an instructor of online courses in sustainable tourism at Harvard Extension's Graduate School of Sustainability, and a Senior Project Associate at the Center for Sustainable Global Enterprise, Samuel Curtis Johnson Graduate School of Management at Cornell University. Her consulting practice EplerWood International fosters sustainable tourism development in Africa, Asia, and Latin America.

Sustainable Tourism on a Finite Planet will give you a comprehensive and deep understanding of all of the multiple actors in the international tourism industry and their environmental impacts and challenges, along with economic, political, and cultural dimensions. The comprehensive analyses of each segment of the industry – tour operators, cruise liners, airlines, hotels, and the emerging online tourism entrants – lay out quantitatively and qualitatively the business models, competitive dynamics, policy, and environmental aspects. The analyses are evidence-based, drawing on a wide range of research and studies. The author's decades of experience in multiple sectors of the industry, including nonprofit, for profit, and governmental organizations, illuminate the book with practical case examples, insightful first-hand experiences, and valuable expert judgments. The presentation is very well organized and exceptionally clearly written. The book is illuminating reading and enriching learning.

James E. Austin, Harvard Business School, USA

Sustainable Tourism on a Finite Planet is both comprehensive and provocative. A must-read for anyone interested in the present and future state of travel and tourism in the face of global climate change, poverty, and ecosystem degradation. Grounded in decades of leadership and experience, Epler Wood methodically presents the comprehensive social, environmental, and financial impacts of what has become one of the most important economic engines of the global economy and introduces new business models and approaches to move us beyond overconsumption of limited tourism assets.

Mark B. Milstein, Cornell University, USA

Epler Wood's book carefully documents why travel and tourism plays a critical role in preserving natural and social capital and its seminal importance to human health, well-being, and compassion in this multicultural world. With equal emphasis, it outlines the importance of replicable measurements of the industry's cumulative impacts, with in-depth analysis of each of the sector's major industrial sectors – hotels, tour operators, cruise lines, airlines, and airports. Read this book to learn how to approach this global industry.

John D. Spengler, Harvard T.H. Chan School of Public Health, USA

How will we protect destinations with projected global population growth and an increase in the demand for travel? Epler Wood lays out an integrated vision for sustainable tourism that effectively addresses the relationship between traveler consumption patterns and their impact on natural resources. For corporate social responsibility and tourism destination managers alike, this book offers indispensable real-world case studies and provides a vision for a pragmatic way forward.

Seleni Matus, George Washington University, USA

SUSTAINABLE TOURISM ON A FINITE PLANET

Environmental, Business and Policy Solutions

Megan Epler Wood

LONDON AND NEW YORK

First published 2017
by Routledge
2 Park Square, Milton Park, Abingdon, Oxon OX14 4RN

and by Routledge
711 Third Avenue, New York, NY 10017

Routledge is an imprint of the Taylor & Francis Group, an informa business

British Library Cataloguing-in-Publication Data
A catalogue record for this book is available from the British Library

Library of Congress Cataloging-in-Publication Data
Names: Epler Wood, Megan, author.
Title: Sustainable tourism on a finite planet : environmental, business and policy solutions / Megan Epler Wood.
Description: Abingdon, Oxon; New York, NY : Routledge, 2017.
Identifiers: LCCN 2016025172 | ISBN 9781138217584 (hbk) | ISBN 9781138217614 (pbk) | ISBN 9781315439808 (ebk)
Subjects: LCSH: Sustainable tourism. | Tourism—Environmental aspects.
Classification: LCC G156.5.S87 E64 2017 | DDC 338.4/791—dc23
LC record available at https://lccn.loc.gov/2016025172

ISBN: 978-1-138-21758-4 (hbk)
ISBN: 978-1-138-21761-4 (pbk)
ISBN: 978-1-315-43980-8 (ebk)

Typeset in Bembo
by Apex CoVantage, LLC

The cover photo is of coffee cooperative Los Pinos in Santa Ana, El Salvador, which is located on the crater Lake Coatepeque. These farmers grow fair trade coffee and in 2010 they began to offer tours of their property with trails that feature spectacular treehouse overlooks. The trail design originated by the author's firm, EplerWood International as part of an ecotourism project funded by USAID. Los Pinos now has a café, cabanas and boat trips on the lake. This project supports dozens of families who were previously living close to the poverty line.

CONTENTS

ILLUSTRATIONS

Figures

Tables

ACKNOWLEDGMENTS

Sustainable Tourism on a Finite Planet was made possible by the class Environmental Management of International Tourism Development, which I teach at Harvard Extension School annually. I began teaching it in 2010 with the encouragement of Dr. John (Jack) Spengler who was the founding Department Director of the Graduate Program for Sustainability and Environmental Management. Jack Spengler is responsible for bringing the field of sustainable tourism to Harvard, and I believe generations of students and professionals will benefit from this. My class immediately attracted high-ranking speakers working within large-scale hotels chains, cruise lines, and tour operators. We also had outstanding lectures on sustainable tourism and economic development from international development institutions and researchers. While all of the speakers for the class are too numerous to credit, I thank them all. Those who came for several years contributed greatly to the content of the course and this book. I thank Rosemarie Andolino, Jane Ashton, Houshang Esmaili, Peter Haase, Arab Hoballah, Kurt Holle, Klaus Lengefeld, Eric Ricaurte, Seleni Matus, Jimmy Smartzis, Jamie Sweeting, Faith Taylor, and Chris Thompson.

The course was also enriched by the participation of the Lenox Hotel sustainability team in Boston. Tedd Saunders is a pioneer in hotel sustainability, and since 2013 he has opened the doors of the Lenox to our class for an annual field trip. Scot Hops, the Chief Sustainability Officer of the Lenox, has been of great support giving lectures annually, supporting the field trip, and encouraging several student projects at the Lenox.

The course has also had outstanding Teaching Assistants (TAs) who have made every year a delight. Alison Hillegeist worked on the class starting in 2010 and continued for six years, and for this I am very grateful. Jessica Blackstock, Laura Carroll, Melanie Ingalls, Mariah Morales, and Jaime Pepper have all contributed as TAs and often guided students in substantive ways on their research. The class has also benefitted from the technical assistance of Harvard Extension's production team, who made quality guest speaker presentations possible from around the world.

I also cannot thank my editorial and research assistants enough. Linda Powers Tomasso, who took my class in its first year, supported me in the early rounds of research. Holly M. Jones, a long-time associate of my firm EplerWood International, provided invaluable editorial assistance. Sara Hall, a doctoral candidate at McGill University, guided me through the final editorial stages of work and offered just the insights required to complete the book's revisions.

I want to thank those who read chapters of the book for comment as I was working on the content. Francis Okello-Okomo of Kenya read the material on economic development and has encouraged me to quickly get the content out. Justin Francis of Responsibletravel.com provided insights on the digital economy and its impacts on tourism development. Vicente Moles of Spain provided insightful input on the cycles of destination growth and decay. Helena Rey of UNEP offered thoughts on the Conclusion, and Dr. Mark Milstein of Cornell worked with me, originally as a Senior Fellow, to foster business student research at the Center for Sustainable Global Enterprise, which set me in a new direction and allowed me to reevaluate what is required to achieve sustainable tourism.

I must also thank all of the students who took the course and contributed research to guide future educational, business, environmental, and policy practices in the field. I also cannot thank my husband, Gregory Epler Wood, enough for his forbearance during the tough times when writing the book interfered with our life. He has patiently listened to all of my moments of frustration, trepidation, and at times plain excitement and joy. I must also thank the entire Wood family who have supported me in every way and given me the pluck and adventurous spirit that has made this work possible.

INTRODUCTION

Dubai attracted 10 million tourists[1] in 2012 to a destination made out of thin air on an arid patch of desert. Just 50 years ago, it was a small village. One of the seven emirates in the United Arab Emirates (UAE), Dubai's foundation is built on the oil economy, but it is the only Emirate without oil. Established as a free trade commercial port in the 1980s, Dubai's airport and duty-free area were still housed under a complex of Saharan tents just 15 years ago. Now, exotic buildings like the Burj al Arab and other architectural wonders built in the economic boom years of the 2000s bring global attention, prestige, and notoriety.

By 2008, Dubai was firmly established as a thriving mecca for business and leisure travelers. On the top of its real estate game for decades, Dubai World, the development authority behind the Dubai miracle, built 5-star hotels, huge duty-free malls, including one with an indoor ski area, a canal for luxury yachts and many other eye-catching developments to entertain the global elite and capture the gaming-crazed public worldwide. But by February 2009, realtors announced Dubai was losing momentum. Foreign workers were laid off, and construction projects were stopped midstream. In November of 2009, Dubai World announced they were likely to default on their debts in the next six months. The Dow dropped 1.3% when the news was released.[2] Major headlines announced, on November 30, 2009, that Dubai was headed for bankruptcy. But by December 15, the slide was arrested. Thanks to the government of Abu Dhabi, the neighboring Emirate, the bankruptcy was averted with a $10 billion lifeline.[3] An oil-rich cousin of the CEO of Dubai World stabilized the market, and Dubai moved forward. The funds came with no strings attached, according to insiders. Dubai had become the first tourism destination to become too big to fail.

In 2012, Dubai's tourism economy grew by 9%,[4] and hotel occupancy rates were reported at 83%.[5] In 2013, Dubai was named the 7th most popular city for international visitors,[6] outranking Hong Kong, Barcelona, and Rome. Only a few journalists have pointed out the incongruity of a desert destination with no water becoming the world capital of conspicuous consumption, entirely built on the unsustainable use of fossil fuels.[7] For those who are concerned, the high end Al Maha desert resort, one of TripAdvisor's top 25 hotels in the world, was built in the desert nearby, which has been awarded for solutions to operating in a desert environment, including water conservation and waste management.[8] The contradiction of an "ecoresort" operating within an ocean of unsustainable practices is not raised as part of award programs of this nature, despite the fact that the celebration of one small example diverts global attention from larger environmental management issues.

The lack of international or national scrutiny may be because travel and tourism has become an essential ingredient to global trade and is central to the development of emerging nations, as the Dubai case vividly demonstrates. The tourism industry represents 30% of the world's exports of services and was worth $1.5 trillion, or 6% of all global exports in 2014. After a significant multiyear recession, the tourism economy has been the first to rebound and is back with 4% growth in 2013 and 2014.[9] It is already one of the world's largest economic activities at over 9% of the global economy, and it is a leading industry in many countries, especially important as a generator of jobs presently responsible for 9% of the world's employment. And these trends are not expected to slow. Having reached the 1 billion traveler mark for international arrivals in 2012, it is projected that there will be 1.8 billion international travelers by 2030.[10]

While there are remarkable economic benefits from tourism growth generating significant revenues, governments largely do not invest in managing environmental, social, or cultural environmental impacts as part of the cost of doing business in destinations around the world. Large-scale developments like Dubai are overseen by development authorities that often provide tax incentives and fast-track approval that sideline even routine impact assessments. In Dubai, water is manufactured via desalinization, using natural gas, making the UAE a net gas importer due to the tripling of water use since 2007.[11] In other parts of the world like Kenya, water services are unavailable or insufficient, leaving locals without clean water at the doorstep of resort complexes.[12] The cost of sewage treatment is not always absorbed either, for tourism facilities or the local workers they depend on. In Dubai, sewage treatment infrastructure was stretched to its limits by 2009, with tanker drivers resorting to illegal dumping.[13] Many resorts around the world simply dump their sewage into the sea.[14] In developing countries, sewage treatment is highly unlikely to be available to hotels, with only 8% of waste water treated and budgetary shortfalls for new treatment facilities chronic.[15] Trash is generated in quantities that exceed local capacity. Experts in Dubai hosted the first sustainable waste management summit in 2013, publically stating that effective solutions are of "vital concern."[16] And 70% of the labor force in the Gulf States is composed of nonresident migrants from South Asia who can be threatened by deportation if they strike for better wages, leaving them without recourse if their work standards are below global norms.[17]

The manufacturing of new destinations with an imported labor force or even the benefit of water resources represents a shift in the delivery and management of tourism, which may increasingly become a global model challenging all the assumptions that tourism is an industry without "smokestacks." The industry has successfully operated without scrutiny because its growth is incremental in many cases and its cumulative impacts are rarely measured. This may be because travel and tourism is a field that has no clear institutional systems for managing its growth, and consumers are fully ready to ignore the environmental contamination or social injustices they may observe and just go on vacation. But as the industry becomes more essential and its footprint larger and larger, its sustainability metrics can no longer be bypassed or disregarded as the world seeks to achieve a more sustainable global economy.

The objectives of this book

The story of how travel and tourism has escaped more oversight is related to its unusual business structures, models that are not investigated or presented as part of the discussion of achieving sustainability in most cases. This book seeks to put the question of business models at the center of the question of improving the travel and tourism industry's sustainability performance. It cannot be ignored that travel and tourism is a service industry that has a wide variety of suppliers working on contract throughout the world, often in dozens of jurisdictions that do not have environmental or social standards that might be typical of Europe, Australia, or North America. It must also be recognized that nations such as the United Arab Emirates are leveraging the tourism economy to build national credibility on a global stage. This often divorces tourism development goals from questions of the sustainable management of natural, cultural, and social capital to benefit the next generation. Because tourism brands are seen as prestige labels to bring higher visibility to new destinations, nations seeking to buff their images often give few fiscal responsibilities to multinational chains, despite the urgent need for their fair contribution toward the vital infrastructure services that are required for the greater good of society. And while tourism policy makers are starting to make note of this issue, they are far too often seen as being chief marketing officers and not chief management officers.

The future of managing tourism on a more strategic, long-term basis will require fundamental changes in governance. Very few nations have centralized permitting systems for environmental approvals, especially for smaller properties, making environmental data on tourism growth scarce and difficult to gather. The industry operates across such a wide spectrum of sociopolitical and geographical systems, there is no unified source of data on its usage of key resources such as energy and water or its solid waste management and waste water treatment performance, even as these metrics become increasingly important for managing the sustainable growth of industry.

This book is being written to lay out how a program of quantitative inquiry might be fostered on a much larger scale within both industry and government and within educational institutions in the future. The structure for this work must be considered carefully and is thoroughly considered in this volume. Every segment of this complex industry is managing essential data about their own institutional impacts but only to a limited degree and with a wide variety of conflicting boundaries, as researchers in the field of tourism have ably demonstrated.[18] Governments have data to move this process forward, but the data is in so many different locations that even the brightest students I have supervised from both Harvard and Cornell have challenges tracking it down. There will have to be a new paradigm of government and academic cooperation to allow the creation of global databases for the purposes of consolidating data sources to support the planning of tourism in future. Cumulative impacts of tourism remain unaccounted for precisely because there is no institutionalized system to gather it at the local, regional, or international level, and this constitutes a serious gap in essential information to guide the development of sustainable tourism worldwide. Baseline information is urgently required in destinations around the world to begin the process of measuring how tourism's growth will affect vital resources in future and guide the process of mitigation.

This book lays out required environmental management approaches for each part of the industry, from airlines to hotels and cruise lines, and discusses how they need to continuously improve and how this will be governed by the business models of each of the industry sectors. It suggests that until environmental and social accounting is incorporated into financial accounting reports, there will remain a false segmentation of effort separating the corporate suite from the Sustainability office and the continued use of anecdotal reporting on isolated achievements to replace genuine environmental and social accountability measures. While environmental sustainability is stressed here, social and cultural planning are also carefully included in Chapter 3, where the role of tourism in economic development is reviewed and systems for building sociocultural assets into planning are stressed. While this volume does not claim to offer a full complement of social sustainability solutions, it does seek to point the way to new areas of required research.

This book seeks to foster a new generation of more prepared sustainability managers for the industry who will be able to work internally and knowledgeably within corporations. This new generation of sustainability managers will be able to guide their companies toward the use of unified reporting systems that will not only help them to benchmark a wide variety of indicators for internal use and improvement but to present data to larger systems for managing the cumulative impacts of tourism. Using talent within corporations and governments will advance the internal dialog and bring expertise to the table where it counts, and this expertise needs to be deployed on the ground, working with operational teams outside of corporate suites to ensure all divisions are taking part.

Advancements in the field of sustainable tourism will benefit from the wisdom of other corporate sectors, many of which have already performed due diligence on their supply chains to meet global standards well beyond what the tourism industry has achieved. The benefits of interchange between sustainability sectors could not be more important, but it is also equally useful for sustainability experts and students who have never worked in the tourism industry to use this book to bone up on the unique aspects of the tourism supply chain before seeking to advise this unusual service industry.

Finally, the book seeks to advance policy analysis and tourism planning by reviewing a wide variety of misguided assumptions that have allowed the tourism economy to be neglected and left without vital financial resources for sustainability management from the budgets of both industry and government. A new innovative balance sheet accounting system is weighed to help identify precisely what investments are necessary to protect the long-term value of tourism in each nation.[19] Tourism and environment ministries may well have to change their brief to ensure that tourism assets are protected with new institutional arrangements that balance the importance of economic development opportunity with the obligation to protect vital ecosystems, monuments, sociocultural capital and local well-being.[20]

The next generation of entrepreneurs and government ministries will need assistance to tackle the growing probability that climate change will undermine the value of billions of dollars of coastal tourism investments worldwide. The industry has only begun to deliberate on how to manage the impact of climate change, which will be unleashed on tourism properties in the next three decades. The Union of Concerned Scientists projects that by 2030 coastal communities and states in the U.S. will see triple the number of high-tide floods each year by 2030, and by 2045 there will be roughly one foot of sea level rise.[21] The Intergovernmental Panel on Climate Change (IPCC) states that extreme sea level rise and changes in storm characteristics are of widespread concern and will have considerable coastal impacts worldwide.[22] This book seeks to demonstrate what approaches will be the most effective for the industry to accelerate its response and take action to measure, report, and mitigate – in order to garner the substantial human and financial resources required to help the industry to decarbonize and deploy proven approaches to mitigation and to protect investment worldwide.

There is a surprising lack of concern about climate change in the tourism industry, which is not discussed in terms of potential losses or the need to anticipate how to protect tourism facilities, even in regions where tourism is the largest industry. The global celebration of the demand side of the tourism business, while the supply side is facing an increasingly urgent threat, is not dissimilar to the subprime mortgages crisis of the period of 2005–2007 – when Wall Street insisted that no financial problems were in the offing, on the eve of the collapse of the unstable subprime economy in 2008. As Figure I.1 confirms, planning for growth is a must, but that should include the protection of assets.

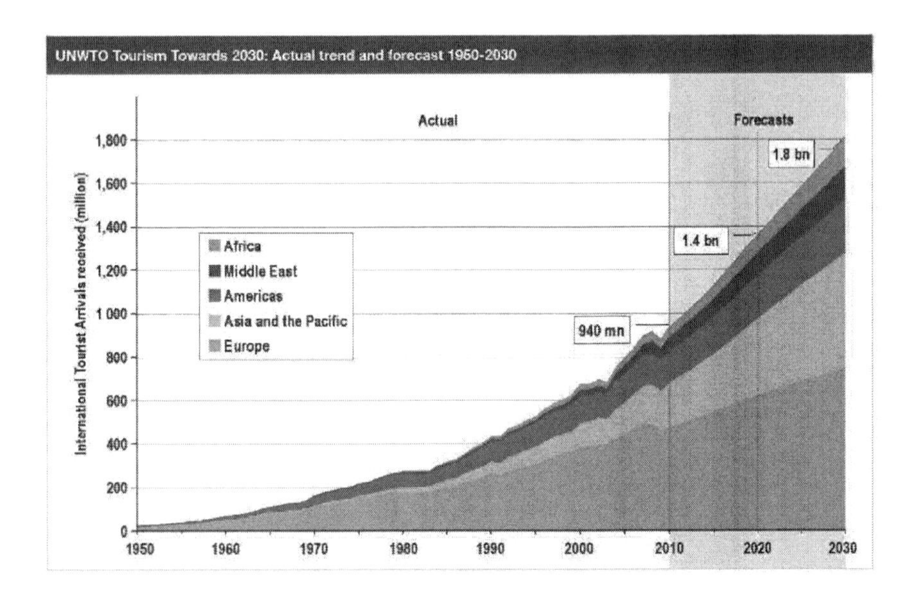

FIGURE I.1 UNWTO 2015 tourism highlights until 2030[23]

Investors, tourism organizations, and global tourism brands rarely discuss the sea change of impacts that such a landslide of visitors will have on global monuments, parks, and coastlines. While discussion transpires, there is only the beginning of action, yet much of the tourism community seems to remain largely unaware of the requirements for the protection of vital tourism assets or the need for a transition to a lower-carbon industry.[24]

While there will be a growing recognition of the requirements for action, and global bodies will lay out a wide variety of fixes, this volume uses student research to a great degree to help build its case. The need for measuring impacts from the ground up is particularly acute in tourism because of the lack of agency oversight and its history of small properties with growing cumulative impacts. The integration of universities at the international and local levels will help gather independent third-party data and guide discussions of sustainability using science-based approaches that will allow monitoring on a scale not seen before. New technologies, such as the latest versions of GIS called geodesign, will help drastically lower the costs of monitoring impacts. GIS (geographic information system) systems will allow even citizen scientists to gather data, provide direct information to established monitoring systems, and inform their governments about points of concern on a wide range of social and environmental issues, using scientific data. They can help pinpoint where vulnerable and compromised systems lie and where investments in alternative energy, clean development infrastructure, and the protection of ecosystems need to be made to better manage the future of destinations. Such citizen scientists can now even monitor real-time data from mobile phones.[25] And to properly utilize these locally gathered data points, the right cloud-based research frameworks need to be established to plug into. A whole universe of live links can begin to light up living maps of organic feedback loops to guide the management of tourism development. Such systems will have to rely on agreed-upon indicators, ideally via global agreements with universities based locally and working via consortium worldwide. These geodesign systems already exist for coastal and participatory planning[26] and need to be deployed to set the research baseline for an accurate assessment of what is required to manage a rapidly growing industrial sector that will need substantial guidance.

Data and research sources

This book uses original research from students who have taken the Environmental Management of International Tourism Development class at Harvard Extension School under my supervision. Replicable and consistent research protocols have been required for 6 years with over 200 students, and the results are revealing. Because this data is difficult to gather without local researchers who can investigate a wide variety of databases, none of which are dedicated to tourism development per se, it has been highly useful to dedicate this globalized student body to the task. Students at Harvard Extension School often study online and thus apply their research time mostly in their home countries and gather data on specific enterprises found in their own backyards. Time and again, they uncover trends that are not reported locally, nationally, or internationally. As this volume reveals, the growing footprint of tourism is causing escalating impacts on ecosystems, human health, and social systems, as well as financial stress for unprepared municipal services worldwide. The student research also delivers strong evidence that the current state of the art of managing corporate impacts is limited to only a very small proportion of the industry, while some 80% of businesses are still not applying any environmental management, despite the fact that cost savings will result in most instances.

What these preliminary unpublished investigations have shown is that there is a vast pool of data on the physical impacts of tourism worldwide that remains uninvestigated. Second, there is a very strong need for consistent data collection protocols, which can help businesses and destinations to make smart investments in green technologies that will save money. Third, there is overwhelming evidence that governments do not have the training, staff, or budgets required to manage the impacts of tourism, causing a growing unreported time bomb that is not reaching the radar of local, regional, national, or international leadership institutions. The student results point to the growing cost of allowing antiquated or poor infrastructure systems to deteriorate even while tourism revenue is skyrocketing, from the Galapagos Islands to Santorini, Greece, leaving precious ecosystems unprotected and local people often paying the bill for emergency fixes via government subsidies that are charged not to the industry but to residents. The longer these problems are hidden or unaddressed, the greater the cost of managing the problem. These students have looked not only at corporate impacts but the impacts that are being placed on municipal systems worldwide to manage the rapid growth of solid waste, the requirements for fresh water far beyond local population needs, and the ever growing physical footprint on coastlines, which require management outside of the ability of local people to deploy. Even the world's most famous beauty spots suffer from poor planning, and coastlines are polluted to the point that bathing is increasingly unsafe. Students learn to gather data that shines a spotlight on the costs of these unmanaged issues for tourism businesses and destinations, and they review how to improve management with a wide range of creative investments in physical planning and infrastructure and in innovative systems of energy production, waste treatment, and waste water management, which all seem to have a reasonable payback period, even without government subsidies.

But the work reported in this volume is not from students alone. A wide range of research is tapped from professionals, NGOs, tourism academics, triple bottom line business experts and academics, multilateral and bilateral institutions, scientists and engineers studying unique forms of managing environmental impacts, tax and policy experts, social scientists, economic development experts, sustainable tourism consultants working worldwide, and biologists and conservationists who seek to protect biodiversity. The concept here is to rely on rigorous academic inquiry but also to tap real-world experience, and seek to ascertain that this experience is rightly and accurately applied to questions of business and governmental management of tourism in future. Additionally, my own experience is applied, with over 25 years of work in more than 25 destinations, as an Instructor at Harvard Extension School, leader of a research program at the Center for Health and the Global Environment at the Harvard T.H. Chan School of Public Health, an NGO leader and founder of a global ecotourism NGO, and a consultant for development agencies who has managed over a dozen international contracts investigating real-world problems with the management of sustainable tourism. There are many innovative approaches to bring more responsible management to the table around the world, and these efforts also inform this book, particularly in the fields of ecotourism and responsible tourism, where lively advocates, practitioners, visionary businesspeople and local community organizers have for years been laying the groundwork for the next phase of sustainable tourism development. These individuals may have more to offer than any other source, as they work in countries around the world developing and testing their own solutions and reinvesting in their local environments and cultures, in ways that can be replicated by the rest of the industry in the future.

Structure of the book

The book is organized to first introduce how the industry's supply chains are connected and why it is such a powerful economic engine for development in emerging economies. It then moves to cover the main industry segments – hotels, airlines and airports, tour operators, and cruise lines – before moving on to discuss how the international industries of golf and skiing are impacting their landscapes, and finally destinations and the challenges that governments face in protecting vital tourism assets. Reviewing the entire supply chain of tourism within one book is unprecedented. It allows both professionals and students to understand the business models of each segment of the industry, to absorb the diverse environmental and sociocultural management challenges for these highly variable business models, and to learn about existing solutions to those challenges. Each chapter provides a glimpse of how the tourism enterprise will morph and change, as the digital economy captures a higher and higher percentage of business, to help readers recognize (1) the risks attached with digitizing tourism products and (2) the measures required to sustainably manage this rapidly transforming industry in the coming decades.

The book presents why tourism – a service industry – does not have an easy way to capture value-added approaches such as fair trade, which guarantees a base price for a commodity product. Unlike other commodity-driven or product-driven enterprises, tourism is not composed of supply chains that accumulate value as the product moves from the factory, production center, or agricultural fields to wholesale and retail distributors in order to reach the end market. On the contrary, tourism's end product is in the country of origin. The difficulty of establishing a better, more accurate set of systems to track the travel industry stems from the fact that this it is not a vertical supply chain model. Travel and tourism companies do not manufacture tourism. Workers are not on production lines. Large companies do not cut costs per unit in factories. Quality control is not managed by sophisticated computerized systems. Service costs must be based upon the economy where the traveler visits and cannot be lowered by sending all travelers to China or Bangladesh. Placing the burden of responsibility on any one company can be a real challenge. Tourists are difficult to control, and the management of the services they buy is not found under any one umbrella. Tour operators may arrange visits, but they are working with dozens of vendors during each trip, most of whom they do not own. Hotels are giving a branded experience, but the owners are usually just franchising the brand for a hefty fee. It is not easy to know where dollars are going, who is responsible for a visit, and certainly not who should be taking care of the air quality, sewage treatment, recycling and other environmental services. It is for this reason that establishing systems of certification based on tourism supply chains will in all likelihood not have the results found in agriculture or forestry. Unlike these commodity-based industries, the end products are rarely owned by one group even in country, making it a difficult challenge to verify if a tourism product is meeting set standards or performing based on verifiable indicators. Only in countries where there is significant political will and strong business support, such as Australia or Costa Rica, has this been achieved with widespread success. It must be kept in mind that many of the active industry players are nonexclusive distributors who legally are not responsible for their "products," only their services.

Tourists are experiencing what the industry calls a product, but it is not produced by any one company or group of suppliers. Rather the tourism product is supplied in pieces, with dozens of vendors involved all along the way, each delivering a part of the travel program in real time. Tracking the delivery of travel through dozens, if not hundreds, of vendors is not the same at all as tracking the delivery of a Nike shoe or a Walmart sweater from the factory to the shelf. While travelers are on the road, they enjoy their purchase in segments – as if a sweater from Walmart can be enjoyed for the cost of the sleeve, then the collar, and so forth until the sweater is complete. Each piece of the travel experience comes from a different factory, which are the destinations travelers visit worldwide. And the shelf is increasingly online with the consumer free to choose without any one company in charge of the overall experience. The diversity of the chain of delivery makes it difficult to put the burden on any one part of the supply chain. And this is why the travel industry has been called a stealth industry.[27]

This book leads the reader on the path of the value chain to lay out how the business of tourism monetizes travel and where it does not and how that "dollar" spent can be applied for better environmental and social management of a trip, or why it may never be reinvested in the local destination, leading to inevitable decline. While the book acknowledges that business and event travel, about 14% of the tourism economy,[28] are an important part of the tourism industry, it investigates leisure tourism as its primary target.

The penultimate chapter investigates the governmental management of destinations and why tourism policies need reform to ensure funds are reinvested in the assets that make tourism possible. Questions are raised about the human capacity to appropriately achieve local management, without more training. Tools to manage the spread of impacts are well-known but little utilized and are shared in this volume via successful cases, well established theory, and concepts from visionary landscape planners.

To conclude, the book seeks to incorporate lessons learned from exploring the supply chain and to review how governments and business can work together to create a new, more efficient, well managed, and sustainable tourism economy that does not overpay for services but does budget in the necessary costs of maintaining essential tourism services and assets. It is proposed that a new framework for tourism governance is required that builds in both the cost of doing business for local authorities, using standardized, audited metrics, and places a higher value on the natural, cultural, and social capital that tourism depends on. The full weight of evidence presented here leads to the conclusion that new institutionalized systems are urgently needed that will protect and finance global tourism assets in the long term. These can be guided by the United Nations Framework Convention on Climate Change and its global COP 21 Agreement and the United Nations Sustainable Development Goals.

No one volume can provide all the answers, but this book seeks to ask most of the pertinent questions and lay out a new framework of investigation, demonstrate how many opportunities there are for cost savings, determine appropriate approaches to mitigation, and help lead to greater consensus on how to guide the sustainable management of tourism in the future. That framework must be based on a full understanding of the industry and how it operates, and it is the author's hope that the book will help policy makers to avoid the many failed experiments of the past, which are detailed here.

Tourism research and statistics

Tourism data is often presented without adequate context, be it economic, social, or environmental. Thus a look at how tourism data is presented will help readers less familiar with the industry to see beneath the surface and achieve more insightful analysis.

Statistics have led to incorrect impressions of the size and scale of the tourism industry since 1993 when the system was first harmonized on an international basis.[29] The numbers of visitors internationally accounted for are only a part of the total impact equation. For example, in 2012, over 1 billion international tourist arrivals were logged.[30] This landmark in global commerce and trade has brought increasing attention to the question of how the travel industry is affecting our planet. But this number only begins to tell the story. International travel statistics are accounted for by the United Nations World Tourism Organization (UNWTO), and they are the best source of data on international travel. However, these statistics only track arrivals across borders. France has long been the number one international destination in the world. In 2014, France received 83 million visits, the U.S. received 74 million, and China 55 million.[31] This horserace, which is frequently published in the press, gives a very skewed impression of how the travel industry is impacting the planet. Why? Because France is not the most heavily traveled country on the planet.

Domestic travel dwarfs these numbers and will continue to do so. In 2011, nearly 2 billion trips were taken within China. In the U.S. in the same year, 1.5 billion trips were taken within its borders. In France, 176 million domestic trips were taken. To understand the possible environmental impacts of the travel industry, the combined domestic and international travel trade must be considered. The following statistics are from 2014 reports:

- **China:** 55 million international and 3.63 billion domestic trips[32] for a total of 3.69 billion trips
- **The U.S.:** 74 million international and 2.13 billion domestic[33] for a total of 2.2 billion trips
- **France:** 83 million international and 199.2 million domestic[34] for a total of 282.2 million trips

For every international traveler, domestic travelers already triple that number in the U.S. and China alone. On an international basis, growth is projected to continue unabated, at 3–4% per year worldwide. In China, domestic travel has grown by 10% on average in the last decade.[35] It is therefore crucial that in future, domestic statistics are used as part of national efforts to account for tourism impacts.

Despite the vast consequences that domestic tourism has on all aspects of tourism management, there has been very little scrutiny of how combined domestic and international travel numbers can be tracked and evaluated. While international visitor numbers are touted and discussed in great detail, domestic travelers are likely to have many more consequences on ecosystems, sacred sites and monuments, and regional development planning. The recent economic growth of Asian economies has produced a vibrant middle class making domestic tourism numbers of enormous consequence in the calculation of tourism impacts. For example, domestic tourism in India grew 11.6% in 2014 alone.[36] Even the poorer segments of populations in countries such as India produce vast footprints, such as during pilgrimages to the Ganges where a gargantuan population of 120 million visitors come every 12 years heavily infecting the river with fecal coliform, causing waterborne illnesses and 5% of the river's pollution.[37]

Tourism's economic impacts

Interpreting statistics is always challenging, and even more so when seeking to review the data on the economic impacts of tourism. The impacts of tourism are not concentrated in urban areas or factory zones. They are spread widely across landscapes and countries. In 1950, the top 15 destinations absorbed 88% of international arrivals; in 1970, they received 75%; in 2010, 55%.[38] The positive news here is that more and more people are benefitting from the tourism economy. The sector directly hires 100 million people worldwide, and job growth prospects are excellent, with projections of steady growth for the foreseeable future.[39] In the developing world, tourism has an impact that far exceeds what we intuitively understand in the developed world. Travel and tourism represents a major opportunity for countries around the world to enter into the global economy in ways that can be spread well beyond overcrowded urban centers, where most workers have gone to join the workforce for a new global society. Tourism has the capacity to bring wealth to rural areas in poor countries. It generates linkages to small and medium businesses in local food and beverage, textiles, handicrafts, and laundry services. It opens up opportunities for local guides and boat operators and builds new markets for local fisheries and agricultural businesses. Interest in culture, music, and dance are revived. The tourism sector is a primary source of foreign exchange for 41 of the world's 50 least developed countries and is known to make a substantial contribution to export earnings[40] And tourism's economic impacts are unusually well distributed. Tourists spread across the landscape and purchase a wide variety of goods and services. It may be true that no other sector spreads wealth and jobs across poor countries in the same way as tourism with 1 in 11 jobs related to travel and tourism globally.[41]

For these reasons, tourism's growth is generally celebrated. It brings opportunity, it brings poverty alleviation, and its environmental impacts are not that noticeable at first, hypothetically a smokestack-free industry. UNWTO and other large tourism corporate industry associations, such as the World Travel and Tourism Council and the Cruise Line Industry Association, are dedicated to gathering statistics that demonstrate the positive economic impacts of the trade. The UNWTO uses the Satellite Account system, whose data some nations gather in order to demonstrate the economic contributions of tourism for governmental management purposes.[42] But, as this volume will demonstrate, all these figures are formulated to demonstrate gross, not net, economic impacts. The power of these gross economic statistics cannot be underestimated. They serve as a not very subtle reminder to governments of the value of attracting private sector foreign tourism investment and the peril if too many barriers are put in the way of the growth of tourism. As a result, governments often offer subsidies to attract foreign developers and certainly do not expect corporations to pay for environmental management or services, as we have seen in Dubai and will be discussed for such locations as Honolulu, Phuket, and Cancun in this text.

It might be assumed that small-scale tourism is more environmentally friendly, but there is little evidence that this is the case. The guest houses, hostels, and low-key backpacker systems attract the type of tourist who loves off-the-beaten-path villages that feel untouched. But countries such as Thailand have seen these types of youth gatherings become so large that there is little local governments can do to properly manage the additional waste, demands on waste water systems, and the growing sprawl and inappropriate behaviors that are part and parcel of the global backpacker party culture. Slowly the bars, the restaurants, and the strip of entertainment, shops, and traveler services sprout along byways and highways. Local people are happy for the business and the types of opportunities tourism represents. They have no idea what will happen in 20 years, and it is believed there is little reason to be concerned. But after 20 years, beautiful beach and mountain destinations around the world are overbuilt, from Goa, India, to Costa Rica. In developing and developed countries alike, citizens are waking up to the need to be proactive, but it is very difficult to introduce planning for tourism. The call for more financial resources to plan tourism or to introduce planning laws is generally met with a lack of concern until it is quite late. Only recently has the country of Costa Rica called for better land-use planning laws in order to help manage their carbon impacts from solid waste, which is the third biggest emission source for the nation and is continuously growing[43] – a problem that is shown to have strong linkages to the tourism economy in this text. Once the process of growth and development takes place, even if it is only small-scale guest houses that are multiplying, this can no longer be assumed to be environmentally benign. A variety of case studies in this book indicate that, in fact, small- and medium-sized tourism establishments have an equally important role in spreading the burden of tourism development across landscapes worldwide.

Travel in the digital age

The travel economy is being boosted substantially by the digital age which is bringing more and more travel experiences into the supply chain. From the top down, the industry is being restructured and is the subject of intense pressures to lower overheads as digital booking engines grab more and more product and price it ever more competitively. At the same time, nearly every beautiful beach and small village is now for sale online for the first time in history. The rapid deployment of[44] travel and tourism experiences via the Internet has accelerated its impacts on fragile destinations, ecosystems, and cultures well beyond what may have been anticipated just 10 years ago. There was once a time when local people could take their time to get organized and learn from nearby destinations. Destinations, like the north coast of Honduras, are well aware they do not want to develop like their Mexican neighbor, Cancun. But destination growth can transpire much more rapidly now. Word of a beautiful beach can spread like wildfire. Thumbs-ups and Likes on Facebook or public review engines like Trip Advisor can lead to very rapid changes in traveler volumes.

The issue at hand is that local people, who live in communities that are the end points of travel and the ultimate source of global demand for quality tourism experiences, have very little pertinent information to help them to gauge how tourism will affect their livelihoods, their public squares, or their coastlines – that were once entirely undeveloped or served as community resources. They have even less idea what the demands on their local services will be. They cannot know that solid waste will arrive in much larger quantity with plastics and batteries they have not seen in such volume before. Many budding tourism developments are simple, rural villages, with dirt roads and poor Internet access. While the recent introduction of mass smart phone coverage in rural areas is changing how people in developing countries access information and communicate, this has not given them the information they need because there is no global effort to inform them of the changes that are likely to transpire or teach them how to account for the costs and benefits of the tourism development process. As this volume will demonstrate, few local authorities would know how to deploy environmental management techniques, even if they knew what techniques would be most timely to consider because there is no established system in most governments to provide input or assistance to help local municipalities to prepare and almost no funding available. Local advocates who seek to protect and conserve their resources must learn without training, via word of mouth. And this can lead to mistakes and deadly divisions between people in communities. While search engines and travel companies instantly commercialize the beauty of places, funding for assistance to organize and protect cultures and environments in fragile destinations is very scarce.

Simultaneously, the pressures of the digital economy and its booking engines are forcing travel suppliers to lower pricing in order to compete. This race to the bottom is pressuring small-scale hotels and tour operators around the world to either drop their commitment to investing in their personnel and sustainability initiatives or lose markets. The transformation of the tourism industry in the digital age is leading to a set of decisions that are reversing the small victories that responsible ecotourism advocates have won, and it is bringing new attention to the question of how tourism markets will preserve both product and tourism assets. Decisions must be made to ensure that the global value of the tourism products and assets are not discounted too drastically by the global digital booking phenomenon.

Sustainable destination management and governments

Governments may hear from their citizens about problems with air quality, water, or waste, but the benefits of foreign exchange from travel and the hefty tax revenue earned gives them little reason to put on the brakes. Every country touts the number of visitors they receive as a point of pride, and, of course, industry promotes the growth of tourism as a very important generator of economic well-being around the planet. Ministries of Tourism generally take responsibility for the effort to bring more travelers to their shores, and they promote their destination brands as a way of showing they are providing solid leadership. Costa Rica successfully sold itself as having "No Artificial Ingredients" from the 1990s through 2009, while at the same time allowing its northwestern coastline in Guanacaste to undergo the greatest growth spurt of uncontrolled, unplanned tourism development in all of Central America.[45]

Ministers partner with industry to create a platform for tourism development that attracts visitors. From a national perspective in much of the world, tourism is an extremely important part of their economic development strategy. As a worldwide export, it ranks only after fuels and chemicals, ahead of food and automobiles, and in developing countries tourism ranks as the first export sector.[46] But Ministries of Tourism are generally not in the highest-ranking bodies of government, such as Treasury or Foreign Affairs. They are often not working directly with Ministries of Environment or Planning. Generally, their mandate is to promote tourism, create positive branding, and develop a positive international profile for the country. They also work closely with business to help promote the country in public–private partnerships.

Ministers of Tourism are members of the United Nations World Tourism Organization (UNWTO). This organization offers very important forums on planning and environmental management in meetings around the world. But at home, Ministries of Tourism are not mandated or given budgets to environmentally manage tourism development. Ministers receive good information and enlightened approaches from the UNWTO, but they face little interest from within their own governments. Travelers pay taxes that are expected to be used primarily to promote tourism through Ministries of Tourism, Destination Marketing Organizations, or Tourism Boards. Ministers who may be concerned about protecting their destinations face an uphill battle. There are no direct budgetary instruments for government to review, plan, or oversee most of the environmental impacts of tourism. Ministries or Tourism Boards do have the power of licensing new facilities and inspection to assure compliance of licenses. But these procedures have not been linked to destination health or well-being. Only recently have such licenses been linked to quality control by countries with real foresight, such as New Zealand.[47] For now, Ministries of Tourism are not mandated to consider how to regulate tourism and its environmental impacts. For this reason, there is global celebration that the growth of tourism is reaching monumental proportions. Conversely, there is almost no effort to point out the extreme danger that this represents to vulnerable societies, monuments, and ecosystems, or the best means to manage this growth to ensure its benefits outweigh its risks.

History of effort in environmental tourism

Global efforts to determine how tourism can be managed sustainably are young indeed. In 1990, I founded The International Ecotourism Society (TIES) with board members from around the world, which defined environmentally managed travel as *responsible travel to natural areas that conserves the environment and sustains the well-being of local people*. It was one of the earliest organizations to link a sustainable development agenda to tourism development. TIES established guidelines, professional training programs, and publications to support sustainability goals for the travel and tourism community with a membership network that supported a staff of 2–5 people from a tiny base office in Vermont in the U.S. Tour operators and ecologically minded hotels backed the effort, as did environmental NGOs and academic institutions, but it never garnered large donations. Nonetheless, it managed to launch international conferences with the support of local organizations in Ecuador, Peru, Brazil, Sri Lanka, Malaysia, Kenya, Thailand, India, the U.S., Costa Rica and Norway, which established ecotourism management protocols throughout the world. The concept of an ecolodge was defined, and sustainable

FIGURE I.2 Hitesh Mehta, co-editor of *Ecolodge Guidelines*, at Sukau Rainforest Lodge, Malaysia

design workshops and business programs to support entrepreneurs were established worldwide. (See Figure I.2.) Renewable energy programs were linked to ecotourism development efforts, and architects and landscape designers worked to take cues from the natural environment to create exciting new designs for lodges in beautiful natural environments around the world. Communities in rural biodiverse parts of the planet organized and created community-owned businesses that have often struggled but are still forging ahead and building improved business models. Efforts to scale up these new paradigms continue to make progress. The concepts of ecotourism, sustainable tourism, responsible tourism, and geotourism all emerged and provide equally valuable direction to the tourism industry worldwide. Hard-working NGOs continue to point out the work that is required. The United Nations Environment Programme has now launched important new guidelines and projects to guide the Sustainable Consumption and Production of tourism. But all of these positive steps have been dwarfed by a reality that needs much greater recognition and a lack of resources to pursue questions of sustainability on a larger scale.

Organizational impediments to change

Tracking the cumulative impacts of the digitized travel experience on earth, its ecosystems, its atmosphere, its cultures and people, and their most essential life support systems in a coordinated effort has never been tried. Because the travel industry is loosely organized, industry members do not perceive the reason to forge alliances to create smarter growth.

One problem is that most reports of the travel industry's environmental impacts are provided by nonspecialists, who are passionate about places but do not carefully account for the travel industry's impacts. Evaluation supported by data can lead to logical solutions and winnable outcomes, while anger and passion rarely do. Small case study examples are often promoted as good information. Data is presented within frameworks that are limited to very special cases. In tourism, it is extremely rare to hear about specific problems, with solid data and discussion of mitigation approaches. Good news examples on a small scale prevail. Award programs help the industry to pronounce itself as proactive and involved in environmental issues, but this rarely translates into larger initiatives from the same organizations that sponsor the awards. The travel industry seems to be escaping the scrutiny that other industries take increasingly for granted. International sourcing of cocoa, coffee, cotton, and palm oil are moving toward 100% sustainability goals with corporate giants such as Unilever and Walmart in the lead. There are no equivalent announcements or efforts in the travel industry.

The model of travel journalism that presently prevails leaves little room for informed investigative journalism. Long ago, most newspapers and magazines relegated travel to a sponsored journalism model, which allows journalists to travel and report on destinations, hotels, and travel facilities of all kinds, with funds from governments, industry, or other sponsors interested in coverage. The large majority of the press is happy to travel and cover the good news about the industry. The press plays a role more like cheerleading, which can only add fuel to the already burning fire of growth. According to senior journalist Elizabeth Becker, a senior investigative journalist who wrote *Overbooked, An Investigation of the Travel Industry*, "To this day, American travel writers are not required to admit in their stories that their travel was paid for by the hotel, restaurant and airline that are the subjects of their articles."[48] Most writers treat destinations like commodities, and they report on what is hot and what is not. Some magazines do not cover the travel world this way; National Geographic and Condé Nast Traveler are among the few. For the most part, journalists and bloggers line up to be underwritten by the travel industry and seem to truly enjoy all the perks of travel they receive from industry, Ministries of Tourism, and Tourism Boards. Furthermore, publications usually do not explain this conflict of interest. The travel sections of newspapers worldwide are tasked with providing good information on where to travel, not investigating how the travel industry impacts destinations.

A new systems approach

There is a broad global consensus that the travel economy is beneficial and that therefore it is best not to discuss its negative side effects, except in terms of general actions and statements of the importance of sustainability. But when it comes to budgets for sustainability, there are only rare cases, such as the Protected Area Conservation Trust (PACT) in Belize, that slice a portion of tourism revenue from visitor taxes and invest the funds directly in tourism's key product, the coral reefs and biodiversity resources of the nation, upon which Belize depends. Within industry, only the largest companies allocate resources to sustainability, and frequently those departments have outsized responsibilities for small, dedicated departments of sustainability or corporate responsibility. For the most part, small-scale businesses do not even perceive the need to demand a rational, carbon-neutral approach to create smarter growth, and the global tour operator community is not supporting the cost of sustainability for their suppliers by carving out enough funds in the total tourism package, a problem that is aggravated by increasingly digital supply chains with discounting as the primary selling point. There is a new opportunity on the horizon, as a result of the global agreement of COP 21 in Paris and the commitment to the new Sustainable Development Goals of the United Nations, which may help to focus on how to restructure sustainable development systems. The conclusion of this book will review how nations and industry might work together to take advantage of these new opportunities to develop a governance framework that finances the cost of sustainability through a reallocation of tourism taxes, green bonds, trust funds, and clean development mechanisms using science-based data to guide decision making and land-use planning powered by geodesign to make certain the process is equitable. But for now the shifting of cost for unsustainable growth is the status quo worldwide. There is a global lack of systems to manage vital resources such as water or to ensure local populations are not left with vast responsibilities for unsustainable growth.

The path ahead

A collective, ground-up comprehensive retooling of travel sustainability systems is required, a conclusion of this book that I would rather not have to make. It is understood that revamping systems and revising how governance and policies transpire is a very bold request, one that seems almost unrealistic, but is being made after numerous revisions and academic reviews of this text. There seems to be little alternative, and this is why the book seeks to bring together a stronger alliance between the travel and tourism industry and universities worldwide, which are a wonderful source of talent and expertise that can bring a new generation of experts to the table. The need for data on the local level is vast, and the importance of cooperation and technical transfer cries out for a system where young people can go and help local governments to get a set of data together and recommended management systems in place. Young people are the most suited to help measure impacts, and they are the most well positioned globally to do so. Involving graduate students in this global task of measuring and monitoring may be the best way to spur reform, but the need for assistance has to be clearly recognized first. It will be the 21st-century generation who can apply advanced digital systems, even from their cell phones, to report on local needs to create bottom-up cloud-based data reports. Only with this data in hand can the global travel and tourism community effectively respond to the fact that tourism is growing far beyond local capacity to manage it.

If beautiful destinations are overrun, which is increasingly a danger, if beaches are overcrowded and often despoiled, if energy prices are skyrocketing due to peak demands from visitors, if nonrenewable water resources are running perilously low due to a lack of basic water efficiency, and if *E. coli* from sewage outfalls is in every major port of call – there is enough evidence to act. The next generation of investment in the growth of tourism must include a more holistic view of what the cost is to maintain a destination in the long term and empower legions of young professionals to manage the data. This book seeks to lay out the question of how to get ahead of the curve and find the right human talent, policies, and management systems to arrest this growing crisis, based on the experiences already compiled, and bring a broader group of experts and policy leaders to the table to consider the most effective means of resolving the issues of managing tourism in the future.

Notes

1 Duncan, Gillian, Dubai welcomed record-breaking 10 million tourists last year, *The National Business*, March 7, http://thenational.ae/business/industry-insights/tourism/dubai-welcomed-record-breaking-10-million-tourists-last-year, Accessed September 13, 2016
2 Financial Crisis: Dubai Bubble Burst http://globalresearch.ca/financial-crisis-dubai-bubble-burst/16328, Accessed September 27, 2016
3 Abu Dhabi Bails Out Dubai with $10B, http://cbsnews.com/news/abu-dhabi-bails-out-dubai-with-10b/, Accessed September 27, 2016
4 UNWTO, 2013, *UNWTO Tourism Highlights*, World Tourism Organization (UNWTO), Madrid, Spain, page 12

5 Dubai hotels close 2012 with 83.6% occupancy, 2012, *Gulf News Tourism*, February 26, http://gulfnews.com/business/tourism/dubai-hotels-close-2012-with-83–6-occupancy-1.1151169, Accessed September 16, 2016
6 Dubai seventh most popular destination in world, 2013, *Breaking Travel News*, May 21, http://breakingtravelnews.com/news/article/dubai-seventh-most-popular-destination-in-world/, Accessed September 16, 2016
7 Becker, Elizabeth, 2013, *Overbooked, the Exploding Business of Travel and Tourism*, Simon & Schuster, New York, and Hickman, Leo, 2007, *The Final Call, Investigating Who Really Pays for Our Holidays*, Transworld Publishers, London
8 Rokou, Tatiana, 2013, International Eco Award for Dubai's no 1 desert resort, *Travel Daily News*, March 6, http://traveldailynews.com/news/article/53630, Accessed September 16, 2016
9 UNWTO, 2013, *UNWTO Tourism Highlights*, 2013 Edition, page 4 offers annual updates, United Nations World Tourism Organization (UNWTO), Madrid, Spain
10 UNWTO, 2015, *UNWTO Tourism Highlights*, 2015 Edition, offers annual updates, United Nations World Tourism Organization (UNWTO), Madrid, Spain
11 As tiny UAE's water tab grows, resources run dry, 2010, *Environment*, June 21, http://reuters.com/article/2010/06/21/us-emirates-water-feature-idUSTRE65K3MK20100621, Accessed September 16, 2016
12 Kassam, Alishah, 2010, *Sustainable Tourism Development, a Closer Look at Water Quality and Resources in Mombasa, Kenya*, Unpublished Manuscript, Environmental Management of International Tourism Development class paper, Sustainability and Environmental Management Program, Harvard Extension School, Cambridge, MA
13 Setrakian, Lara, 2009, Filthy rich: Dubai choking on sewage, ABC News, February 1, http://abcnews.go.com/International/story?id=6781673&page=1, Accessed September 2016
14 De Stefano, Lucia, 2004, *Freshwater and Tourism in the Mediterranean*, WWF – Mediterranean, Rome, Italy, page 7
15 UN Water, 2015, *Wastewater Management, a UN-Water Analytical Brief*, UN-Water, Geneva, Switzerland
16 CW Staff, 2012, First sustainable waste management summit in Dubai, *Construction Week Online.com*, August 15, http://constructionweekonline.com/article-18088-first-sustainable-waste-management-summit-in-dubai/, Accessed September 16, 2016
17 The states of forced labor, n.d., *The New York Times*, http://paidpost.nytimes.com/aware/the-states-of-forced-labor.html, Accessed September 27, 2016, and UAE: Repression on all fronts, crackdowns on dissidents; blind eye to abuse of migrant workers, January 29, 2015, *Human Rights Watch*, https://hrw.org/news/2015/01/29/uae-repression-all-fronts, Accessed September 27, 2016
18 Holcomb, Judy L. Randall S. Upchurch, and Fevzi Okumus, 2007, Corporate social responsibility, what are top hotel companies reporting? *International Journal of Contemporary Hospitality Management*, Vol. 19, No. 6, 461–475, and De Grosbois, Danuta, 2015, Corporate social responsibility reporting in the cruise tourism industry: a performance evaluation using a new institutional theory based model, *Journal of Sustainable Tourism*, Vol. 24, No. 2, 1–25, DOI:10.1080/09669582.2015.1076827
19 Ambrosie, Linda M., September 2012, *Tourism: Sacred Cow or Silver Bullet?* PhD Dissertation, Haskayne School of Business, Calgary, Alberta, Canada
20 Ibid., page 169
21 Spanger-Siegfried, Erika, Melanie Fizpatrick, and Kristina Dahl., October 2014, *Encroaching Tides: How Sea Level Rise and Tidal Flooding Threaten U.S. East and Gulf Coast Communities over the Next 30 Years*, Union of Concerned Scientists, Cambridge, MA, http://ucsusa.org/sites/default/files/attach/2014/10/encroaching-tides-full-report.pdf, Accessed October 9, 2016
22 Climate Change 2007: Working Group II: Impacts, Adaptation and Vulnerability 6.3.2 Climate and Sea Level Scenarios https://ipcc.ch/publications_and_data/ar4/wg2/en/ch6s6-3-2.html, Accessed September 27, 2016

23 UN World Tourism Organization, 2015 Edition, *UNWTO Tourism Highlights*, page 14, Offers annual updates http://e-unwto.org/doi/pdf/10.18111/9789284416899, Accessed September 27, 2016
24 Scott, Daniel, Peeters, P. and Gössling, S., April 2010, Can tourism deliver its "aspirational" greenhouse gas emission reduction targets? *Journal of Sustainable Tourism*, Vol. 18, No. 3, 393–408
25 ARCNews, Winter 2016, *ArcGIS Is a System of Engagement and a System of Record, Applying the Benefits of Geography Everywhere*, ESRI, Redlands, CA
26 Ibid.
27 Becker, Elizabeth, 2013, *Overbooked*, The Exploding Business of Travel and Tourism, Simon & Shuster, New York, NY, page 17
28 UN World Tourism Organization, 2015 Edition, *UNWTO Tourism Highlights* offers annual updates http://e-unwto.org/doi/pdf/10.18111/9789284416899, Accessed September 28, 2016
29 UNWTO, December 2010, *The System of Tourism Statistics: Basic References*, Madrid, Spain
30 Tjolle, Valere, 2013, Tourism's billion opportunities get bigger, *Travel Mole*, March 5, http://travelmole.com/news_feature.php?c=setreg®ion=1&m_id=_rs~s~T_~Ad&w_id=8755&news_id=2005436, Accessed July 16, 2013
31 UNWTO Tourism Highlights, 2015 Edition offers annual updates, http://e-unwto.org/doi/pdf/10.18111/9789284416899, Accessed September 27, 2016
32 Sector Report, The Tourism Market in China, http://ccilc.pt/sites/default/files/eusme_centre_report_-_tourism_market_in_china_update_-_sept_2015.pdf page 11
33 U.S. Travel and Tourism Overview 2015 Offers annual updates https://ustravel.org/system/files/Media%20Root/Document/Travel_Economic_Impact_Overview.pdf, Accessed September 27, 2016
34 Travel and Tourism to France to 2019 http://prnewswire.com/news-releases/travel-and-tourism-in-france-to-2019-300164182.html, Accessed September 27, 2016
35 Sector Report, The Tourism Market in China, http://ccilc.pt/sites/default/files/eusme_centre_report_-_tourism_market_in_china_update_-_sept_2015.pdf page 11
36 Ministry of Tourism, Government of India, 2015, *India Tourism Statistics at a Glance*, Offers annual updates, Market Research Division, Ministry of Tourism, Government of India, New Delhi, page 1
37 Vuppuluri, Richa S., December 2015, *Solid Waste Management Systems at Religious Pilgrimage Destination in India*, Unpublished Manuscript, Environmental Management of International Tourism Development class paper, Department of Sustainability and Environmental Management, Harvard Extension School, Cambridge, MA
38 UNWTO, 2012, *Challenges and Opportunities for Tourism Development in Small Island Developing States*, Madrid, Spain, page 5
39 Economic Impact Analysis, offers annual updates, http://wttc.org/research/economic-research/economic-impact-analysis/, Accessed September 28, 2016
40 Roe, Dilys, Caroline Ashley, Sheila Page and Dorothea Meyer, 2004, Tourism and the Poor: Analyzing and Interpreting Tourism Statistics, PPT Working Paper No 16, page 22 http://195.130.87.21:8080/dspace/bitstream/123456789/441/1/Tourism%20and%20the%20poor%20analysing%20and%20interpreting%20tourism%20statistics%20from%20a%20poverty%20perspective.pdf, Accessed September 28, 2016
41 Economic Impact Analysis, offers annual updates, http://wttc.org/research/economic-research/economic-impact-analysis/, Accessed September 28, 2016
42 Ibid.
43 Government of Costa Rica, Ministry of Environment and Energy, September 2015, *Costa Rica's Intended Nationally Determined Contribution*, San Jose, CR, page 13
44 Qualmark, http://qualmark.co.nz/quality_experiences.html, Accessed September 13, 2016
45 Honey, Martha, Erick Vargas and William H. Durham., 2010, *Impact of Tourism Related Development on the Pacific Coast of Costa Rica*, Summary Report, Center for Responsible Travel, Stanford University and Washington, DC

46 UNWTO Tourism Highlights, 2016 Edition, offers annual updates, page 6, http://e-unwto.org/doi/pdf/10.18111/9789284418145, Accessed October 10, 2016
47 Qualmark, http://qualmark.co.nz/quality_experiences.html, Accessed September 13, 2016
48 Becker, Elizabeth, *Overbooked, The Exploding Business of Travel and Tourism*, Simon & Schuster, New York, pages 27–28

1

THE CHALLENGE OF SUSTAINABLY MANAGING TOURISM ON A FINITE PLANET

In 2010, I stepped in front of a digital classroom at Harvard for the first time, with 75 students ready and waiting to learn about how to manage the global impacts of the international tourism industry. I had a skilled producer of online content to help set up the class and the support of outstanding teaching assistants. As of 2015, over 250 students have taken the class from nearly every corner of the globe, often learning about its unique supply chains and influence on global commerce for the first time. The course lays out the significance of the international tourism industry, which represents nearly 10% of the global economy, from economic and environment management viewpoints. It provides students with an understanding of how mainstream tourism operates, its supply chains, and how each sector of the business approaches environmental management. Students learn how the industry is presently managing air, water, waste water, solid waste, sprawl, and ecosystem impacts. And they are asked to undertake final research projects that scrutinize how one sector of the industry can effectively approach the environmental management of water pollution – sewage, solid waste, toxics and other runoff, air pollution including carbon emissions, noise, solid waste, destruction of natural habitats, competition for natural resources, and most recently public health problems transpiring in local populations.

It never occurred to me that this course was not taught elsewhere. I had not realized that neither hospitality curricula nor travel and tourism schools have classes on environmental management that cover each sector of the tourism industry. When initiating the class at Harvard, our overriding goal was to introduce students to the business model of each sector of the tourism industry and their supply chains first and then review the best examples of environmental management. For the first 3 years, guest speakers composed the majority of the course, offering their own personal experiences with initiating sustainability departments in such major corporations as Wyndham, Royal Caribbean, Sabre, and TUI Travel. In the following years, guest speakers were added from Chicago and San Francisco Airport Authorities and United Airlines. The venerable Lenox Hotel and their team in Boston began to hold an annual field trip. And we now host representatives of destinations with government leaders from such countries as Belize and Mexico. The United Nations Environment Programme offered their expertise, as did global NGOs and experienced international consultants.

These speakers brought the very latest information on environmental management of tourism to Harvard's campus. Thanks to Harvard Extension School's premium online classroom program, not only could speakers be effectively brought to the classroom from different offices in North America and Europe, they could be live-streamed to a wide variety of students who live in destinations across the globe.

A new and exciting set of facts began to emerge from the corporate perspective, airport authorities, and destination leaders. Lessons learned were built into the learning program. While the content received excellent reviews from the student body, the textbook began to seem insufficient because it did not reflect the material or its approach. I ultimately decided to write this book to give the course the right text and offer the information and research gathered in the class to inform students beyond Harvard of the benefits of learning about the industry and its environmental management practices before entering into the profession of tourism or hospitality.

There is a sense of excitement about teaching this course every year. One of the biggest payoffs of this teaching program consists of the outcomes of research from the diverse and talented student body. Using a standardized methodology for every research paper written, our class began to accumulate very important data on questions of environmental impacts not accounted for by corporations, global multilateral institutions, or governments. I found myself reading student papers with a mixture of foreboding, excitement, and gratification. Any researcher might understand my reaction. A process of collecting data had been launched, which was bringing in results that were not available from any other source. The data's significance appeared to be not only important but of vital interest to governments and industry around the world. While there is certainly a vibrant NGO community seeking to report on questions of tourism development impact, they do not have the funding or scope to do in-depth research or reporting on the many cases of impacts that can be found only by investigating government databases related to the management of water, waste, and pollution. I found myself reading every paper with growing concern and attention. I did not want conclusions that were biased or unfounded or based on student opinion. I often stressed the importance of accurate data and the use of unbiased sources to be sure we could rely on the findings.

Students reported on the tourism industry and destinations with an eye on what local statistical information could tell us. If corporations reported on lowering waste, my students looked at local landfills. If governments stated they had improved waste water treatment, they checked public health reports for coastal water quality. They were encouraged to use only objective data to support their findings and graded down if they did not follow this guideline. Their case studies were broad ranging. Questions of how well solid waste is managed by airports was undertaken by numerous students, who found that most airports do not holistically report or manage waste for their concessionaires or for the airlines. They investigated sustainability practices of local small town hotels with an equally balanced eye to understanding their costs and what government services were available to help them become more efficient. Local strips of bars and restaurants and even delicatessens got the eagle eye in resort communities. One student even traveled by fishing boat to gather data on solid waste management in a cruise port that had not posted data online or in any corporate reports, and despite having participated in our classroom, the sustainability office would not return her phone calls. Chinese environmental management was studied by Chinese students, who could read local reports and determine how well international reports jibed with local statistics. Hundreds of student reports flowed in.

These students have continued to plug into the process of gathering data from both business concerns and destinations, providing a steady source of ground-up data that has never been available before. This has helped to create a new source of information that has guided the author's thinking about how global data could be gathered in future.

Without baseline information or standardized methodologies, there is little basis to make recommendations for the environmental management of industry. This is not to say that corporations are not reporting. Corporate reports using the Global Reporting Initiative and the Carbon Disclosure Project have been highly valuable. But these respond to internal benchmarks and improvements within corporations. And according to in-depth academic investigations of corporate reporting from the cruise line industry, which is the fastest growing leisure travel market segment, there is a consistent failure to use third-party verification, problems with unclear presentation of information, poor reporting on scope and sources of information, and few performance assessments.[1] Another review finds that the reports offer little beyond basic compliance to preserve cruise lines from risk, but show little commitment to managing business sustainably.[2] The patchy, anecdotal patterns of information reporting are not unique to the cruise industry. While certainly reporting is improving, there is still very little data that can meet the needs of developing a global benchmarking system for tourism impacts worldwide. Students have investigated specific hotels, for example, that are part of corporate benchmarking systems, only to find time and again that local hotels are not part of global corporate reporting systems.

The issue of managing such a complex industry with so many local tentacles, which are often not owned or operated by the larger brand, time and again proves to be one of the greater challenges of accounting for tourism's impacts. Cruise lines are unusual for the tourism industry because they do own much of their "footprint" but also because of their history of divorcing themselves from any legal liabilities even in their home ports, as will be discussed in Chapter 7; the question of capturing their local impact data is similar to the issue with hotels, tour operators, and airlines – which often cannot manage issues related to waste management, sewage treatment, and water and energy efficiency without much more participation from local bodies. Responsibility for reporting on environmental goals at the local level is avoided for this reason. There is little cooperation with local governments on this problem. It is essentially a legal and policy standoff, with corporations taking responsibility for their own metrics within the facility but avoiding any engagement on the question of cumulative impacts at the destination level. There are high-quality NGO reports and a wide variety of academic articles in journals, but these investigations have largely not informed decision making by government or industry. And what reports have been done on certifying destinations are still few and far between, with no science-based system to gather results, a lack of measurable indicators, and surprisingly little intention to publish what is gathered.[3]

NGO approaches to destination management

NGOs have picked up the slack in many destinations, but their scope to achieve real policy breakthroughs has been limited to bringing together all of the parties to discuss how to cooperate. By and large, the gold standard has been to hold stakeholder meetings with constituents on a regional basis to discuss best practice and review local guidelines and policies. This has led to a set of subjective goals that are difficult to measure or validate.

I myself, as a former NGO leader and founder of The International Ecotourism Society, spent 12 years leading such meetings. I understand their value and the importance of creating common ground where local people, their governments, NGOs, and industry can sit at the table and discuss how to combine efforts to achieve enlightened approaches. But this technique has not lent itself to fostering rigorous inquiry, nor has it produced a baseline of quantitative information upon which the world can now act.

Having sat in hundreds of local stakeholder meetings, it is clear that the field of sustainable tourism has been led by actors that seek to demand that their opinions are the most educated and informed and that their labels are the best labels to apply to the process. Instead of discussions based on data that all can share and comprehend, angry and upset stakeholders seek to sway public meetings with strongly held viewpoints based on their own problems with government, while government seeks to assuage business but in reality cannot offer legitimate solutions without budgets to manage tourism's impacts. Neither the gurus nor the protesters are advancing approaches that are genuinely constructive because they are based on opinion and anecdotal information.

On the grander stage, global leadership of the United Nations World Tourism Organization (UNWTO) has been highly valuable to organize tourism's governmental leaders to focus on sustainability and its benefits. Large organizations such as the World Travel and Tourism Council (WTTC) bring strong resources to bear in order to point out the industry's value economically, socially, and environmentally. Conferences have been held to discuss how tourism brings peaceful interchange, human development, and personal understanding of our place on the planet. There have been in-depth investigations of how tourism benefits the poorest sectors of the global economy, more than most commodity-driven export industries. The tourism industry has gone a long way to prove its worth and to investigate how it can improve. But despite all the effort to improve local benefits, channel funds toward the protection of natural resources, and point out the potential of the strong economic results of the industry, dialog remains in forums where world leaders are largely not present, and little attention is paid to the industry's goals to play a larger role on the policy stages of the world.

What is needed is a very realistic and transparent process of looking at the cost side of managing tourism development and growth with serious actors paying close attention to the potential consequences. Fortunately, there are intentions to develop a Sustainable Tourism dashboard that go beyond economic indicators announced by the World Travel and Tourism Council in 2015. But the WTTC readily admits in its 2015 report that travel and tourism's footprint is likely to increase, and the challenge remains to accurately quantify it and achieve significant reductions.[4]

There are many scientific, data-driven options to account for impacts, and trained staffers can bring more simple efficiencies to this process. For small businesses, basic spreadsheets are often all that is required to guide investments that result in quick paybacks and save on valuable nonrenewable resources. But there is a lack of trained members of the hotel, tour operator, and cruise community to carry out these tasks, especially within middle management. And while some exceptional corporate leaders are making sustainability a core business strategy, too often these strategies are not enabled beyond a central corporate team.[5] While all industries are driven by rate of return on investment, and every industry will review whether environmental management pays off, the approach of the tourism industry and government is consistently to review the value of investment in the marketing and not the maintenance of tourism products.

Primary steps to unify sustainable tourism

In 2011, sustainable tourism as a field of study was advanced considerably by the Green Economy initiative. While statements of values and aspirations have dominated the field for 20 years, the Green Economy report,[6] developed by the United Nations Environment Programme (UNEP), set out to quantify the impacts of tourism across the planet for the first time in order to prepare for the Rio+20 event held in 2012. Unlike any other UN environmental summit in history, the tourism industry was included in the dialog on the main stage and was part of a global effort to review how industry can advance sustainability through investment in a transition to a greener economy. Academic researchers received funding for a preparatory report that became a baseline of valuable data on tourism's global impacts that had never been calculated and summarized in the past.

The report found that tourism has (1) energy-intensive transportation with growing greenhouse gas emissions, (2) excessive water consumption, (3) discharge of untreated water and sewage, (4) generation of solid waste well beyond the capacity of local economies to manage, (5) growing damage to marine and terrestrial biodiversity, and (6) growing impacts on the survival of local cultures, built heritage, and traditions.[7]

The Green Economy Tourism report,[8] which is now published separately, is essential reading. The report reconfirms that there are no systematic international country data sets on environmental impacts from tourism. The facts gathered in this report are valuable indeed.

Energy: Tourism's growing consumption of energy and dependence on fossil fuels has important impacts on global GHG emissions. Tourism creates about 5% of GHG emissions. Air transportation contributes 40% of these impacts and accommodations 21%. Tourist trips are estimated to continue to grow by 180% in the next 20 years, with distances traveled to increase by over 200%. Emissions from air transport are set to grow by 161% with some efficiency gains. With these growth scenarios, aviation will become a key source of GHG emissions by 2035.[9]

> Given the rising global trend for travel and the growing energy intensity of most trips, future emissions from the tourism sector are expected to increase substantially, even considering current trends in technological energy-efficiency gains in transport (air and ground) and accommodation.[10]

Water: Water consumption by tourists can be directly competitive with local needs. Tourists tend to use much more water than local people, with luxury tourists consuming more than 4 times what an average European user would consume at home.[11] Water use is significantly increased by golf courses, extensive gardens, swimming pools and fountains, spas, and water parks. Golf courses frequently require more water for irrigation per day than the sum total of daily needs of 80% of the global population.[12]

Sewage: Discharge of untreated water, both gray water and sewage, is surprisingly pervasive from tourism facilities. UNEP reports that even in the Mediterranean region, it is commonplace for hotels to discharge untreated sewage directly into the sea with only 30% of municipal waste water from coastal towns receiving treatment before discharge.[13]

Solid Waste: Solid waste is an exceptionally difficult problem for tourism, with waste management and recycling still lacking in airports, on airlines, in tourism destinations, and even in hotels and bars and restaurants that are only beginning to manage food waste through composting.

Biodiversity: Damage to habitats from unchecked tourism development is widespread. Beachfront developments have the most aggressive patterns of development. Many tropical beachfront destinations hug islands or shorelines that are also fringed by coastal mangroves. To make way for facilities on the beach, mangroves are often sacrificed. Coral reef systems are also at great risk from runoff caused by large-scale tourism and port development. The Caribbean Sea is threatened by the density of tourism along coastlines, with an astounding 25,000 visitors per square kilometer per year in St. Thomas and a median of 1,500 visitors per year for the region, according to the scientists reviewing coral reef health for Global Coral Reef Monitoring Network.[14] The Global Coral Reef Monitoring Network scientists were able to statistically correlate the drastic decline in coral reef cover with the density of visitors per year. Locations with more than 2,635 visitors per square kilometer per year have only 6–13% coral cover, compared to islands such as Bonaire, which has 37% coral reef cover and only 253 visitors per square kilometer per year.[15]

Threats to Local Cultures and Heritage: While tourism can contribute to the protection and even the restoration of cultures and cultural heritage according to UNESCO, maintaining sites to meet growing tourism presents difficult challenges for heritage site managers.[16] The planning and management of cultural heritage sites is inadequate in many of the world's most significant global heritage sites and continues to receive inadequate consideration despite years of discussion about the need to improve and manage tourism growth. The challenge is highest in emerging economies where there are problems of training and a lack of policies that take into account the need to preserve sites. UNESCO has expressed its growing concern regarding unplanned uncontrolled tourism in many heritage sites.[17]

Cultural erosion is a well-known side effect of tourism growth. Indigenous groups that have had little contact with the outside world are understandably at greatest risk from the introduction of foreign, Westernized practices. External influences can transform some cultures in one or two generations and weaken interest in traditional customs and skills. There is also the danger of commodification of culture where dances, stories, and crafts are put up for sale without proper understanding, respect, or benefit to local people. As local people start to perform rituals exclusively or frequently for tourists, the meaning of these rituals begins to fade.[18] Examples of this type of commodification can be found on every continent, from the hula of Hawaii to the *adumu* (jump dancing) of the Maasai. Efforts to restore true cultural sanctity and respect to traditional dances and ceremonies can take generations.

Green, inclusive growth

The Green Economy report focuses on how the world can transition to a more equitable, culturally respectful global economy that allows for scaled-up economic growth. The report predicts that greater efficiencies in resource use will bring great benefits to society at large. In tourism, careful monitoring of natural and cultural resources has many benefits. Residents of tropical islands depend on vibrant fisheries for their livelihoods. If coral reef systems are sacrificed to wasteful construction practices, locals lose their source of protein and their way of life. Local governments can no longer afford to allow the wholesale development of their coastlines without protection of the water and their reefs.

Because environmental and cultural impacts and assessment reports are managed through agencies that have responsibility for a wide range of projects on a national or international basis, sifting through government databases is the only way to come to an understanding of how tourism is impacting energy, water, biodiversity, cultures, and heritage. Documenting these impacts in a coordinated fashion will allow researchers to advance solutions that can be measurable and worked on by researchers all over the world using the same frameworks – not a set of proprietary systems that cannot be revealed for fear of losing funding or brand equity. While sustainable tourism experts have sought to set out guidelines and principles, the lack of measurable standards has led to a fractured set of methodologies and standards that are aspirational and not constructed to be based on measurable fact that can be internationally shared.

This book seeks to tease out where it is possible to measure results and set unified standards for local implementation. It will take local researchers and students to bring together the data for this approach, and it is best if they do so based on data that is available through local sources and via new approaches to building measurable databases in the cloud for researchers worldwide. There is an urgent need to have sustainable tourism research to become globally in sync and transparent, using measurable standards that can guide local decision makers and have real impacts on local lives and well-being.

Measuring the cost of tourism for delivering sustainability

As tourism growth continues in the next 30 years, many questions remain unanswered regarding how global organizations, governments, NGOs, and industry can manage the development of tourism in a coordinated fashion that will meet sustainability criteria. While tools and suggestions abound, there are few studies that review the actual cost of this process.

The Center for Sustainable Global Enterprise at the Samuel Curtis Johnson School of Management of Cornell University undertook a project in Belize in 2011 to do just that. A team of MBA students I supervised with Dr. Mark Milstein were tasked to look at revenues earned from tourism and the cost of "doing business" on Ambergris Caye, Belize, the country's most popular tourism destination that accounts for over 12% of Belize's GDP. The study, written by Cornell graduate students, concluded that:

> The rapid growth of tourism has occurred in a largely unplanned manner without an understanding of the effects of this development on environmental, economic and social factors. The infrastructure required to support this development is not being financed adequately.[19]

This study opens up a set of primary questions that remain unaddressed in the field of sustainable tourism. What are the underlying costs of delivering sustainable or responsible tourism? While the global dialog has focused on best practice, guidelines, sustainability reporting, and certification, there are few studies of the real cost basis of delivering sustainable tourism on the ground at the local level.

Around the world, as tourism escalates, there will be an increasing amount of pressure on beaches, monuments, and invaluable parks and protected areas. These places operate by and large without recompense from the industry or certainly without adequate resources to respond to the growing number of visitors that are now visiting and will visit in future. The value of public resources has generally been understood to be the burden of local citizens to pay for and protect. At best, the parks and protected area system may charge a special fee to foreigners to help cover the cost of their visit. But the majority of the global system does not value the tourism product offer or the cost of building necessary public water, sewer, and waste treatment systems to protect the underpinnings of sustainability at the most fundamental level.[20] Tax systems for tourism are not designed to reinvest in the tourism product; they are designed largely to market tourism. As a result, tourism policy makers lack adequate funds to cover the cost of planning and public services. This causes an inevitable decline of destinations that lack basic funds for maintenance, protection, and upgrade.

This conclusion was the result of the original study in Belize by the Cornell team, and it took 6 years of study to validate it because it seemed too simple to be right. Consequently, I have been asking this simple question to both faculty and students at both Harvard and Cornell ever since the Belize study: "If a factory were to be built that is marketed extensively but never receives any of the profits or operating costs to maintain it, what will happen?" Unfortunately, the conclusion seems clear. One faculty member at the Harvard Kennedy School said, "You have a big problem there." And indeed we do. Throughout the world, there is a billion-dollar system to market tourism and no clear budget or management system to pay for the maintenance of some of our world's most treasured cultural and natural sites. While protected areas and monuments have more experience and authority, they are also under tremendous stress. And outside of these monuments, there is virtually no governmental or industry system in place to create a quantified system that covers the cost of protecting regions that tourists are flooding in growing waves.

Case examples of lack of maintenance

The Caribbean Basin, once a prestigious port of call for luxury travelers, is a good example of how an outstanding destination can be devalued. Countries such as Jamaica, once a premiere Caribbean port of call, began to recognize that their main tourism centers were becoming degraded. Despite educated efforts to attract more visitors and diversify their markets, occupancy levels in Jamaica began to fall below 50% in the 1990s.[21] Left with a declining tax base and dependency on tourism, their only options were to increase tourism through additional marketing and seek to improve occupancy, often by lowering pricing even further. This spiral of promotion and deflation left many costs uncovered to maintain their destination and provide for local residents. By 2002, their main resort centers, Negril, Montego Bay, and Ocho Rios had exceeded the local governmental capacity to provide for affordable local housing, necessary schools, health care, recreation, community centers, water supply, sewage, and solid waste treatment. Coral reefs were threatened and beaches eroded, and there were regular episodes of fecal contamination of the swimming waters.[22] Investment in the necessary infrastructure fell in Jamaica. And competition quickly began to deflate costs even further, as the Dominican Republic and Cuba began to offer a high-quality, cheaper alternative.

While the Caribbean basin, known to be one of the most tourism-dependent regions in the world, continues to grapple with its future, this cautionary tale (covered further in Chapter 8) has one important message for governments throughout the world, which is to anticipate the problem of deflation and cover the costs of remaining sustainable over the long term – or face inevitable decline.

In the case of Ambergris Caye of Belize, the Cornell student team sought to gather information on the costs for the destination to manage water, waste treatment, road infrastructure, electricity, land use, education, town revenues and expenses, health and crime and safety, and habitat conservation. The data was gathered in an effort to develop a new system for the government that would enable it to consolidate the data from a wide variety of government databases and create a set of predictive and analytic tools that automatically draw conclusions based on real statistics and trends.

The students met with a wide variety of agencies, both at the national level and in San Pedro, the main village on Ambergris Caye. While government officials were willing to help, each agency had data squirreled away in different systems using different monitoring approaches. The effort to measure the actual costs of managing tourism versus the benefits could not be completed due to the inability to gather data in a prompt and efficient manner. Nonetheless, some of the facts were clear. For example, Ambergris Caye was not offering adequate sewage treatment for the islanders. "Currently the (sewer) system services 3,400 customers and treats 160,000 gallons of wastewater per day. Roughly 25–30% of the island's population currently have access to the sewer system. Due to a lack of funding, the sewer lines have not been expanded."[23]

For the country of Belize, their number one tourism destination was proceeding with tourism development without adequate sewer systems, a direct threat to their local swimming waters, coral reef, and sea life. For the most part, Belize is not different from most tourism destinations. Not only do they lack sufficient sanitation and public services that can serve tourism as well as local residents, most destination and local authorities lack information on the cost of developing tourism sustainably at the local level. Without more data, the exercise of discussing principles at the national or international level may be doing more harm than good. If countries such as Belize do not have budgets to offer sewage treatment for all residents on their primary island destination, there is a clear information and policy gap. Until the cost of making tourism sustainable is understood, the larger effort to introduce best practices and guidelines or certification for sustainable tourism becomes moot.

Next steps to measure tourism's global impacts

Tourism is part of the transformation of the world economy, both as a business and as a leisure industry. It is becoming an increasingly essential component of foreign trade. It is the circulatory system for the modern global economy. There is little chance it will grow smaller. The only question is how quickly it will continue to expand.

It is time to review the cost of tourism development across the globe with a fresh set of perspectives. Business is expanding at rates that have never been seen in the past. And destinations are growing in ways that are not anticipating how tourism impacts can be managed.

The increasingly competitive costs for air travel both long haul and short haul, along with the extraordinary availability of affordable flights due to the global emergence of discount short-haul airlines, have made air travel the mode of transportation for over 50% of travelers crossing international borders worldwide. Air travel has become 2.5% of all carbon emissions from all human activity worldwide and has the potential to outstrip almost all other sources of carbon pollution if business as usual reigns. (See Chapter 5 of this volume.) The aviation industry thrives on the growth of the global economy and is essential to the growth of emerging economies such as Brazil, Russia, India, and China (BRIC). These BRIC countries have high rates of tourism growth and are symbols for how the global economy is being transformed.

- China is the 3rd largest destination for international travel, with 57.6 million arrivals in 2011, growth of 3.4%.
- The Russian Federation received 22.8 million international visitors with a growth of 11.9%.
- India received 6.2 million international visitors with a growth of 8.9%.
- Brazil received 5.4 million international visitors with a growth of 5.3%.[24]

Since Rio+20, researchers have continued to examine how tourism growth will affect mankind's ability to exist on the planet, as the competition for valuable natural resources, land, and food increases.[25] Although energy per guest night and per trip is extremely difficult to estimate because of the variation of trip lengths, levels of luxury, and the type of activities included, broad estimates have now been made. These estimates also review CO_2 emissions, fresh water use both direct and indirect, and land use and food per traveler.

Each of the projections are modeled in three scenarios, *economic slowdown*, *business as usual*, and *global growth* covering the period between 2010 and 2050. This groundbreaking report by Stefan Gössling and Paul Peeters comes to the conclusion that the global growth of tourism will likely outstrip all efforts to achieve efficiencies. The global tourism system is in a phase of exponential growth, and impacts on resources are accelerating rapidly. Conservative business-as-usual scenarios for tourism anticipate the following growth pattern by 2050:

- **Energy use:** Growth by a factor of 2.64
- **CO_2 emissions:** Growth by a factor of 2.64
- **Fresh water use:** Growth by a factor of 1.92
- **Land use:** Growth by a factor of 2.89
- **Food use:** Growth by a factor of 2.08[26]

With this report now completed in 2015, there is less and less doubt that the environmental impacts of tourism will roughly double or nearly triple by 2050. How much the impact on global resources can be mitigated will be greatly influenced by the steps taken to environmentally manage the industry.

The tools required

Every industry has tools to measure its environmental impacts, and the tourism industry is no different. The largest hotel chains in the world, such as Hilton and Accor Hotels, are now using global systems to gather and manage their data to help each hotel to lower its total impacts. This data is guiding their operations, and helping their corporations to see a greater return on investment. Tour operators, such as TUI, have gone to great effort to measure their global footprint, lower their carbon emissions, and develop strategies to ensure that their supply chains are not harming the planet. These are important advances. But newer tools are now needed, which allow industry and government to work together to manage data overtime and understand the trends of how their cumulative impacts will impact the long-term cost of doing business. GIS systems can now be deployed to track indicators on a regional level to create a global data-driven mapping system that could be open source for governments and industry worldwide to monitor the sprawling footprint of tourism together.[27]

Municipalities that are seeing heavy impacts from the growth of tourism must begin to manage data on a more advanced level to understand the costs of providing services to their citizens and to the tourism industry on an equitable basis. But most municipalities lack the data, expertise, and revenue to do so.

Several student case studies, written by Harvard Extension School students, illuminate the importance of this process.

Water issues in Mombasa, Kenya: case study

Individuals around the world suffer from water scarcity with droughts increasing in frequency due to climate change. East Africa has been subject to increasingly serious droughts. In 2011–2012, a human exodus from Somalia into Kenya was caused by severe drought, with 9.5 million people desperately leaving home in search of water, mostly in Kenya where services were provided in the worst drought in 60 years. Kenya itself was severely affected by drought and continues to suffer from increasingly arid conditions, some of which are caused by overgrazing and farming practices, but some conditions have been brought on by a climate change. This life-and-death drama for survival puts the use of water by tourism facilities in particular relief.

Harvard Extension student Alishah Kassam, from Kenya, looked at how tourism facilities in Mombasa are using water, the issues with overuse, the lack of proper maintenance of water infrastructure, and the problem with inequity of access to water.

According to Kassam, the growing problems with accessing clean water in East Africa make the overuse of water or the pollution of water by travelers particularly egregious. East Africa's moderate droughts and floods occur on a regional basis every three to four years. Increases in the frequency and severity of,droughts and flooding will likely put pressures on the economy and tourism development, impacting future water resources. These impacts will soon, in all probability, disrupt the supply of water.[28]

The importance of tourism to economies in Africa is growing, with growth rates recorded at 5% in Sub-Saharan Africa in 2012. Nonetheless, this economic growth is not translating into better environmental services for local people, for example clean water in Mombasa, Kenya, as reported by Kassam.

> Tourism development in Mombasa, Kenya has led to increased prosperity, however, a number of environmental problems have threatened sustainable tourism development and have imposed constraints on its future growth. The systems that support the tourism industry are under strain, in particular, access to freshwater. Tourism is Kenya's leading foreign exchange earner with Mombasa, being the second largest city in Kenya, having the highest concentration of hospitality facilities and infrastructure in Eastern Africa. During peak periods, basic water management and resources prove to be inadequate to meet high season demands causing adverse environmental consequences. At the same time, the impacts of climate change such as high rainfall variability, frequent droughts and increase in global average temperatures are imposing additional pressures on water availability and accessibility.
>
> The issue however is that the poor are unable to access sufficient water to meet their daily basic needs in Mombasa yet, the hotel industry uses significant amounts of water to meet tourist demand such as daily showering, clean sheets and bath towels in addition to maintaining the beauty and greenery surrounding the hotel.
>
> Results from a survey reported that 80% of hotels in Africa are faced with water shortages that were affecting their businesses.[29] Eastern Africa has the lowest safe drinking water and sanitation coverage in Africa and domestic water demands will increase due to increased tourism, rapid population

growth and urbanization. Mombasa has had a 44% increase in population in the last 10 years, with a daily water demand of 200,000 cubic meters.[30]

Mombasa is dependent on water sources from outside the district and is supplemented by groundwater sources. In some cases, groundwater is a major source of water available in addition to wells and boreholes. However, groundwater sources are being polluted because of contamination by sewage, through surface runoff and cross-contamination from septic tank, soak pit systems and pit latrines from hotels, seriously degrading the water quality and safe drinking water for local residents. Mombasa's water supply has been further harmed by high-suspended sediment loads damaging pumps and the silting of the weir supplying the water.

Degraded water quality caused by discharge of wastewater from hotels is having a serious effect on these water sources due to sewage contamination through surface runoff and cross-contamination from septic tank, soak pit systems and pit latrines. People therefore lack adequate and safe drinking water and are exposed to preventable water-related diseases. Similarly, ground waters in certain areas have high mineral content that are not suitable for drinking. In areas where groundwater provides a large part of the water supply, movement of water is usually slow and its pollution may go undetected for a long time. Health effects such as diarrheal and other diseases caused by biological or chemical contaminants, cholera and other water-borne and vector-borne diseases, have led to major health problems. These problems are likely to be exacerbated with rising temperatures and humidity in Mombasa.

Water shortage is partly due to poor maintenance of existing infrastructure and insufficient investment in new dams and boreholes to keep up with tourism, population increases and climate change mitigation.[31] Kenya's failure to develop surface and groundwater resources has intensified its vulnerability and consequently, very little stored water per capita is available. Therefore, surface and groundwater resource storage areas are rapidly depleted during severe droughts. In addition, this issue is intensified due to large amounts of illegal and uncontrolled abstraction of water from rivers, water allocation problems during droughts, and inadequate storage availability.

The poor in Mombasa are more susceptible to environmental problems and hazards including water management, water pollution due to lack of funding sources, and knowledge of the importance of water. While the hotel industry aims to meet the requirements of water by tourists in Mombasa, many local people are being deprived of water to meet their basic needs. Large resorts are requiring significant amounts of water that is being extracted from the local environment. Increases in tourism are having a domino effect as many other related industries are also increasing and being spread in the region such as oil refineries, cement factories, and textile industries. This increase is attracting a large labor force from outside the region and increasing population growth which is exerting high pressure on water sources which in turn is affecting the quality and quantity of water supplied.[32]

What becomes clear in the Mombasa, Kenya, study is that tourism is coming into direct conflict with local water needs in a region that is poor and plagued by droughts. While tourism is meeting economic growth and development needs, there are no efforts to square how this type of development is affecting needs for local water. How much water can be spared from local needs to provide for the needs of tourism growth and development? In Kenya and countries around the world, there is a need to equitably and clearly do the math, or conflict may be the inevitable result.

Benefits

Enter Type of Benefits Below	Enter Benefits for Hotels	Enter Benefits for Condos
Condo fees	—	3,000
Taxes from residents	—	10,000
Taxes from construction jobs	—	500
Taxes from jobs	50,000	—
Foreign direct investment	200,000	200,000
Increased property taxes	50,000	50,000
Increased economic activity	500,000	50,000
Tourism funds to government	200,000	2,000

Present Value of Benefits			13,866,913	4,375,011

Enter Discount Rate:	10%	
Enter Interest Rate:	6%	
Enter Time Horizon:	20	years

	Hotels	Condos
Total Benefits (Costs)	13,825,313	4,250,209

FIGURE 1.1 Mock-up of cost benefit tool for Ambergris Caye, Belize (dollars)[33]

The students at Cornell proposed a Knowledge Management System for Belize that would help the country to account for the costs of managing tourism on Ambergris Caye. They built a model, which was tested and presented in Belize. They based this tool on Excel spreadsheets, with direct links to data in key areas of the island and key indicators for social, environmental, health, and local citizen well-being. What is particularly important is that they explored town revenues on the island to understand the capital costs of tourism development. The final model prepared also utilized predictive tools, including simple cost–benefit analysis, to allow town government to judge whether hotels or condos might be of better economic benefit to them; an example is shown in Figure 1.1. These efforts were ultimately not pursued by the Belize Tourism Board, which funded the project. But they clearly demonstrated the feasibility of developing simple knowledge management and predictive tools that could assist municipalities in future in making critical decisions.

The perils of underestimating tourism's requirements

In the Introduction of this book, the question of energy and water use in the destination of Dubai is discussed. If Kenya is dry, the United Arab Emirates, where Dubai is located is much drier still. And yet it is frequently difficult to fairly evaluate the use of water in such a destination because it is being manufactured through desalinization in a country that is blessed with outsized fossil fuel resources. Why should the United Arab Emirates not spend their fossil fuel equity on the development of tourism? Or for that matter, why shouldn't Saudi Arabia, as a model of sustainable tourism investment, use equity from oil profits? These are not always easy metrics to discuss because every country deserves a shot at the global benefits of tourism. But perhaps there is another way of viewing the investment in tourism and evaluating its efficiency in the long term in order to be certain that in 50 years' time such an investment still makes sense.

To test the validity of future investments in tourism, in very arid regions, one excellent reference is the Global Footprint Network. This program measures how much biologically productive land and water a population requires to produce the resources its consumers need and absorb the wastes it generates.[34]

For example, as fossil fuel resources become increasingly scarce, the value of every energy unit explored is rising. The metric, called Energy Return on Energy Invested (EROEI) helps to put the inefficient use of increasingly expensive energy development into perspective. Originally, oil in the U.S. returned 100 to 1 when Texas oil workers were pumping for gushers in the 1930s. Now with deep oil wells in the Gulf of Mexico and off the coast of Brazil, the EROEI is declining rapidly, with its current rate at 15 to 1 in the U.S. Understanding the EROEI of tourism's use of fossil fuels could be one useful measure of evaluating how destinations are using their nonrenewable resources to provide for water, for example.

When considering the case of tourism development in Dubai, or for that matter Saudi Arabia, efforts to tout new uses of renewable energy need to be carefully evaluated and reviewed. So, often, destinations can easily tout the anecdotal use of renewables, when in fact they are completely dependent on fossil fuels for their water and energy, and that dependency is not being factored into the presentation. The United Arab Emirates' own managers have stated that there needs to be a reevaluation. "We need to convince them that water here isn't a free resource. It's not even a natural resource. It's manmade. It is costly, and it has big environmental impacts," said Mahamed Daoud of the state-run Environment Agency of Abu Dhabi.[35]

Abu Dhabi provides Dubai with nearly all its energy and water resources at increasing cost because its vast natural gas reserves, which account for 98% of its power generation, are insufficient to meet the needs of the region. Few pressures have been placed on Dubai to mandate more efficient uses of energy.[36] As a result, Dubai might have the lowest possible EROEI imaginable. While Abu Dhabi has announced a renewable energy buildup, scholars note that it will represent only a token reduction in carbon omissions growth and is mainly positioned to improve the UAE's reputation.[37] One of the problems with the current approach to sustainable tourism is that no metrics are being applied that allow researchers to compare global footprints or Energy Return on Energy Investment as a criteria for rating how well nations are using their diminishing resources to develop and attract foreign exchange.

Rhodes Island, Greece: case study

Dubai is not the only destination that has failed to review its return on energy investment in tourism development. Harvard Extension School graduate student Sofia Fotiadou studied Rhodes Island, where 85% of the electric supply is generated by outdated oil-fired power stations with highly subsidized fuel paid for not by tourists but by Greek citizens in her paper on electricity production in the hotels of Rhodes Island, Greece.[38] Fotiadou found the following issues.

> Tourism has long been a major driver for the Greek economy. In particular, the last two years Greece has been challenged by unprecedented tourism growth. As a result of a strategic reinforcement of its profile as a tourism destination, the country has seen a 26% increase in international tourism arrivals compared with 2012.[39] Of course, no one can blame a country that is striving to overcome the worst financial crisis in its history for using tourism as one of its fastest and most reliable growth factors. However, tourism hardly ever comes solely with gains; it entails costs, sometimes ignored and/or skillfully unacknowledged.
>
> South Aegean Islands are the second most popular Greek tourism destination after the capital Athens. In these islands, tourism is directly or indirectly the main economic activity. Many of the Aegean islands are not connected with the mainland electricity grid and use autonomous grids, fed mainly by local oil-fired

power plants. Renewable energy takes up only a minor percentage of the total supply which averages at 16% for these islands.[40] The remaining 84% is covered by diesel or heavy oil (mazut) generators. Reliance on petroleum oil triples the cost of electricity production as compared to the respective cost on the mainland. In order for all consumers in Greece to have the same charge per electricity unit, the extra cost generated by petroleum oil use in the Islands is allocated to all Greek consumers through an extra charge in their electricity bills.

Although it is difficult to precisely screen out tourism's consumption as locals also increase their consumption during tourism season due to the rise of ambient temperature, the share of locals in the increase of electricity demand during this period is most likely to be insignificant. Indeed, an extensive statistical analysis throughout many of the Greek islands with isolated grids by Gravouniotis[41] found that summer tourism demand is to blame for most of the financial burden on local power generation.

The tourism season in Rhodes lasts from late April to end of October, peaking during July and August. Considering international tourism arrivals in peak season and taking into account that tourists in South Aegean spent on average 9.6 days on destination,[42] it is possible to estimate approximately the population increase on the island in any day of the tourism peak months. Tourists more than double the population during tourism peak from a population of 115,000 to a total population of 234,000 on any given day.[43]

This vertical rise in population translates into a respective increase in electricity consumption. During tourism season, the doubling of population results in more than doubling of electricity demand. A co-driver for increase in electricity demand is the coincident rise of temperature due to intensification of air-conditioning. See Figure 1.2.

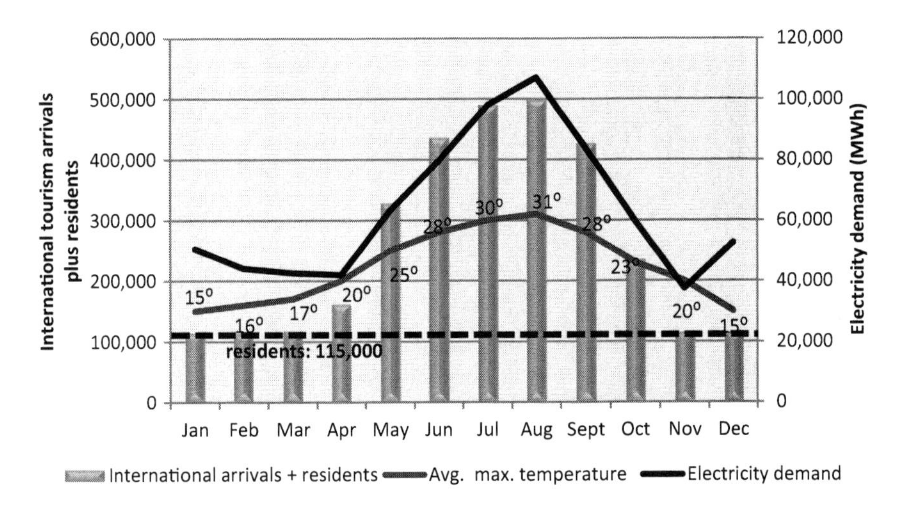

FIGURE 1.2 International tourism arrivals in 2013 in relation to monthly electricity demand and monthly average maximum temperature, Rhodes, Greece[44]

TABLE 1.1 Oil subsidization for tourism's electricity consumption, Rhodes, Greece[46]

*Electricity generation per month**					
Maximum of non-tourism season (MWh)	*Average of tourism season (MWh)*	*Average increase in electricity generation per month of tourism season (MWh)*	*Total for 6-month tourism period (MWh)*	*Average (2012–2013) cost to society per MWh (€)*	*Total cost to Greek society per year (million €)*
46,121.18	72,725.80	26,604.62	159,627.70	124.66	19.899

*The table refers only to electricity that was generated in thermal power plants and does not include electricity generated from renewable resources.

In 2013, total international tourism arrivals were 1.782 million and considering the average stay per tourist to 9.6 days, the total nights spent by foreigner tourists were roughly 17 million. If the external cost of tourism's electricity consumption was to be allocated to each tourist then each night spent should be charged with an extra 1.17€. Otherwise the total cost to Greek society annually is nearly 20 million Euros; see Table 1.1. Considering that tourists visiting the South Aegean spent on average 842.9€, this adds only a minor fraction and appears quite feasible with the right government policies in place. Indubitably, such a policy would avert the social discontentment and radically improve the reputation of hotel enterprises.[45]

Value of investment in environmental management and sustainability

There are significant gains to be made by investing in energy efficiency and the consequent lowering of greenhouse gases (GHG) emissions, as well as water consumption, better waste management practices, and biodiversity conservation. As the Greece study demonstrates, the cost of the service of energy to the tourism industry is presently being supported through additional expense to Greek society and is perversely creating a system where more fossil fuel energy systems are being built to support the tourism economy. Clearly, this system needs reevaluation.

The United Nations Environment Green Tourism Report undertook a modeling exercise to look at the savings from investment in efficiency and good practices. With basic investments in water savings devices, for example, the report concluded there could be an 18% improvement in the use of water on a global level.[47] Be it water, energy, waste, or waste water treatment, every government in the world is now facing a question of how to manage their resources to pay for their citizens' needs while offering appropriate incentives for economic growth and development. What this volume presents is that many poor emerging economies can ill afford the expense of subsidizing energy generation, waste water, and solid waste treatment without participation from the tourism industry. A fair allocation of public and private expenses is required.

One example of a nation that needs economic development, has outstanding tourism resources, but can ill afford the additional burden of unaccounted-for costs to the economy is Haiti. Harvard Extension student Nathalie Brunet studied Haiti's tourism development strategies after the devastating earthquake of 2010 and reviewed how the nation will manage the increased solid waste the expansion of tourism will generate. While Brunet, who is a native of Haiti, was unable to gain access to all government planning documents, she determined that Haiti plans to double the number of hotel rooms in the country, from 10,000 in 2015 to 20,000 by 2030.[48] Her paper documented how much additional solid waste might be generated by doubling the hotel rooms and the problems this might cause given the low capacity of the government of Haiti to manage solid waste for its own citizens. Not surprisingly, Haiti is unprepared for the additional burden on their solid waste system.

> Indeed, 37% of the population lives in the West department,[49] in the periphery of Port-au-Prince where all of the political offices and public services are concentrated, including the Metropolitan Municipal Solid Waste Collection Service (SMCRS), which is the only public entity mandated to collect municipal solid waste. Due to SMCRS' insufficient resources, solid waste collection is sparse in the city and residents frequently burn piles of trash in open air, releasing toxic fumes. Approximately 640 tons of trash per year (40% of the quantity produced in the region)[50] is collected and sent to the legal landfill – considered an open dumpsite by international standards – serving the greater metropolis; it is located 3.5 km northwest of the Port-au-Prince international airport. In secondary cities and rural regions of the country, there is no public infrastructure for waste collection and treatment, leaving municipal governments with limited to no capacity for provision of such services.[51]

Brunet found that the expansion of hotels will place a heavy burden on local waste management systems. For example, her case study of the Arcadins Coast of Haiti gives a vivid example of the problem.

> Seven hotels with 675 rooms, private vacation homes and approximately 50,000 residents share the resources of the Arcadins Coast, stretching around a 19km strip of beach north of Port-au-Prince. Amenities and services generating waste include multiple restaurants, a marina, a spa, and motorized vehicles for water sports and on land (ATV). Plastic containers and aluminum cans are the most prolific ones generated on site, added to tons of floating plastic arriving on the beaches from the capital region during the rainy season via 17 rivers. Members of the civil society attempted to reach an agreement in the past couple of years with the local authority for the implementation

> of a waste collection system in the area, but were not successful. Meanwhile, private companies are not inclined to collect waste on the coast due to the cost burdens imposed by the long-distance travel from Port-au-Prince, and the additional equipment and vehicles that it will require from them. Some hotels lack proper documentation and reporting tools to measure their waste management practices, they simply apply triage, and transport recyclables to buyers in Port-au-Prince. However, for the most part, hotels dispose of their waste in empty lots they own at a distance from the resorts, with no government intervention or control.[52]

Brunet concludes that there is an absence of an adequate legal framework for linking the expansion of tourism to a more advanced waste management strategy for Haiti. She reviews options and recommends an analysis of the additional cost of managing tourism's solid waste management requirements. She suggests that the Center of Facilitation of Investments in Haiti calculate how much additional revenue will be needed for each new tourism facility to not only provide local employment and revenue to the state but cover the cost of solid waste management. There is no system in Haiti or in many other countries to review these additional costs, which result in an increasing burden on local economies with no solutions in sight. Haiti is the poorest country in the Western hemisphere and clearly cannot raise the revenue from its own citizens. There is an obvious problem here, but to date Brunet has found no discussion of how to solve it.

Solutions can be developed that rely on financing that can enable a transition to greener technology and more efficient services that save money for all involved. Fotiadou finds that small hotels in Greece must first focus on energy efficiency with incentives from government. Second, private sector investment in renewable energy is required to lower the total burden on a small tax base and to offer good potential returns while lowering costs in the long term. Prospects for return on investment in many regions where tourism is popular are very good. She writes: "South Aegean islands have abundant natural energy resources and constitute ideal regions for implementation of these technologies. Based on real data from the installed photovoltaics on Rhodes their annual performance is 1745KWh per KW of installed PV power, one of the highest in the Mediterranean region.[53]

While the volume of tourism escalates in destinations around the world, the time is now to invest in lowering the total cost basis of all services, electricity, waste, and waste water while investment is in play.

It should not be ignored that in many parts of the world, there is a very low tax base. For example, I visited the municipality of Puerto Lopez, Ecuador, in 2015 – an up-and-coming destination on the coast with a thriving whale-watching industry and a growing number of hotels. The tax base of the municipality is highly limited, as it is surrounded by protected areas and is restricted from collecting taxes from certain community areas that live at the bottom of the pyramid. As a result, the municipality can collect tax from only 20% of its total area. I met with the municipality's team. They were smart and eager. They wanted training. But they had never studied the cost of managing the burden of growing needs for waste treatment, recycling, demands for more fresh water, and the need to put in better roads. Such a cost analysis needs to be done, not only in Puerto Lopez but in municipalities around the world.

Investment in the necessary infrastructure is at the core of the question of how the globe can absorb the growth of tourism sustainably. One problem that will need much more investigation is how international tourism business pays for this infrastructure and how much is covered by residents. Investment is needed to offset the impacts of the growing footprint of business as usual. It is difficult to understand how the sustainable tourism global effort has been able to discuss instituting sustainability practices for the past 25 years, without beginning to resolve how to manage the excess costs for destinations to manage solid waste and waste water or how to focus attention on how energy is generated, how local dirty grids can be either avoided or improved, and how local water resources must not only be protected but managed for the local population in such a way as to ensure that it does not lose its own sources of clean water.

Conclusion

The future of managing tourism will depend on the collection of destination-related data that allows for local people to understand the true cost of managing tourism. While the travel and tourism industry has shown that it has important economic impacts and seeks to prove its value to local economies worldwide, there has been a lack of transparency of the cost of managing vital local infrastructure required by tourism. As a result, there has been a growing problem with the lack of management of waste and waste water. Global studies on the footprint of tourism demonstrate that its growing use of energy will require greater investments in alternative sources of energy, such as in the case of Rhodes Greece, and the excess production of solid waste should not be foisted upon poor countries, such as in Haiti. Corporations have improved their reporting systems through globally accepted systems such as GRI and CDP, but academic research shows that reports for cruise lines lack third-party verification, are only for the most limited required scope, and generally show that the corporations do not seek to use sustainability as a core business strategy. Destinations often seek to attract tourism as a key source of foreign exchange and therefore do not charge the industry what is required to cover local clean energy, waste, and waste treatment services. Local people are left without protection and at times are even presented the bill, through tax subsidies, for tourism's costs. The net benefits of tourism are therefore not presented or understood.

In the future, better systems to integrate public and private accounting of the key investment required to make tourism a sustainable industry are essential. The lack of human resources or educational programs in tourism destinations suggests that much more capacity building will be required both to assist municipalities with measuring and accounting for local tourism impacts and to bolster the capacity of middle management in the tourism industry to develop the impact measurements systems that can contribute to a public–private dialog on the investment required for sustainable infrastructure in tourism destinations.

Notes

1 De Grosbois, Danuta, 2015, Corporate social responsibility reporting in the cruise tourism industry: a performance evaluation using a new institutional theory based model, *Journal of Sustainable Tourism*, Vol. 24, No. 2, 245–269, http://dx.doi.org/10.1080/09669582.2015.1076827, Accessed September 13, 2016
2 Bonilla-Priego, Jesus, Font, X., and del Rosario Pacheco-Olivares, M., 2014, Corporate sustainability reporting index and baseline data for the cruise industry, *Tourism Management*, Vol. 44, 149–160
3 Personal communications, Louise Twining Ward to Megan Epler Wood by email 2015
4 World Travel and Tourism Council, 2015, *Travel and Tourism 2015, Connecting Global Climate Action*, World Travel and Tourism Council, London, page 14
5 Epler Wood, Megan, June 12, 2015, *Global Expert Survey of Education and Training Programs & Feasibility Study for Certificate Training in Sustainable Tourism*, Unpublished Report, Center for Sustainable Global Enterprise, Samuel Curtis Johnson School of Management, Cornell University, Ithaca, NY
6 UNEP, 2011, *Towards a Green Economy: Pathways to Sustainable Development and Poverty Eradication*, unep.org/greeneconomy, Accessed September 13, 2016
7 UNEP, *Green Economy, Tourism Investing in Energy and Resource Efficiency*, page 414, http://unep.org/resourceefficiency/Portals/24147/scp/business/tourism/greeneconomy_tourism.pdf, Accessed September 13, 2016
8 Ibid.
9 Gössling, Stefan, *Carbon Management in Tourism, Mitigating the Impacts on Climate Change*, Routledge, International Series in Tourism, Oxon
10 UNEP, *Green Economy*, page 417
11 Ibid., page 418
12 Ibid.
13 Ibid., page 418
14 Global Coral Reef Monitoring Network, International Union for the Conservation of Nature, 2014, *Status and Trends of Caribbean Coral Reefs*, 1970–2012, IUCN, Gland, Switzerland, page 16
15 Ibid., pages 80–81
16 Pedersen, Arthur, 2002, *Managing Tourism at World Heritage Sites, a Practical Manual for World Heritage Site Managers*, UNESCO World Heritage Centre, Paris, France, page 11
17 Ghanem, Marwa Magdy and Samar Kamel Saad, 2015, Enhancing sustainable heritage tourism in Egypt: challenges and framework for action, *Journal of Heritage Tourism*, ID: 1029489, DOI:10.1080/1743873X.2015.1029489
18 Pedersen, *Managing Tourism at World Heritage Sites*, page 33
19 Benton, Celia, Robert Matelski, and Jacob Shirmer, 2011, *Ambergris Caye Knowledge Management System Project, 2011*, Unpublished Manuscript, Sustainable Global Enterprise Immersion, Center for Sustainable Global Enterprise, Johnson School of Management, Cornell University
20 UNEP, 2011, *Green Economy Report*, page 577
21 Special Advisory Services Division, Commonwealth Secretariat, 2002, *Master Plan for Sustainable Tourism Development of Jamaica*, Commonwealth Secretariat, London, page 65
22 Ibid., page 152
23 Ibid., page 15
24 UNWTO, 2014, *UNWTO Tourism Highlights*, Madrid, Spain

25 Gössling, Stefan and Paul Peeters, 2015, Assessing tourism's global environmental impact 1900–2050, *Journal of Sustainable Tourism*, Vol. 23, No. 5, 1–21, DOI: 10.1080/09669582.2015.1008500
26 Ibid.
27 The author's research program at the Center for Health and the Global Environment is proposing this system for tourism planning. Harvard T.H. Chan, School of Public Health, Center for Health and the Global Environment, Accessed September 13, 2016, http://chgeharvard.org/about-this-program/projects-1
28 Kassam, Alishah, 2010, *Sustainable Tourism Development, a Closer Look at Water Quality and Resources in Mombasa, Kenya*, Unpublished Manuscript, Environmental Management of International Tourism Development class paper, Sustainability and Environmental Management Program, Harvard Extension School, Cambridge, MA
29 Goodwin, H., 2007, *No Water, No Future*, International Centre for Responsible Tourism: Occasional Paper No. 9, https://academia.edu/2985954/No_Water_No_Future_-_Tourism_Drinking_Destinations_Dry_World_Travel_Market, Accessed September 28, 2016
30 Munga, D. and S. Mwangi, 2004, *Pollution and Vulnerability of Water Supply Aquifers in Mombasa Kenya: Interim Progress Report*, http://unep.org/groundwaterproject/Archives/Kenya-midReport.pdf, Accessed September 13, 2016
31 Mogaka, Hezron, Samuel Gidhere, and Richard Davis,2006, *Climate Variability and Water Resources in Kenya: Improving Water Resources Development and Management*, The World Bank, Washington, DC
32 Kassam, Alishah *Sustainable Tourism Development* a Closer Look at Water Quality and Resources in Mombasa, Kenya, Unpublished Manuscript, Sustainability and Environmental Management Program, Harvard University Extension School
33 Benton, Celia, 2012, *Knowledge Management System Project in Ambergris Caye, Belize: Incorporating a Predictive Tool into the Original KMS Prototype, Final Report Phase 2*, Unpublished Manuscript, Center for Sustainable Global Enterprise, Johnson School of Management, Cornell University, Ithaca, NY
34 Global Footprint Network, Standards and Research Department, 2009, *Ecological Footprint Atlas 2009*, Global Footprint Network, Oakland, CA
35 Solomon, Erika, 2010, *As Tiny UAE's Water Tab Grows, Resources Run Dry*, US Edition, Reuters, http://reuters.com/article/2010/06/21/us-emirates-water-feature-idUSTRE65K3MK20100621, Accessed September 13, 2016
36 Dargin, Justin, Addressing the UAE Natural Gas Crisis: Strategies for a Rational Energy Policy, The Dubai Initiative, Policy Brief, Belfer Center for Science and International Affairs, Harvard Kennedy School, Cambridge, MA, page 3
37 Krane, Jim, February 11, 2014, An Expensive Diversion: Abu Dhabi's Renewable Energy Investments Amid a Context of Challenging Demand, James A. Baker III Institute for Public Policy, Rice University, Houston, Texas, page 3
38 Fotiadou, Sofia, December 2014, *Cooperative for Sustainable Electricity Production in Hotels of Rhodes Island, in Greece*, Unpublished Manuscript, Environmental Management of International Tourism Development class paper, Department of Sustainability and Environmental Management, Harvard Extension School, Cambridge, MA
39 SETE, 2014, *International Tourist Arrivals at the Main Airports, October 2014/2013 (Provisional Data)*, SETE, Athens, Greece, sete.gr, Accessed September 13, 2016
40 Hellenic Electricity Distribution Network Operator, 2014, *Photovoltaics–Statistics*, http://deddie.gr/el/themata-tou-diaxeiristi-mi-diasundedemenwn-nisiwn/sundeseis-stathmwn-ananewsimwn-pigwn-energeias/fwtovoltaika-kai-alles-ape/, Accessed September 13, 2016
41 Gravouniotis, P., 2004, *Controlling the Electricity Demand Surge in the Greek Islands with Solar Air-Conditioning: A Model of Market Penetration*, London, http://sidsgg.webs.com/2012/proceedings/Paraskevas%20Gravouniotis_Controlling%20the%20electricity%20demand%20surge%20in%, Accessed September 13, 2016

42 Research Institute for Tourism, 2014, *Development of Tourism and Basic Figures of Greek Hotel Accommodation Sector 2013*, Hellenic Chamber of Hoteliers, Athens, Greece, http://grhotels.gr/GR/xee/ITEP/DocLib2/Tourism-Hotel_2013_gr.pdf, Accessed September 13, 2016
43 SETE, 2014, *International Tourist Arrivals at the Main Airports*
44 Fotiadou, Sofia *Cooperative for Sustainable Electricity Production*. Elaborated data from SETE, 2014, weathersparc.com, Hellenic Independent Power Transmission Operator, 2013
45 Fotiadou, Sofia, Cooperative for Sustainable Electricity Production. Author note: This includes (1) pollutant-specific emissions, for the EU, from oil fuel/technology used for electricity production (average emissions per unit of electricity generation include emissions from the operation of the power plant and the rest of the energy chain); (2) pollutant-specific (SO2, NOX, NMVOCs and PM2.5), country-specific, damage cost factors, and (3) damage cost factors for CO_2, common PM2.5 to all countries
46 Fotiadou, Sofia, *Cooperative for Sustainable Electricity Production*. Author note: To compensate for the uncertainty of the share of locals in electricity demand during tourism season, the maximum and not the average monthly consumption during nontourism season is considered here. Moreover, the share of renewable energy is maximized during tourism season (total renewable energy contribution during nontourism period is 44,861 megawatts and during tourism season is 54,816 megawatts); however, the calculations consider only thermal power generation, and this also partly compensates for the uncertainty of locals' share.
47 UNEP, 2011, *Towards a Green Economy*
48 Durosier, Andy, Director of Special Programs, Ministry of Tourism and Creative Industries, personal communication with Nathalie Brunet, December 2, 2015
49 Source: Institut Haitien de Statistiques et d'Informatique, 2015, http://ihsi.ht/produit_demo_soc.htm, Accessed September 13, 2016
50 Source: Groupe BURGEAP, LGL S.A., June 2014, *Rapport final Programme d'inclusion du recyclage informel du site de Truitier à Port-au-Prince*
51 Brunet, Nathalie, December 2015, *Trash and Tourism in Haiti*, Unpublished Manuscript, Environmental Management of International Tourism Development class paper, Department of Sustainability and Environmental Management, Harvard Extension School, Cambridge, MA
52 Ibid.
53 Fotiadou, Sofia, *Cooperative for Sustainable Electricity Production*

2

MANAGING A SPIDER WEB

The tourism industry supply chains and sustainability

> To me the new generation travel industry is a knowledge based industry akin to what we might want to call as a Knowledge Processes Outsourcing (KPO). All you have to do is stay focused on what today's traveler buying travel services will continue to need in greater abundance – specialized knowledge and expertise. And acquire competencies in sourcing and bundling various travel components to build a travel itinerary for the discerning traveler.
>
> – *The Luxury Indian Traveler*[1]

The travel industry was once a brick-and-mortar industry that featured lively offices on main streets with tempting brochures and posters of Paris and London, Tokyo, and Rio de Janeiro. My mother was a travel agent for a spell, and our entire family were travel bugs who relied on her and the agency to book all of our trips to Europe and my father's business travel to Brazil, Asia, and Australia. But the main street agency our entire family once enjoyed was transformed to a lackluster factory by the 1970s. By then, travel agents were simply booking tickets for most of their day or fronting mediocre tours at discount prices without the guidance or inspiration many customers deserved. Arthur Frommer, one of the deans of travel writing, noted: "After 30 years of travel writing standard guide books, I began to see that most of the vacation journeys undertaken by Americans were trivial and bland, devoid of important content, cheaply commercial and unworthy of our better instincts and ideals."[2]

Travel supply chains were driven by the assumption that every tourist needed to see the exact same sights, causing the Louvre, the Sistine Chapel, and other European "musts" to be overpriced and overcrowded. Customers were subjected to mob scenes at every museum, and European capitals were clogged with buses all stopping at the exact same locations daily. While the old-fashioned package tour still exists, the paradigm started to shift in the 1980s when the so-called New Age of Travel began to emerge, described in Frommer's excellent classic book.[3] Alternative travel gave life to the package tour, and a wide variety reinvented travel, offering creative customized trips to meet the needs of the next generation.

Between 2000 and 2010, a new transformation in packaging travel took over, and this trend is now well under way. This time the industry is transforming how travel is booked via the Internet, and they are creating a wide variety of customized choices, allowing consumers to literally piece together their own supply chain, with advice, but not necessarily that much intervention. To meet this new demand, travel companies produce websites that feature an array of enticing photos from every corner of the globe with videos to transport the imaginations of their consumers. It is the obligation of this new generation of travel providers to process knowledge and expertise and to offer the traveler insights into having fresh and unique experiences. They must anticipate every possible request for information both in advance and while the visitor is on the road. A digital menu is laid out that can be customized instantly with local experts. Travelers are offered a panorama of ideas, which will entice their imaginations and help them plug into local trends, and see the best of the locale they are visiting without crowds. Instead of being carted in buses to tourism "shows," they receive tips on music, food, and culture, as well as updates on local events with tickets at the ready. Even the cruise vacation has become a fabulous fantasy land of choices between shows, water parks, restaurants, and spas, all available for booking on an iPad in every room. And the new generation of traveler is ready to rate their experiences on the fly for millions of others to consume and help them to find what best suits them and post their pictures on social media.

Travel agendas are now easily put together and optimized according to client interests. Just as books or movies are targeted to our tastes on Amazon and Netflix, online travel agencies are now ready to offer us customized trips on the spot, booked with our interests prescreened, while we are on the road running from meetings, to transport, to airport gate with mobile device in hand.

As the 21st century progresses, travel will increasingly be managed in the digital sphere with our digital assistants guiding us as we transport ourselves around the planet. The "etherization" of our world has created a market of dreams that can be delivered increasingly without human interaction. In 2012, 40% of all travel bookings were made online in the U.S. and 36% in Europe. (See Chapter 6.) Now those bookings are quickly migrating to our mobile phones. As of 2014, 40% of all travel bookings were on mobiles, doubling the rate of booking by mobile in one year – and quickly replacing the desktop.[4]

The transformation of the travel industry is still under way. The digital commerce revolution has consequences at every level of the travel and tourism supply chain, and it is making environmental management and sustainability planning more challenging and urgent. This chapter cannot predict accurately how these trends will change the marketing of travel, but it is clear that the industry is restructuring at a rapid pace and that this will continue for at least the next decade. To understand how the digital transformation of travel is driving change, every student of sustainability in tourism must first understand its supply chains from the bottom up and the top down. This chapter seeks to introduce these issues, summarize them, and then allow readers to explore them in more detail in Chapters 4–9.

What makes travel supply chains unique

How travel is purchased very much dictates how the supply chain is organized. As our vacation dreams are packaged and launched into digital orbit, the engines behind our selection process become more like choosing a movie than a place. The retailers, in the form of agencies and tour operators, focus on the marketable attributes of each country and emphasize the key words that travelers respond to online. The hotels, airlines, and cruise lines all offer a supply of rooms, airline seats, and/or cruise cabins, which can be booked directly or through a retailer. The destination offers attractions, such as coral reefs or human-made monuments, or entertainment centers. Each of these offerings can be combined on a customized basis for the traveler buying either directly online or through a wholesaler and retailer.

The customized nature of travel today is a great improvement over the days when bland, packaged tourism was the norm. But the explosion of digital sales points has created a new problem that is more serious. We have arrived in the 21st century at a point where travelers have instant access to any ecosystem, tribal culture, or historic monument on earth without limits. A web of digital sales sites all offer similar trips down the Rhine, near the Pyramids, or on the Great Barrier Reef. These incomparable products are offered by travel agencies online and offline; by cruise lines, hotels, tour operators both internationally and domestically; and by local guides with brochures often outside the gates. The number of vendors for even one world monument would easily be in the thousands. This is the challenge of the established product, monument, or must-see in the travel world: to manage the growing number of tourists who can reach their gates and expect entry without limit and without paying a great deal to the destination itself.

There is an equally large challenge for the newly discovered destination. As each new product is uncovered and sold, it is well-known that it ultimately is despoiled. The film *Gringo Trails*, a feature-length documentary, portrays how backpackers discover a pristine beach in Thailand only to watch it be transformed to a youth rave party hangout, where high-powered drinks are pumped out in volume and mass drunkenness is a regular affair.[5]

To understand what is transpiring, the student of the industry must review travel supply chains. Travel supply chains are also called tourism value chains, tourism industry chains, or global tourism value chains almost interchangeably. Despite the clear potential benefits of tourism value chains, existing studies have not fully defined or measured the value created by the tourism value chain, due in part to highly fragmented research.[6] There is often confusion as a result because tourism's value chain is not vertical, and suppliers, wholesalers, and retailers all overlap a great deal.

Unlike other industries, the travel industry's supply chains are a spiderweb of small and medium businesses that accumulate value for the entities marketing travel but largely not for the products they sell. (See Figure 2.1.) There is a retail and wholesale structure that represents the product both locally and internationally. And there is an equal or larger number of suppliers locally who deliver the product on arrival. They include hotels, restaurants, local attractions, shops, and local food providers. These providers do not work consistently together for the delivery of each product. The supply chain is not vertical by any means and, in fact, is largely horizontal, with each of the offerings available represented by a different chain of suppliers. All of these players are combined in a wide variety of combinations to create the experience each consumer now purchases.

While the old days of mass travel kept supply chains fairly simple, the new digital travel universe creates a unique supply chain for almost every traveler. Mapping these supply chains is thus a unique challenge that has become increasingly complex.

The supply chains of tourism need more analysis to come to a global understanding of how the protection of global assets, such as protected areas, can be paid for within this complex spiderweb of offerings. At present, there is no established system for the industry to incorporate the cost of preserving valuable tourism assets into supply chains, and this will only become a more critical problem in the next 20–30 years.

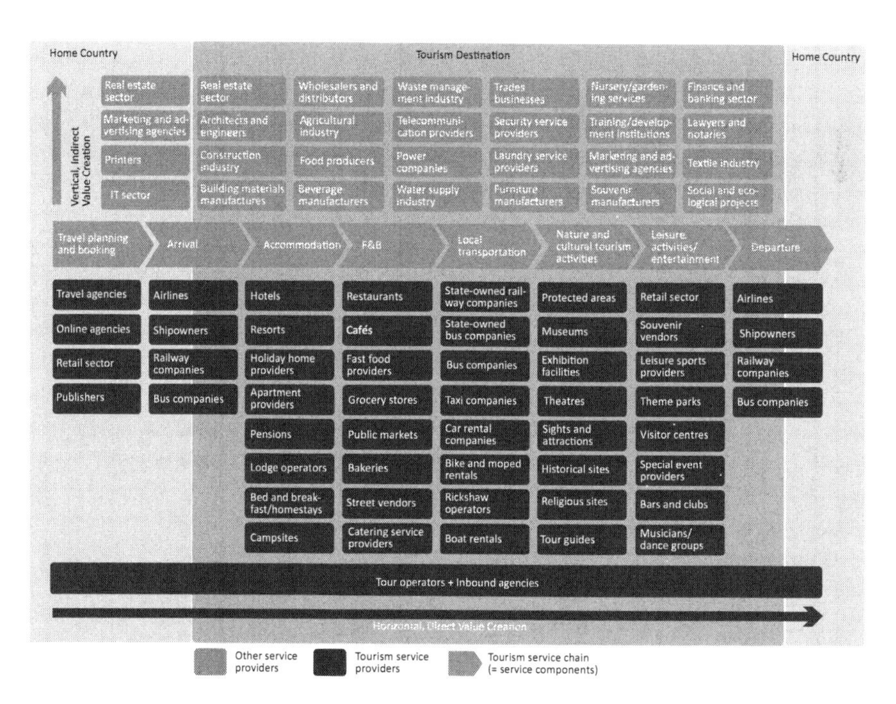

FIGURE 2.1 Travel and tourism supply chains[7]

Supply chains and sustainability management

In November 2013 at the Harvard On-Line Forum on Tourism and the Environment, Tensie Whelan, president of Rainforest Alliance, laid out a challenge. Why, she asked, are so many large tourism companies touting philanthropic goals and not incorporating those goals into their supply chains. She suggested it was surprising that the industry is not "proactively reducing the impacts of their supply chains," considering the huge advances in other corporate sectors.

Globally recognized companies around the world are using their supply chains to protect business value as global competition continues to undermine even the most secure brand reputations. As brands globalize, they are increasingly dependent on their suppliers to protect not only their product but their product's reputation. In the old days, the most effective means to gain competitive advantage was to vertically integrate and take ownership of all of the stages of production. Branding was applied to a system of quality control that inspected and secured each level of the chain of sale. The international buyer could use its purchasing power to control prices and quality and dictate terms from the top. But for most firms, those days are gone. Outsourcing is now on such a grand scale, with globalized products being produced at such a low cost, that companies can no longer dictate terms from the top. The globalization of supply chains has lowered costs but raised risks. Companies such as Apple or Levis have been caught condoning practices that would be unacceptable at home. Human rights abuses and environmental degradation are side effects of a system that is no longer entirely under the control of the buyer.[8] But times are changing.

In the past 20 years, activists have sought to expose how companies such as Walmart have sourced leather and beef from cattle ranches in threatened regions of the Amazon. Nestlé, Mars, and Hershey have suffered public protests for sourcing chocolate from suppliers responsible for degraded habitats with endangered species and using child labor.[9] To protect their global image, large companies now routinely seek to manage their supply chains by calculating how much environmental and social risk they are incurring with each sale from production to marketplace. To distance themselves from risk, companies such as Unilever have committed to 100% sustainable agriculture sourcing by 2020.[10] When Tensie Whelan of Rainforest Alliance spoke of her disappointment that the travel industry is not proactively building sustainability into their supply chain management, she spoke as a leader of the movement to engage with corporations and help them to deliver on ambitious brand promises of the type Unilever has made to stop the abuses once typical of globalized supply chains.

Why is the travel industry not caught up in the same ambitious effort to reduce risk and raise the profile of the industry's responsibility? Certainly there have been concerted efforts to help the industry to do so. A large part of the barrier to achieving supply chain sustainable management in tourism can be traced to the fact that a tourism experience is not a physical commodity. Each tourism experience is unique and cannot be effectively traced back to the producer to be certain it is being sourced sustainably. Companies in commodity-driven businesses, such as agriculture, are using supply chain tracing, supplier codes, and auditing to document the source of their products and the practices associated with the commodities they purchase.[11] Even in this case, research by leading authorities at the Committee on Sustainability Assessment found real gaps in delivering measurable results on the ground for producers.[12] Well-meaning reports lacked good replicable protocols for measurements, examples of outcomes under a variety of differing conditions, statistical significance, and an ability to compare results between studies.

These authorities recommended that all certifying systems include:

- Reliable measurements that can be compared between studies looking at key factors and
- Science-based mechanisms to understand which initiatives and interventions improve sustainability and which do not.[13]

In the case of tourism, there is a wide variety of standards that are often not comparable and frequently not measurable. The overall effort to create a standardized system under the Global Sustainable Tourism Council (GSTC) has not overcome this problem. The GSTC only provides "guidance in measuring compliance that is not intended to be the definitive set of criteria."[14] The indicators are largely not measurable and not comparable under different conditions and have no statistical significance. To achieve more accountability, a different kind of accounting will be required. It may be possible to measure impacts by geotagging tourists who are willing to be tracked as part of a global sample of the impacts of tourism – but this is still in the realm of futuristic thinking. A more immediate method will rely on the regular measurement of science-based indicators of tourism's cumulative impacts on destinations using modernized GIS systems. Because every tourism experience is unique and every trip has its own footprint, the most accurate measures of tourism impacts will be found where travelers converge, in destinations. Impact measurement in destinations using GIS technology is still rare, but it is the ideal technology to record impacts and encourage participation. There are now "ready to use apps, to allow stakeholders to visualize, analyze and collaborate anytime, anywhere and on any device.[15] The future of this type of collaboration, using maps and science-based indicators, is on our doorstep and ready to apply.

Basic principles of travel industry organization

The travel industry has a retail, wholesale structure that revolves around a nested set of experiences and product that cannot be owned. This distinguishes tourism from most other industries. Even other industries that rely on natural capital, such as mining and forestry, count on reserving access to the timber or minerals they intend to exploit over a set period of years. In the case of tourism, no company can own, reserve, or license the Pyramids or the Great Barrier Reef. As a result the industry revolves around selling a product that can be sold by thousands of competitors. Some attractions are owned by corporations, such as Disney parks, which are part of a supply chain that is owned by the Disney Corporation. But most of the travel and tourism world depends on public attractions, such as the Taj Mahal, where few effective systems are in place to manage the growing demand.

A master thesis on the Taj Mahal lays out a set of challenges that are not atypical of world heritage sites worldwide.[16] In 2010, the Taj Mahal was the most popular tourist destination in Asia with 4.1 million domestic visitors and 0.6 million international visitors, for a total of 4.7 million a year. There is inadequate water, little medical aid, poor restrooms, and no safe storage. The heat is intense, and the lines are long. These observations were backed by survey assessments of visitor satisfaction against an expected mean for tourism sites. The Taj falls well below standard, and visitors are quite dissatisfied after their visit.

Why? Because the effort to manage 60,000–65,000 visitors a day is woefully understaffed and underfunded. An organizational chart demonstrates that approximately 17 different agencies have responsibility for the Taj. There is no trust and no coordination among regional, local, national, and international agencies, and corruption is high. There is a lack of leadership that leaves the local officials to protect the monument from hawkers, phony guides, and growing air pollution without proper organization or morale. Sustainable practices are simply not in the cards."[17]

Demand at locations like the Taj continues to rise. In 2014, 5.37 million Indian visitors went to the Taj, an increase of 31% in 4 years.[18] Tourism will continue to grow dramatically. The new global middle class is spreading its wings. As more consumers come online, online sales engines will multiply, and more and more operators and hotels will sell the Taj. But no member of this system will have influence on the management of the Taj or take responsibility for it.

To avoid such tragic results, researchers suggest that tourism supply chains must embrace a perspective that operates on a wider level of value production than simply for business. A new "ecology of cooperation" needs to be achieved where value is created for all sectors of the tourism economy, including the environment and essential cultural monuments called tourism value ecology.[19] (See Figure 2.2.)

If the genuine value of the location is not embraced through a more holistic process, such as tourism value ecology, heritage sites like the Taj Mahal will be unbearable to visit. It may seem essential to address this problem, but the industry has already found an alternate solution to the loss of heritage and ecosystems: private attractions, urban entertainment, underwater sculpture gardens, and other manufactured tourism experiences. Casinos are one of the fastest growing examples of the manufactured tourism destination, and humorously they are achieving iconic status in their own right. Las Vegas, which was built in the desert without access to any of the normal attractions of tourism, was made possible by the introduction of commercial air service and a new highway network, which attracted gamblers to a growing honky-tonk district of casinos and bars on Fremont Street, which was enhanced by the introduction of neon lights. This strip is now considered a monument of important heritage value.[21]

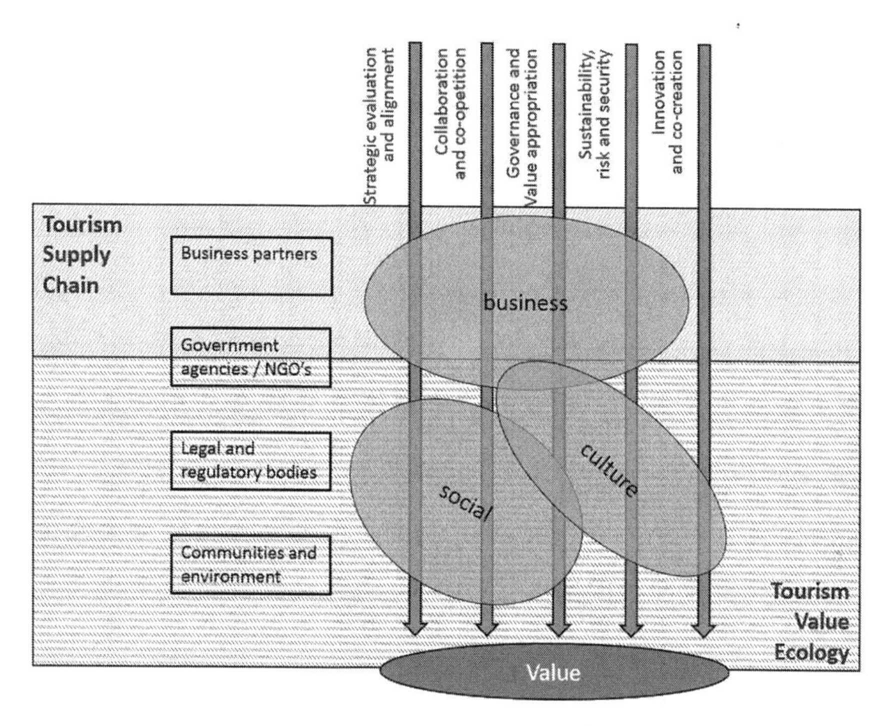

FIGURE 2.2 From tourism supply chain to tourism value ecology[20]

The creation of a destination out of a transportation hub in the desert enhanced by entertainment and hotels is now a much imitated strategy. Combining office buildings, convention and exhibition centers, cultural attractions, leisure, and spas, medical, and wellness facilities offers a human-made alternative to natural attractions. Dubai, which was once a desert with few facilities, is an indoor entertainment capital where millions come primarily to shop. "Dubai is full of tourists, but they aren't the tourists we know. These are economic tourists, and they're coming from the Middle East and eastern Africa for the express purpose of shopping."[22] No authenticity of culture or natural settings is required, just goods for sale and entertainment in the form of luxury shopping malls.

If the travel and tourism industry builds entertainment and business centers on a mass scale, there is no requirement for tourism value ecologies, which include local people, governments, ecosystems, and culture in the equation. Industry can own and manage the product being delivered and control their supply chains to a large degree – a huge competitive advantage. But a significant percentage of the public still demands the real world. They want volcanoes, beaches, rain forests, and snowcapped peaks. There is a hunger to see our most beautiful ports, such as Rio or San Francisco. These cannot be replaced by artificial capitals of pleasure and shopping. The innumerable valuable locations of the world will be endangered if supply chains are not coordinated to discourage congestion, develop value for the protection of sites, and create an ongoing set of maintenance programs that allow tourists and the industry to protect the very locations they are depending on.

Retail structure of the industry

The travel and tourism industry reaches the consumer through direct marketing increasingly online. Online travel agencies (OTAs) sell tickets, hotels, and attractions direct to the consumer. They take a commission on sales to support their operations but of course do not own any of these items. In 2014, Expedia became the largest retail tour operator in the world.[23] While travel agencies still exist in physical form, they are increasingly a virtual experience. Business travel agencies are still very competitive because they can manage volume travel for large corporations and offer excellent service and reduced costs.

Meta-search engines are now leveling the digital playing field between OTAs and their suppliers, making every part of the chain capable of direct digital retail sales, in effect pumping more and more product into digital pipelines. This type of key word search engine process already fuels significant profits for online booking agents, while placing a strain on their suppliers to maintain profit margins if they cannot play the same game. Accessing meta-search engines can be viewed as an opportunity for small and medium-sized tourism businesses as well, but it entails driving more revenue through online search and keeping operating costs very low. The OTAs have already benefitted from meta-search, and their model clearly makes it a logical place to drive a great deal of their marketing investments. But now META is becoming essential to all players in the supply chain who are seeking to drive business directly to their portals.[24] If they do not, they lose potential profits. This all leads to an ever growing competitive environment to capture customers online and lower margins. The new digital tourism supply chain is pictured in Figure 2.3. This graphic demonstrates how all segments of the industry are now routed through digital reservation systems, which are then routed through global distribution systems. Only then are they sold through either wholesale tour operators or online travel agents. Buyers are accessing this information through the variety of channels pictured at the bottom of the figure.

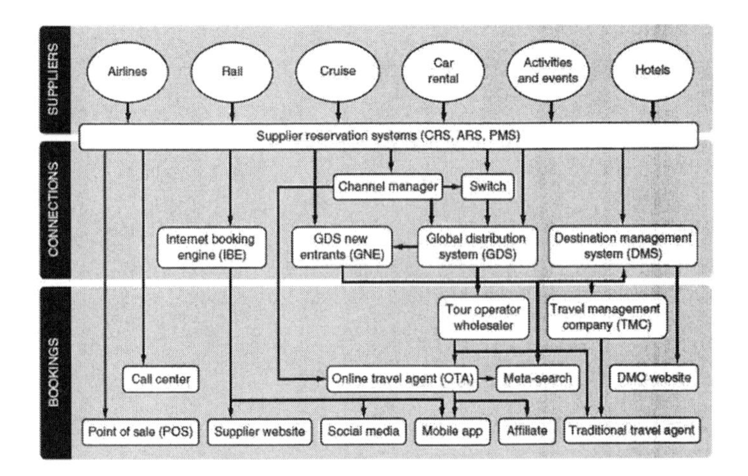

FIGURE 2.3 The digital travel distribution system[25]

Wholesale structure of the industry

The wholesalers of the tourism industry are called tour operators. For a century and a half, they have put together packages for travelers, which include transportation, hotels, and guides. Outbound tour operators are found in source markets that export tourists, and inbound tour operators are located in destination markets that import tourists. Tour operators are in an interesting/precarious position in terms of balancing their business needs with those of other stakeholders, clients, and the like, which has serious implications for how to successfully implement environmental management strategies.

The outbound tour operators were once very powerful in the market because they managed a good portion of the supply chain. They now manage client information needs via websites and 24-hour call centers, offering a range of customized itineraries that can be constructed with clients in online chats or on the phone. They charge a higher overhead rate for the customer service they provide, which exceeds what OTAs can offer. Their agents are now called specialists who guide the visitor and advise them based on their own extensive travel experience. Trips are touted to be 100% tailor-made with the best guides and real insights into the country you are visiting.

Inbound tour operators are the lesser known partners in the tour operating business. They manage all of the bookings for the outbound operators in the destination country. They book hotels, hire transportation, and manage guides on the ground and often handle the majority of this work for their international outbound operator customers in a business-to-business wholesale transaction.

This part of the industry works closely with the reality of the destination. These participants are responsible for all of the visitor customer service in country, which gives them a close relationship to the clients and their needs. These are the real travel experts who manage our visits to attractions, including parks, protected areas, and cultural heritage sites. They help us avoid traffic congestion, guide us to the clean restrooms, and evacuate us in rough security situations. Their work puts them in contact daily with both the situation on the ground and the needs of the client.

Inbound tour operating is above all highly dependent on location and quality of service. In these circumstances, the tour operators can be the most supportive players in the world for tourism sustainability. And with the rise of the digital economy, inbound companies now effectively manage tours for their own international customers increasingly without the help of outbound tour operators. A global spider web of marketing and packaging organizations are trading on how well they know the site and how effectively they can deliver the travel experience from small towns and capitals across the globe. If they reach customers effectively and capture their buying power, there could be benefits for the traveler and the destination.

Transportation

All travel is connected by transportation companies, a major part of the travel and tourism supply chain. The integration of transportation into the study of managing sustainable tourism in the 21st century is essential. Airlines, trains, buses, cars, and ferries offer mobility, without which the travel and tourism industry could not exist. Forty percent of carbon emissions comes from cars on the road, 2% by rail, and 6% from cruise and boating, but the majority of travel and tourism's carbon impacts come from flying in airplanes.[26] Air transport is used by 53% of travelers during their business or leisure journeys, making air travel the dominant form of travel on the planet. Approximately 2 billion air travelers are taking flight annually in the current decade, and that number is expected to triple by 2026[27]. The challenge of environmentally managing the aviation industry cannot be overestimated, with air travel currently contributing 2.5% of all carbon emissions from human activity worldwide and estimated to reach as high as 15%.

Today's transportation environment is increasingly congested, and this is taking a toll on the travel economy. Traffic jams burden a large proportion of major cities, and security controls create backups for those seeking to cross borders or travel through airports. Creating more efficient means of transport and developing more well managed transportation corridors, which service borders and airports, will lower the overall environmental impacts of travel and improve the well-being of travelers. But these goals should not be conflated with the goal of removing restrictions for all travel and opening all borders, which will only magnify the current question of how the impacts of travel can be managed.

Next-generation research on improving transport

The transportation industry is a unique sector, with extraordinary resources and some visionary leaders who see the future of how our planet may operate in ways that average citizens may not yet be considering.

Research has yielded a fascinating treasure trove of innovative concepts that promise to transform the currently congested travel transportation experience into a more customized, GPS-driven system. There may come a day when local transport providers can eliminate the difficulties of congested arteries and smoothly connect travelers to international air corridors, avoiding the current nightmarish system of constant delays. A wide variety of new technologies are on the horizon to ease the transportation burden of travel, offering less stress and more comfortable systems, which also will lower travel's growing carbon footprint.

These systems are not only in the design stage, they are being tested and piloted. Integrated intermodal mobility providers give passengers in Germany smooth access to different forms of transport all on one account, including trains, car and bicycle sharing, and air transport. New centralized traffic management systems are being deployed that help cities plagued by clogged traffic corridors. In Rio de Janeiro, a command center now integrates city management, emergency response, road traffic, and public transport management, allowing city managers to detect traffic and reroute vehicles. Automotive manufacturers will need to integrate vehicle navigation systems and voice controls to enable drivers to receive communications from central traffic management, but this type of interactive management is fully available to public emergency and security vehicles and could be easily privatized.[28]

Traffic management systems reduce congestion and CO_2 emissions by rerouting traffic away from overburdened corridors. In a hypothetical case, central traffic control systems in Mexico City will be able to anticipate traffic increases and project when CO_2 limits will be reached. Systems like Google Transit will automatically provide travelers with alternate car, rail, public transport, bicycle, and pedestrian routes with an estimation of costs based on their choice of alternate routing and transport system.[29] This will lower commute times, raise the livability of growing urban areas, and reduce the growing emissions of the transportation sector.

These ideas are not futuristic. They do involve creating personal portfolios of preferences tracked by large corporations such as Google and allowing traffic control centers to govern where we drive. But the transportation experience may be greatly improved. Already travelers have happily embraced Uber cars, permitting private cars to zip to our locations with the use of mobile device GPS systems. And NextBus has already made it possible to arrive at a bus stop on a bicycle within minutes of the arrival of our own bus and pop it on a rack to transport us to the next intermodal site, which may be a subway. This eliminates wasted time and fuel. With systems like this expanded in future, personal transport advisors working with our GPS location will be able to transmit our best transportation options to us, not only to get home from work but to get from home to our airport gateway. And this same system might help to manage how many travelers arrive at security lines simultaneously, instead allowing us to have a bite in a restaurant to avoid security line congestion. And GPS will be the essential ingredient to improving the problem with air traffic delays, which is essential in order to improve the travel experience and cut down on CO_2 emissions associated with airports. While the next steps for transportation may well improve the travel experience, the smoother linkages achieved will only make travel even more popular and increase the numbers who choose to travel.

Hotels and cruise lines

The hotel and cruise line industry, while very different, are in comparable positions within the global travel and tourism supply chain because they offer accommodations be it on land or sea. They are the link in the chain that offers a place to sleep and relax with air cooling and heating, water, lighting, and food frequently on a 24-hour basis. They manage this infrastructure according to local legal requirements on land or according to the international laws of the sea. They design their facilities according to in-depth market research exercises that are constantly updated in order to be certain they are meeting the needs of their clients in all the different facilities they offer.

Hotels and ships have separate divisions to manage their restaurants, kitchens, conference rooms, spas, grounds, and activities, including golf courses and swimming pools. Efforts to environmentally benchmark ships or hotels can be stymied by the diverse guest uses they accommodate, which require highly different systems for benchmarking and monitoring. This is made even more complex because of the different levels of luxury, rated by stars, that these firms sell and the type of facility they seek to offer, from palatial private islands with private chefs and personal grounds to tents with small hot plates and hiking trails nearby.

Hotels

The business model of the hotel world has an important impact on how the travel industry views its supply chain obligations. Franchising, a business practice that began in the U.S., has become the dominant business model for major hotel brands throughout the world. In the next few decades, every major brand in the world will employ the franchise system to manage risk in a fast moving global expansion period. Franchising, explored fully in the Chapter 4, does not favor sustainability benchmarking and management. While certain hotel groups, such as Hilton, have now installed sustainability management systems that are required as part of a branding agreement, on the whole hotel groups cannot take responsibility for their franchised properties. Local hotel owners, who are the most subject to risk and potential downturns in the economy, are frequently loath to make long-term investments to improve the efficiency systems of their hotels because it is the owners, not the brand, that must invest in the upfront costs.

What has emerged as a countertrend is the digital sharing economy – with the San Francisco–based Airbnb attracting the lion's share of this rapidly evolving market. Airbnb and its competitors fill a niche that will be difficult to outcompete. Large brands with their expensive franchise agreements are causing the price of hotels to escalate. Small and medium-sized properties frequently have small margins and limited marketing power. The booking engine Airbnb solves these problems. It provides enormous visibility for each owner and demands only a small service fee in return. It places every home owner in charge of customer service and eliminates the frustration of corporate insensitivity to individual needs. This billion-dollar global room and home sharing software system could help slow the rapid growth of new hotels and their infrastructure to a degree and take advantage of an enormous stock of housing, apartments, and homes that have space for more visitors. And to double the benefits, the system puts extra cash into the hands of local people who can use it to improve their livelihoods and the sustainability of their homes. But even with these benefits, for cities with housing shortages or central cores that are becoming dominated by tourist apartments, regulation will be needed.

In 2015, the *San Francisco Chronicle* found 5,000 Airbnb listings in the center of the city,[30] an excess of offerings that is driving out local residents and making the city ever more unaffordable. In many cities, landlords have converted rental properties to unhosted Airbnb apartments, forcing affordable housing residents out of their homes. Despite the growing controversy, San Franciscans gave a thumbs-up to short-term rentals in 2015 via referendum.[31] An Office of Short-Term Rental Administration and Enforcement was established, which requires all hosts to apply in person. And legislation has expanded the right of nearby owners to complain if they observe a violation of the short-term rental law.[32] San Francisco's law limits rentals where the host is not present for 90 days, and only primary residences can be rented in order to prevent landlords from evicting tenants to create full-time hotels.[33] Each city with outsized short-term rental demands will need to determine how to regulate this market, but the sharing economy has innumerable benefits for both hosts and guests alike.

Cruise

The cruise industry is a vertically integrated industry, which gives it greater reason to invest in efficiencies on board. Each investment can be amortized and lower the cost of operations of the ships. Ships must be managed with the utmost concern for operational efficiency, and the crew is required to deliver not only on comfort but on passenger safety. Ship designers and operators generally have an exceptional understanding of how to deliver a high-quality experience at the least cost with the most efficiency possible. The cruise ship industry innately seeks to lower its use of fuel and other resources, a goal that is built into the DNA of the shipping industry.

But the history of cruise ship environmental management is young. Despite their culture of efficiency and service, cruise industry motivations for environmental management are lowered by a legal system that does not hold them entirely accountable, as is fully explored in Chapter 7. Like all shipping operations, cruise ships are governed by international laws which permit them to be governed under foreign flags. This allows them to avoid strict environmental and labor regulations typical on land and gives freedom to operate with an unusual amount of sovereignty. The cruise ship industry is not beholden to any port and can change ports at will. Their accommodations are mobile, and this permits them to deal with small countries and islands in a manner of a powerful Goliath forcing weak governments (Davids) to submit to long-term business arrangements that are not adequately to their benefit.

While this industry wants to improve its relationships with local governments, its supply chains need to reflect and incorporate more of the common good of destinations and the increasingly beleaguered oceans and seas where cruise ships operate. To date, they have offered small grants, supported destination management initiatives led by NGOs, and encouraged the certification of land-based suppliers.

Destinations and attractions

The bottom of the supply chain is the destination. All of the other links in the chain are designed to take advantage of the destination, and destinations nurture the entire industry. The destination is the location or site "that is central to the decision of the traveler to take the trip."[34] The destination is a classic "common pool" resource, which is not owned by any member of the supply chain, unless it is a privately owned attraction such as Disney World.

Most destinations come under either local, regional, or federal management. But the goals of the managers are often not directly related to the management of tourism. For example, protected areas across the planet are now seen as vital reserves to protect biological diversity. Park authorities often see tourism as a threat to their goals of preserving ecosystems because park resources can be degraded when traveler numbers increase and budgets remain stagnant. For many parks around the world, it is increasingly difficult to demonstrate that it is possible to equitably manage tourism and create a "virtuous cycle" of benefit.[35] But the option is there to create a system that requires valuable payment to preserve the park and reinvest in these invaluable assets. The U.S. National Park System receives nearly 293 million visitors annually with a budget of approximately $10 per visitor. The service continues to manage tourism while also protecting resources, acting as an environmental advocate, partnering in community revitalizations, acting as a world leader in the parks and preservation community and pioneering the drive to protect America's open space.[36] If every tourist were to generate an additional $10 for the management of the parks, its backlogs of projects would be decreased and its ability to manage the U.S. natural and cultural heritage would be greatly increased. But the system, nonetheless, represents a valuable model for capturing the value of tourism for the protection of world-class tourism assets.

It is certain that tourism can be very damaging to global heritage and natural sites unless there are good management systems in place and budgets to protect these areas. If the budgets are available and the personnel is trained, tourism is one of the best means of protecting valuable ecosystems and cultural heritage. Tourism raises interest in the preservation of these areas and has tremendous potential to pay for the cost of protection.

There has yet to be a major coordinated effort by the global protected area community to seek to generate revenue from tourism that could drive revenues toward the management of the earth's most important biological and cultural resources, in fact their global membership organization – the International Union for the Conservation of Nature (IUCN), the world's foremost organization leading the parks and protected area community – has no staff people with funding to work on tourism at all at the international level and instead depends on a voluntary international working group to develop knowledge and demonstrate best practices.[37] But while national parks and protected areas could certainly benefit from more protection and more budgetary resources, they are well resourced and benefit from coordinated oversight and governance compared to regional destinations outside park and protected area systems.

The cycle of growth and inability to manage destinations

Most destinations fall outside the governance of one authority. This causes confusion and often allows vital natural and cultural resources to remain unprotected from tourism growth. In destinations around the world, even when there is considerable effort to plan tourism infrastructure and hotels, the surrounding areas do not benefit from this foresight, as in the famous case of Cancun, Mexico.

Cancun, Mexico: case study

Cancun is located on a small peninsula on the Caribbean coast of the Yucatan peninsula. In the late 1960s, the Mexican federal government developed a master plan for Cancun that allowed hotel development along a dune strip that was undeveloped at the time of construction. The Yucatan was once a region with low populations and a tranquil Mayan community living near incomparable Mayan archeological sites. Its center of commerce was the elegant colonial city, Merida. After the creation of Cancun, the region quickly attracted hundreds of thousands of immigrant workers, all of whom moved to a sprawling town with insufficient urban infrastructure. The success of Cancun caused a cascade of new resort complexes that quickly crawled down the coast, now stretching 100 miles to the border of Belize.

Mexican researchers have documented that in the Cancun region only 30% of the waste water (aka sewage) generated by tourists and locals is treated, allowing contaminated water to flow directly into the Yucatan's peninsula's groundwater, underground rivers, cenotes, and the sea without treatment. And because of the many immigrants to the area and the sprawling worker city that is located just over the bridge from the peninsula where the luxurious resorts are found, hundreds of thousands of tons of solid waste were transported annually to illegal garbage dumps through the 1990s and 2000s.[38]

The cycles of tourism development generally follow a pattern of early rapid growth, long plateaus, and deterioration. It is very difficult to maintain destinations in a steady operational state that allows local stakeholders to deliver quality tourism experiences over a 20- to 40-year period. In Chapter 8 the question of how to apply regional planning principles is explored to respond to the continuing privatization of real estate and natural and cultural resources that are traded off for foreign exchange receipts, just as any other commodity for sale, and not seen as invaluable resources that have real value in the marketplace if protected in the long term.

The process of creating a common understanding of how a destination will change and become valuable to those who own valuable natural assets is rarely introduced at all or not in time. I have been in the position of having to inform local people of the negative cycle of development that might transpire if they do not cooperate and hold on to their resources. In my work in countries such as El Salvador and Bangladesh, local residents have no idea why speculators are at their door and why their beach resources are so valuable to outsiders. Without more information, these local people are vulnerable to selling out far too early in the destination development cycle. These cycles are difficult to arrest unless local people become more aware of the value of their local undeveloped locations and begin to work to protect them.

Common pool resource management

Destinations are common pool resources that often have no recognized value in communities until the development cycle begins. A common pool resource (CPR) is a natural or human-made system that has many common benefits to society and commercial value for many users. CPRs are often equated with the tale of the Tragedy of the Commons, which in historical England left grasslands denuded due to an inability to agree on limits on grazing rights. Fishing grounds, watersheds, grazing areas, bridges, parking garages, streams, lakes, oceans, and other bodies of water are all CPRs. For years the field of economics has favored the dynamics of resource privatization, arguing that more reliable benefits can be financed, built, and distributed through capital markets. As income inequality continues to grow worldwide and resources become less available to a growing percentage of the poor, privatization is proving to be a perilous and flawed solution for deciding upon a common future. David Sloan Wilson's fascinating book on the evolution of altruistic human behaviors suggests that both private gain and public management must coexist in human societal evolution in order to obtain sustainable outcomes for the planet. Competition hones quality decision making by individuals and their businesses, but it also favors exploitation of the losers, allowing a winner-take-all mentality to flourish.[39] As global societies grow on an unprecedented scale and stake out real estate with increasingly unlimited resources, some will gain access to the most beautiful places, and others will lose out entirely, unless a more savvy process is developed and soon. Most would agree that parks or waterfalls or beaches should not be privatized because this is unlikely to produce equal access to these locations. The fundamental academic text for understanding these dynamics is the Nobel Prize–winning book and research by Elinor Ostrom, *Governing the Commons*.[40]

When destinations grow, real estate sales quickly change the value of land and the density of development and the type of commercial uses. Outside real estate buyers arrive to convert cheap or almost worthless land into valuable property. Once real estate is sold to outsiders, locals have few options. Locals sell quickly because they don't perceive the financial benefits they might receive in future from their destination and simply opt for the short-term cash benefits they will receive immediately. This is a textbook CPR problem.[41]

Without a system to protect destination resources, locals generally do not receive a reasonable percentage of the benefits from their native lands. At times, there can be residents who intuitively understand the importance of collectively managing and preserving their destination. Monteverde, a mountaintop destination in Costa Rica, is one such example.

Monteverde, Costa Rica: case study

In the 1950s, a Quaker community bought nearly 3,000 acres in the central mountains of Costa Rica and developed a "strong, self-reliant system of governance that blended Quaker values of simplicity democracy and non-violence with personal and community responsibility."[42] They established a town meeting system as a means of giving all residents a voice and have used consensus decision making to foster local services including crafts, galleries, restaurants, and a woman's craft cooperative, and they created the Monteverde Cloud Forest Reserve, which is recognized as an important biological treasure that benefits the community and the nation of Costa Rica. The reserve protects the watershed of the region and has become a center for scientific investigation.

When tourism began to grow in Costa Rica in the 1980s and 1990s from small numbers of scientists to tens of thousands of tourists, Monteverde was on every itinerary and frequently still is. To maintain the reserve's biological integrity and to pay for management, the Tropical Science Center, which manages the reserve, raised entrance fees and began to restrict visitor numbers to 100 at a time. Visitor trails and interpretation areas were restricted to just 2% of the reserve.[43]

The Monteverde community continues to work through many questions of balancing the common good with benefits for individuals. The price of land did shoot up, and many locals did sell their land. But efforts to slow the growth and keep the benefits local have been constant. Hundreds of thousands now visit annually. But it remains a mixing ground where tourists, scientists, and locals can enjoy this domain and share a certain ethic of preservation. Recent research confirms that Monteverde has maintained its ethic of preservation with norms that remain strong and with a common goal to be a sustainable tourism destination in the long term.[44]

Monteverde evolved with local ethics and homegrown governance systems. As tourism grows, more organized initiatives to encourage and build strong consensus and sustainable destination governance are just emerging. The latest thinking on the governance of common pool resources suggests that entirely new forms of ecological governance are required to provide a more locally based, well educated, and prepared democratic system for managing local resources. Local citizens must first understand the value of what they have. Elinor Ostrom's research on managing shared wealth has supported thinking that not all development must be for private economic gain and that there can be successful systems to manage shared wealth.

The essential elements of Ostrom's work calls for knowledge, cooperation, and leadership as follows:

- There is a need for well documented information on the costs and benefits of actions.
- Individuals must understand that the resources have value for them in the long term.
- Local organizers must establish a trusting relationship with regular communications among stakeholders.
- A respected system of monitoring is needed that guides sanctioned uses of commonly valued resources.
- Leadership is essential; it fosters community-level problem solving.
- Collective rules must be in place that are perceived as fair and respectful of local needs.[45]

The human right to a clean and healthy environment could become a form of common law that allows local people to protect themselves from overdevelopment. New legal strategies might help protect civil society in the face of rampant development and allow them to establish new forms of governance that can protect their common pool resources. A stakeholder trust is one new form of governance that allows communities to gain legal rights on a community basis over their natural and cultural assets. This could help prevent the classic development cycles that remove local citizens from the process of governing their futures. Instead, they might become the trustees and guardians of their own common pool resources.[46]

These are novel ideas, but they draw upon the value of collective wisdom. In the case of tourism destination development, such creative thinking may apply and allow communities to claim local trustee rights over valuable beaches and waterfalls, as well as other historical assets that once belonged to their ancestors. Some of these systems are already working on a small scale. In Kerala, India, the government gave the management rights of a local scenic waterfall to the local community in a small ecotourism development area called Thenmala, which had become overrun, eroded, and littered with trash. The community was permitted to charge a small entrance fee to the waterfall. They used the funds to pay local women to manage the site and keep it clean, replanted vegetation that had been lost, and offered villager-led guide services to keep locals on trails.[47]

Indigenous communities around the world seek to manage their resources and benefit from tourism in a more collective manner. This has resulted in new styles of governance to protect ecotourism resources. In the case of the community of Infierno, Peru, in the Amazonian province of Tambopata, the development of a lodge prompted discussion and collective planning for how wildlife, habitats, and even cultural traditions should be used, showcased (or not), and protected. What is quite inspiring is that the community defines the "good life" primarily with words of peace, unity, and getting along with family and community. While entrepreneurship and gaining the most for oneself is growing as an ethic in Infierno, the community has reconfirmed their goal to invest communal profits from the lodge into purely communal endeavors through a community fund.[48]

Indigenous societies offer a fascinating view into how communal norms and management could preserve destinations in future. Other examples of collective land management systems in Botswana and Namibia are explored in Chapter 3. In the next 20–30 years, destinations will require collaborative local systems that allow residents to share a vision for their future and to put these systems into action.

The future of sustainable supply chain management for tourism

Supply chains build value with each transaction. Tapping that value and institutionalizing an investment in the local destination will be the task of local people and their governments. They will have to work very hard to protect the value of their places and to preserve their culture and environment in a win-win proposition that they lay out with future investors, tour operators, and hotels. The more an area is overdeveloped, the sooner the value of the destination will fall.

If the global industry invests in the value of sustainable destinations by securing a portion of tourism revenue and invests in historic preservation, environmental management and conservation, training, education, and management skills at the local level, the entire tourism system will retain higher value. If the standard destination cycles of growth are allowed to transpire without destination management, real estate sell-offs, sprawl, and environmental and cultural degradation will devalue precious natural and cultural assets. The capacity to manage a positive cycle of destination management is not part of our educational lexicon yet and is not found in most policy manuals. Local people have few operational models to work from to guide decisions. The tools available will be discussed in detail in Chapter 8. But there is still a great deal of room for creative thinking and innovation. At Harvard there are students already gestating models for large-scale tourism investments that help retain local value and that foster investment in renewable cycles of building mutually supportive systems of tourism with local food, culture, and land conservation strategies. These positive cycles are more than possible to design and can be fostered quickly.

For example, a proposed large-scale Chinese investment in tourism in Iceland could have been transformed from an insensitive landgrab to a positive contribution to local society with steps not that difficult to undertake. A study, performed by Magni Magnason,[49] for our class at Harvard on International Development of Sustainable Economies, on Grimmsstadir, Iceland, looks at alternatives for the Zhongkun Investment Group (ZIG) of China to develop a sustainable resort complex in northern Iceland.[50]

Grimmsstadir, Iceland: case study

Property tycoon Huang Nubo, of ZIG, launched an effort to invest $200 million in 2011 to create a resort and mountain park complex in Iceland, which was blocked by Iceland's foreign investment laws. The site was near Vatanajokull National Park in the area known as the Diamond Circle, a well-known tourism route with views of spectacular glaciers and ice sheets. The local people farm, hunt, and fish on land that is highly sensitive to erosion and not easily developed without serious environmental and cultural impacts. Magnason reviewed the developers' own plans and proposed that Nobu set aside 99% of the 300-square-kilometer parcel he was ready to procure and place it in a land trust, while retaining the 300-hectare site he required for his resort. He gave the envisioned trust organization a legal obligation to protect the environmental and cultural heritage of the region, suggested financing with a one-time start-up cash infusion diverted from the original proposed investment in the land, and found that operational costs would be covered with a small percentage of tourism annual revenues from the resort.

Such creative approaches to the environmental and cultural conservation of sites not within protected areas are essential to the next generation of tourism development. Land trusts and other vehicles to preserve cultural and environmental assets not under federal or local protection could revolutionize how governments both attract investment and secure tourism assets. If strategic, unprotected areas are defined by law, nations could require an investment in land trusts to secure a percentage of revenue in cultural and environmental conservation, while making certain that revenues are there in perpetuity to manage the sites locally via an active and empowered civil society.

Conclusion

In the next generation of planning for the growth of tourism, a wide variety of innovations will be required that allow for business investment, government cooperation, and local involvement in the management and protection of tourism assets. The changing dynamics of the tourism business, the globalization of markets, and the demand for more efficiency and lower overhead, driven by the digitization of travel markets, needs to be fully integrated into the process of developing solutions. The boom in digital marketing is becoming a larger part of the tourism industry every day and will inevitably have large impacts on the entire supply chain.

The tourism supply chain must begin to incorporate a larger, more holistic set of values that includes the responsible management of destinations. Yes, there are successful examples of ecotourism and responsible travel around the world, and there are good cases of large-scale industry taking the right approach and investing in lowering their footprint and contributing to the well-being of local people. But the business model of tourism in general does not favor this. There are rare cases such as responsibletravel.com, which fosters a growing ledger of good examples and best practice of travelers online. Justin Francis, CEO of responsibletravel.com, states, "[T]ourists are realistically the only people who can scrutinize the full supply chain of destinations globally, expose issues and create change."[51] But despite the undeniable value of conscious consumers, the pressure these consumers have brought to bear on major questions of international tourism development has been small. In the case of tourism, sourcing the provenance and sustainability of tourism offerings will never be an option. Rather the sustainability of the sites visited will have to be measured, and that will require much more due diligence on the ground than has ever transpired in the past.

The solutions will be developed only if the next generation of pioneers has a thorough knowledge of how the tourism business operates and the capacity to view tourism not in terms of isolated, anecdotal examples of the past that feature small-scale tourism development solutions. Instead, this new generation must look at the global picture and bring knowledge management systems to the local level, while plugging data on cumulative impacts into the most advanced data systems. These students must also understand that land outside of protected areas needs immediate attention. There is a strong need to focus on collaborative systems for involving local people in the governance of their common pool resources, at the same time beginning the process of legislating protection of invaluable areas, through land trusts and other land preservation systems that also protect local livelihoods. The Iceland study by Magni Magnason demonstrates that such goals can be achieved while still allowing substantial investment in new tourism development to transpire.

The digital marketing community must also test and vet options for driving more value through the digital supply chains to local products, which have no immediate source of revenue. The essential fact is that more investment in destination management and preservation is needed. If local governments are not in the position to manage the use of their land or heritage sites in the near term due to a lack of capacity and budgetary resources, and if the number of travelers continues to grow at global rates of 4–5% a year with domestic travel often doubling these figures, the value of the most precious destinations on earth will plummet. Overcrowded public cultural and natural resources, such as the Taj Mahal, are already starting to lose value.[52] Arresting this and securing a common good require planning and cooperation, such as what was achieved in Monteverde, Costa Rica. Each of the remaining chapters in this book describes a clear set of solutions that will help to manage each major link in the supply chain more effectively in future.

Notes

1 Kumar Goel, About the Luxury Indian Traveler https://linkedin.com/pulse/luxury-travel-indiai-ravi-goel, Accessed September 28, 2016
2 Frommer, Arthur, 1996, *The New World of Travel*, 5th Edition, Simon Shuster Macmillan Co., New York, Preface, page xiii
3 Ibid.
4 Rezdy, n.d., *Travel Statistics for Tour Operators*, https://rezdy.com/resource/travel-statistics-for-tour-operators/, Accessed September 13, 2016
5 *Gringo Trails*, n.d., Are tourists destroying the planet – or saving it?, http://gringotrails.com/, Accessed September 13, 2016
6 Tham, Aaron, Robert Ogulin and Willen Selen, 2015, From Tourism supply chains to tourism value ecology, *Journal of New Business Ideas and Trends*, Vol. 13, No. 1, 1–65
7 Beyer, Matthias, March 2014, *Tourism Planning in Development Cooperation, a Handbook, Challenges, Consulting Approaches, Practical Examples, Tools*, GIZ, Boon and Eschborn, Germany, page 27
8 Dauvergne, Peter and Jane Lister, 2013, *Eco-Business, a Big-Brand Takeover of Sustainability*, MIT Press, Cambridge, MA
9 Ibid., page 91
10 Ibid.
11 Ibid., page 93
12 COSA, 2013, *The COSA Measuring Sustainability Report: Coffee and Cocoa in 12 Countries*, The Committee on Sustainability Assessment, Philadelphia, PA
13 Ibid.
14 Global Sustainable Tourism Council, http://gstcouncil.org/, Accessed September 13, 2016
15 ARCNews, Winter 2016, *ARCGIS Is a System of Engagement and a System of Record*, ESRI, Redlands, CA, pages 1 and 4
16 Bains, S.K.B., 2011, *Can International Tourists Have a Better Experience at the Taj Mahal*, Thesis Submitted to Auckland University of Technology for Master of Tourism Studies, Faculty of Applied Humanities, School of Hospitality and Tourism
17 Ibid., page 166

18 Devi, Aditya, Number of Foreign visitors to Taj Mahal on wane, January 2, 2015, *The Times of India*, http://timesofindia.indiatimes.com/india/Number-of-foreign-visitors-to-Taj-Mahal-on-wane/articleshow/45723601.cms, Accessed September 29, 2016

19 Thamet al., From Tourism supply chains to tourism value ecology

20 Ibid., Figure 3, page 58

21 Weaver, David B., 2010, Contemporary tourism heritage as heritage tourism, evidence from Las Vegas and Gold Coast, *Annals of Tourism Research*, Vol. 38, No. 1, 249–267

22 Kasarda, John D. and Greg Lindsay, The Aerotropolis Emirates, in *Aerotropolis: The Way We'll Live Next*, Farrar, Straus & Giroux, New York, page 298

23 Sun, sea and surfing, 2014, *The Economist*, June 21, http://economist.com/news/business/21604598-market-booking-travel-online-rapidly-consolidating-sun-sea-and-surfing, Accessed September 13, 2016

24 SiteMinder, 2014, Q&Q with SiteMinder: metasearch for travel and tourism, May 6, http://siteminder.com/blog/meta-search-for-travel-tourism/, Accessed September 13, 2016

25 Benckendorff, Pierre J., Pauline J. Sheldon and Daniel R. Fesenmaier, 2014, *Tourism Information Technology*, 2nd Edition, CABI, Oxfordshire, UK, page 44

26 UN World Tourism Organization, 2015 Edition, UNWTO Tourism Highlights, page 14, offers annual updates, http://e-unwto.org/doi/pdf/10.18111/9789284416899, page 5, Accessed September 29, 2016

27 Oxford Economics, n.d., Aviation the Real World Wide Web, Overview, Oxford Economics, Oxford, UK http://oxfordeconomics.com/my-oxford/projects/128832, Accessed October 1, 2016

28 World Economic Forum, 2013, *Connected World, Transforming Travel, Transportation and Supply Chains*, in collaboration with the Boston Consulting Group, Geneva, Switzerland

29 Ibid.

30 Cutler, Kim-Mai, 2015, San Francisco Supervisor, Mayor Proposes Tighter Regulation on Short-Term Rentals, Airbnb, TC, April 14,, http://social.techcrunch.com/2015/04/14/airbnb-new-regulations/, Accessed March 17, 2016

31 Romney, Lee Tracy Lien and Matt Hamilton, November 4, 2015, Airbnb wins the vote in San Francisco, but city's housing debate rages on, *Los Angeles Times*, http://latimes.com/local/california/la-me-ln-airbnb-san-francisco-vote-housing-debate2-20151104-html-story.html, Accessed September 29, 2016

32 Cutler, Kim-Mai, *Mayor Proposes Tighter Regulation*

33 Stephen, J.D. Fishman, n.d., *Overview of Airbnb Law in San Francisco*, http://nolo.com/legal-encyclopedia/overview-airbnb-law-san-francisco.html, Accessed September 13, 2016

34 UNWTO, *Indicators for Sustainable Development for Tourism Destinations; A Guidebook*, 2004, UNWTO, Madrid, Spain, page 8, box 1.2

35 Bushell, Robyn and Paul Eagles, 2007, *Tourism and Protected Areas: Benefits beyond Boundaries*, The Vth IUCN World Parks Congress, Gland, Switzerland, page 10

36 Happy 100th Birthday, National Park Service, http://nps.gov/aboutus/history.htm, Accessed April 17, 2017

37 World Commission on Protected Areas, n.d., Tourism-TAPAS, https://iucn.org/protected-areas/world-commission-protected-areas/wcpa/what-we-do/tourism-tapas, Accessed October 11, 2016. Epler Wood sits on the Executive Committee of this voluntary organization

38 Martinez, Elva Esther Vargas, Castillo Nechar, M., and Viesca González, F.C., 2013, Ending a touristic destination in four decades: Cancun's creation, peak and agony, *International Journal of Humanities and Social Science*, Vol. 3, No. 8, Special Issue, page 23

39 Taken from Wilson, David Sloan, 2015, *Does Altruism Exist? Culture, Genes and the Welfare of Others*, Yale University Press, New Haven, CT

40 Ostrom, Elinor, 1990, *Governing the Commons: The Evolution of Institutions for Collective Action*, Cambridge University Press, Cambridge

41 Adapted from ibid., page 34
42 Honey, Martha, 1999, *Ecotourism and Sustainable Development, Who Owns Paradise?* Island Press, Washington, DC, pages 150–156
43 Ibid.
44 Gora, Ashley, 2013, *Sustainable Tourism Norm Transfer and the Case of Monteverde*, Lake Forest College Senior Theses, Costa Rica
45 Bollier, David and Burns Weston, 2015, Advancing Ecological Stewardship via the Commons and Human Rights, in *State of the World 2014: Governing for Sustainability*, eds. Renner, Michael and Thomas Prugh, The Worldwatch Institute, Washington, DC, Chapter 9, pages 91–104
46 Ibid., 101–102
47 Personal observations made by Epler Wood during visit in 2004–2005
48 Stronza, Amanda Lee, 2010, Commons management and ecotourism: ethnographic evidence from the Amazon, *International Journal of the Commons*, Vol. 4, No. 1, 56–77
49 Magnasan, Magni, 2014, *Procurement of Land for $7 Million to Support Future Luxury Hotel and Golf Course Development by the Zhongkun Investment Groups (ZIG) in Grimsstadir, Iceland*, Unpublished Manuscript, Environmental Management of International Tourism Development class paper, Department of Sustainability and Environmental Management, Harvard Extension School, Cambridge, MA
50 Chinese property tycoon eyes Norway as Iceland project on hold, Bloomberg News, http://bloomberg.com/news/articles/2014-02-13/chinese-property-tycoon-eyes-norway-as-iceland-project-on-hold Accessed September 29, 2016
51 Francis, Justin, February 2, 2015, personal communication
52 Devi, Aditya, 2015, Number of foreign visitors to Taj Mahal on wane, *The Times of India*, January 2, http://timesofindia.indiatimes.com/india/Number-of-foreign-visitors-to-Taj-Mahal-on-wane/articleshow/45723601.cms, Accessed September 13, 2016

3

ECONOMIC DEVELOPMENT OF TOURISM IN EMERGING ECONOMIES

Introduction

Bangladesh is a country with a large population of over 160 million people living in a delta region roughly the size of the U.S. state of Louisiana. Over 30% of Bangladeshis live below the poverty line, the largest percentage of whom live and work on small farms.[1] The textile industry has raised the country to a middle level of poverty, and this has fueled a surprisingly rapid growth of domestic tourism.

Bangladesh has beautiful tourism assets, including the longest beach in the world on the Teknaf Peninsula, a protected upland hill tribe region in the Chittagong District, parks that shelter biodiversity in the Lawachara area, and St. Martin's Island, a coral island that is a mecca for travelers. Domestic tourism is thriving and has the potential to be a source of revenue for sustainable development. Ecotourism has an especially important role to play in helping to conserve the country's protected ecosystems, with numerous opportunities to involve local communities; however, there are many barriers to success, including the rapid expansion of the mainstream tourism trade. While tourism is growing on the island, local communities are not enjoying the benefits that they could if tourism were managed with a carefully planned pro-poor development strategy. As part of my work on a USAID development project, I prepared a Community-Based Ecotourism Strategy that recommends a wide range of measures to manage tourism sustainably and contribute to economic growth.[2] Some of the most pressing issues can be captured in my early reports.

> There is an active ecotourism development and awareness program on St. Martin's Island. In particular, they have developed a tourism zoning plan with technical assistance from the International Union for the Conservation of Nature (IUCN.) Attempts to institute zoning or tourism regulation

> have not produced outcomes as the Department of Environment does not have the ability to enforce regulations. There is evidence that tourism development is proceeding apace without restraint. The community seeks relief as they are receiving few benefits from tourism. They are losing their land, and their natural resources are being destroyed. Fresh water is disappearing. The development project could approach this problem in two ways, support more eco-commerce and train the villagers to offer independent services – or develop the villagers' capacity to become a trained part of the growing mainstream tourism workforce.
>
> It is unlikely that the development project will be able to arrest the growth of tourism on the island, and villagers have reportedly lost 90% of their land to outside developers. It is therefore not entirely logical to encourage more village-based hotels, which will not be competitive with mainstream tourism. Village based eco-commerce will take a great deal of training over a long period of time to create a limited tourism economy for the villagers. Rather it is recommended that the project use "pro-poor" tourism development tactics on St. Martins Island. This entails helping villagers to work within the growing mainstream tourism trade not seek to compete with it. This entails developing community capacity to work for the hotel corporations as trained employees, and to provide needed services to the growing hotel industry, from cleaning, to food, to guiding, to handicrafts.[3]

Such a set of recommendations did not seem ideal. But St. Martin's island had already passed a tipping point where the opportunity to introduce a more positive economic development strategy that achieves both conservation and poverty alleviation was already disappearing. Despite such challenges, there is no question that tourism is a source of economic growth and job creation around the world. It is directly or indirectly responsible for 8.8% of the world's jobs, 5.8% of world exports, and 4.5% of global investments.[4] It is the only service sector that provides concrete trading opportunities for all nations regardless of their level of development, and it is the number one source of foreign exchange for a growing percentage of lesser developed countries.[5] The tourism economy is of more consequence in lesser developed economies. Countries in Africa and Asia are projected to grow at double the rate of developed nations and garner a market share of 57% of all international tourism arrivals worldwide by 2030, equivalent to over 1 billion foreign visitors.[6]

Economic development and tourism: key issues

The role of tourism in global emerging economies could be transformative. While the potential is great, the delivery of tourism as an economic development tool has been weak. Projects have lacked a strong platform for donor and private sector investment. Yet there is more-than-adequate evidence that tourism can contribute to clean development infrastructure improvements; build labor forces that are pro-poor; empower women, young people, and marginalized populations; and stimulate cultural heritage and environmental conservation.[7] While many good case studies exist to prove this potential, as will be presented here, there are very few substantiated research-based models demonstrating that sustainable tourism can function as a fully inclusive and sustainable form of economic development. According to economic development experts, tourism has too often been unable to increase the value of the sector for the poor.[8] While tourism is positioned to end poverty and hunger, to ensure healthy lives, and to create more inclusive and higher-quality education,[9] there is only slight evidence that this is being achieved on any scale. This chapter investigates evidence of how tourism's economic development benefits could be improved based on well researched methodologies. The evidence presented in this volume will be based as much as possible on longer-term longitudinal studies that use statistical evidence and third-party academic review.

While the international sustainable development community has invested substantial resources in tourism as a source of economic development in emerging economies, there have been a wide variety of critiques and analyses that legitimately point out that these investments have been far too oriented to supply-side product development in rural poor and biodiverse areas and thus fail to succeed due to inadequate attention to connecting local, rural suppliers to the existing tourism marketplace. Too often, governments and donors have invested in small-scale local tourism products without adequate attention to connecting these products to the tourism supply chain. Future investments must be predicated upon existing supply chains that both bring international markets to the local level and build the capacity of local suppliers to meet market demand. Such work needs to be measured via a process that is best known as contributing to sustainable livelihoods. The sustainable livelihoods framework measures financial and human capital, as well as natural, cultural, and social capital. This framework is fully detailed in this chapter and is recommended as the essential methodology for incorporating a triple bottom line approach to sustainable tourism economic development.

The process of developing sustainable tourism as an economic development tool would also greatly benefit from the use of value chain analysis to determine where weak links lie and how to reinforce the linkages between local products, regional buyers willing to invest in local well-being, and international buyers driven increasingly by markets that demand social responsibility. At present, though tourism is often discussed as economically beneficial to local people, there is very little investment in local capacity to develop or manage tourism from national or local governments, though there is evidence that some countries are moving in this direction for the first time.[10] Instead nearly all pro-poor tourism investment has come from the international donor community, which has sought, though inconsistently, to invest in tourism as a form of sustainable development. The history of these efforts provides important guidance to future practitioners on the pitfalls to avoid in this field.

History of tourism as an economic development tool

Global efforts to determine how tourism can be managed sustainably while fostering economic growth are young indeed. The concept of sustainable tourism was formally established only after the United Nations Earth Summit in Rio de Janeiro in 1992, which brought together 117 heads of state to discuss how conservation could be reconciled with the need for economic development. The Summit was the most ambitious, well attended, high-ranking conference ever held on the environment, and at the time it was the largest gathering of world leaders in history. Biodiversity conservation took center stage with a binding treaty requiring nations to inventory their species. Climate change was addressed for the first time on a global scale, and a declaration of the importance of environmentally sound development was outlined to create a global strategy, which became known as Agenda 21.

While the Earth Summit is generally recognized as the launch point for the field of sustainable tourism,[11] the idea of linking tourism management to the principles of Agenda 21 came several years after the event. Agenda 21 became one of the central frameworks for leveraging sustainable tourism as an economic development tool, and it was used in Europe in particular, where local authorities were given the mandate to apply zoning, environmental regulations, licensing, and other tools to achieve sustainable tourism development as part of the Agenda 21 process.[12] Recognizing the important groundwork laid by the Earth Summit and Agenda 21, the Cape Town Declaration of 2002 called upon multilateral donor agencies to include sustainable tourism in their programmatic efforts, stressing the following points regarding economic responsibility:

- Exercise preference for those forms of development that benefit local communities and minimize negative impacts on local livelihoods.
- Maximize local economic benefits by increasing linkages and reducing leakages, by ensuring that communities are involved in and benefit from tourism.
- Adopt equitable business practices and pay and charge fair prices.
- Provide appropriate sufficient support to small, medium and micro enterprises to ensure tourism-related enterprises thrive and are sustainable.[13]

Subsequently, the United Kingdom's Department for International Development, the Netherlands Development Organization, and the UNWTO began to focus on sustainable tourism as an economic development tool that could directly serve to alleviate poverty. These efforts were tied to the concept of responsible tourism, which became a globally accepted term for sustainable tourism in the marketplace, particularly in Europe. At the same time, however, U.S.-based donors were working primarily with the concept of ecotourism and community-based tourism through their own agencies. Throughout the world in this period, investment of approximately $7 billion was spent on ecotourism and community-based tourism over

5 years, with only 17% of the projects identified as tourism in donor frameworks. Because the projects were managed under funds largely related to the conservation of ecosystems, conservation scientists were frequently tasked with experimenting on improving livelihoods using ecotourism.[14] Few of these individuals had ever worked in tourism, and they failed to properly account for tourism market realities in the development of small community-level projects. A significant proportion of these projects failed, and there were legitimate articles pointing out these problems from Europe,[15] and I personally devoted much time as an NGO leader at the time to constructing a framework for donors that would take into account the many factors required to make ecotourism a success at the community level, which was originally presented in 2001 at a Stanford University conference.[16]

In October of 2005, global leaders gathered in Washington, DC, for the World Tourism Organization (now United Nations World Tourism Organization) Tourism Policy Forum, which brought together the highest leadership of the organizations that have helped to support the analysis of tourism as an economic development tool: the U.S. Agency for International Development (USAID), the World Bank, and the Inter-American Development Bank. Andrew Natsios, Administrator of USAID stated:

> Sustainable tourism does not simply happen. It requires an overall strategy, detailed planning, with a host of supporting mechanisms including public–private partnerships, appropriate legislative and institutional reforms, training and public education, infrastructure and technology, finance and credit systems that reach down to the poor, and continued monitoring and evaluation.[17]

Since 1998, I have researched extensively how donors capitalized community ecotourism projects in ways that did not have long-term benefits for conservation NGOs and donors.[18] This was a difficult period in the field of ecotourism because there were so many failed projects that were so entirely disconnected from market realities. More than once I saw local community members sitting outside fully built lodges, financed by donors or governments, waiting for customers to arrive without any knowledge of how to attract them. A great deal of investment was lost, and mistakes were made. But these errors should not lead to the conclusion that community-based tourism cannot work. There is every reason to believe that tourism can be of real value to local, rural economies, as will be presented here. Investments must simply be predicated on involving preexisting members of the supply chain, companies that already are attracting tourists to the region.

Thorough statistical analysis has now shown that, "the expansion of tourism may contribute to the economic prosperity of a country, but the social and environmental benefits are not spontaneous. Policies and actions must channel tourism growth into the improvement of the socio-economic conditions of local populations."[19] In fact, tourism growth has not been shown to significantly improve the Human Development Index in a large-scale study looking at three groups of countries in different economic development categories. The conclusion of this research was that tourism "cannot be considered as a tool to significantly improve the living conditions of the population in the least developed countries."[20] Instead, researchers suggest that international organizations assess more "prudently than before whether their investments in tourism will give the desired results."[21]

Funds to create a deeper analytical system upon which all economic development experts can draw have generally been lacking. Lessons learned continue to be put on the shelf without further study. Investments come and go. While sustainability is in every speech, it is very difficult to find good data on how much investments in sustainable tourism have benefitted local economies. How we define the core questions in the future must first relate to legitimate research on the outcomes of investment in sustainable tourism, particularly at the community level where much of the investment has gone and has been lost.

Research on tourism and local economic development

Although research on the relationship between tourism and local economic development was not guiding investments adequately at the donor level, excellent work was being done on the ground, especially in Africa, which has been particularly fertile ground for research on sustainable tourism as a tool for economic development. A pro-poor research program was launched by the Overseas Development Institute (ODI) and the International Institute for Environment and Development (IIED), together with the Centre for Responsible Tourism, to identify how the private sector, NGOs, and donors could spread the distribution of benefits to the poor from tourism's heavy growth in developing countries.[22] Pro-poor researchers found that there are many reasons to invest in the potential of tourism to alleviate poverty:

1 Scope for wide participation from the informal sector of economies
2 Ability to attract consumers to the products offered by the poor
3 Tourism's reliance on natural capital, such as wildlife and scenery, which does not require capitalization and allows easy entry to the business
4 Tourism's proportion of benefits that often go to women[23]

Tourism is understood to be "pro-poor" if it generates net benefits for the poor.[24] From 2000 to 2010, a wide variety of articles were published, providing more and more solid evidence that tourism is a unique economic tool to reach poorer segments of society if appropriate measures are taken.[25]

In *Responsible Tourism, Critical Issues for Conservation and Development*, detailed case studies demonstrate where tourism projects have succeeded in creating real economic contributions to local livelihoods in South Africa, Botswana, and Tanzania, among other countries. These successes are achieved by the private sector working with governments and communities in order to create policies that do result in contributions to rural, poor people.[26] Joint ventures between private wildlife companies and government have had an excellent record. For instance, after the breakup of the apartheid government in South Africa in 1993, many proactive policies were put in place by SANParks, the South African parks agenda, to put titles of the land back into the hands of the local black communities that had been excluded and removed from protected areas in the country. This paved the way for communities to achieve joint ventures, with up-and-coming companies such as Wilderness Safaris, which brought investment and opportunity to local people.[27] In the period of this book's documentation, there was a changing landscape of opportunity, where governments were fine-tuning their programs, and the private sector was increasingly proactive in seeking to help communities to benefit.

Partnerships with tour operators that are dedicated to working with communities is a tested solution that has reduced risks for communities and that has had proven social and environmental benefits. A study of the socioeconomic benefits of camps run by Wilderness Safaris in Malawi, Namibia, and Botswana, with 618 surveys in 25 rural communities, covering 13 different ethnic groups, found that the employment opportunities provided to educated youth in rural areas lessened the likelihood of young people moving to urban areas, encouraged the support for the protection of natural resources, improved access to schools and clinics, and offered better opportunities for education and training locally. The study reminded readers that community projects and ecotourism operations must be aligned with the expectations of the community and "an understanding of the cultural, economic as well as non-economic characteristics of the communities concerned."[28] Another long-term longitudinal study in Botswana demonstrated that communities benefit from safari tourism if they have proper governance systems allowing them to manage tourism on their own terms.

Okavango Delta, Botswana: case example

Botswana is a highly regarded safari tourism destination that receives about 100,000 visitors a year.[29] Historically, foreign companies dominated Botswana's well developed tourism economy, leaving little opportunity for local people, but now the country receives praise for its proactive policies that favor the poor. A study carried out in the Okavango Delta of northwestern Botswana[30] reviewed the specific economic development impacts that have benefitted local communities. Ten years of data was gathered, enabling researchers Mbaiwa and Stronza to measure tourism's long-term economic, social, and political impacts on three villages, over a longer period than most researchers are able to investigate. While exclusive tourist companies once sought to remove the villagers from areas their clients visited in the Delta, in the 1990s the Botswana government gave the land to the villagers to prevent further privatization of wildlife resources by safari companies. This enabled the villages to manage their own destiny and lease the opportunity to operate on their land to experienced safari companies that could work compatibly with local people.

According to the study, the residents recalled when "there were no jobs and poverty was very serious."[31] Now community-based tourism is the primary source of livelihood for 60% of the village residents. They thatch lodges, produce and sell crafts, and lease community land to safari companies in the Delta. This has given families more security and has allowed children to stay in the area. Most are employed as cooks, cleaners, storekeepers, and guides. The main use of their earned income is to buy food, build houses, support aging parents, and pay expenses for children's schooling. Local residents were quoted as saying, "You no longer fear that you might find your people having relocated elsewhere. There are benefits here."[32] Over 93% of households reported an increase in income over the 10-year period of the study.

Collective benefits were also studied as part of a review of how tourism income builds social capital, a form of communal benefit that speaks to how communities make decisions for the long-term benefit of their region. Through a consensus process, the three villages in the study used their community-based tourism organization to agree on funding social services and community development projects for all three villages. There were contentious issues, where at times funds were mismanaged or misappropriated. In response, better governance systems were put in place during the 10-year period, and records are now kept to ensure benefits are more evenly distributed among all families.

While community members once feared they would be expelled from wildlife areas that tourists visited in favor of wealthy safari companies, now there is a new balance and quality of life for local people. This research "contradicts claims by scholars that community conservation and development projects are failing to achieve rural development."[33]

There are reasons the Delta residents have fared so well. According to author Mbaiwa, "there is no doubt that the adoption of ecotourism has increased social capital in many villages in Botswana."[34] In particular, the formation of 91 community-based organizations (CBOs) has given rural communities a structure for managing their affairs. Operating legally as trusts, as of 2007 these CBOs involved 135,000 participating community members, or 10% of the country's population.[35] These trusts act as intermediaries between the government, NGOs, and the communities. They have developed land use management plans for their ecotourism areas, and they manage community commerce, maintain records and financial accounts, and report at annual meetings. The CBOs legally manage the land and have de facto oversight of the wildlife on the land, which represents a vital piece of equity upon which communities can finance their future. The CBOs have minimized unfair revenue distribution, and they have leased rights to operate safaris to experienced companies. Overall, the organization of the system is clearly working on behalf of local people, an excellent state of affairs that could be duplicated.

New metrics to measure sustainable tourism's contribution to economic development

Tourism as a national policy for governments is undergoing rapid change in order to acknowledge the importance it has as a powerful generator of foreign exchange. Many countries are updating and reorienting their plans to develop more cohesive integrated policies in response. Governments have not placed tourism policy makers in the most optimum positions within the cabinet, and policy roles are almost always fragmented between many different government ministries and departments. This undermines the effectiveness of managing tourism as a genuine economic development tool, with too many fingers in the pot and too many decision makers with their own diverging goals. Tourism revenue and economic impacts come from many sources as a result, including border security, aviation, national parks, hotels, tour operators, cruise lines, and all of their suppliers. The complexity of measuring these impacts is legend.

The most widely used economic impact accounting system is the Tourism Satellite Accounting System (TSA), which captures tourism expenditures and consumption. This accounting system, overseen by the UNWTO, requires the review of national statistical tables related to characteristic tourism products, such as hotels, food and beverage, transport by air or by land, cultural services, and sport on domestic and international levels.[36] Gross revenues from members of the supply chain that sell products lower down, such as hotels being sold by tour operators, must be carefully subtracted from the total to avoid double counting. While TSA can be an effective tool, the manual for managing TSA accounting was only completed in 2008, and many countries still do not use the system.[37] Additionally, while TSA has led to many improvements in the monitoring of the economic impacts of tourism, it does not seek to measure net economic impacts, or the net benefit to a host country from international visitors, known as yield.[38] The ability for tourism to become a genuine sustainable economic development tool must be weighed not by gross impacts but by net receipts, and that process is still in its infancy. Reports on how countries reinvest tourism revenues into economic development goals should be the wave of the future, but at present it is exceedingly rare, according to the 2016 OECD review of tourism policies.[39]

The primary duty of government across the world is to license facilities, tax them, and use these funds to identify new products, nurture investment in new tourism products, disperse tourism as widely as possible, and internationally and nationally market and promote tourism for the country.[40] This results in more economic development, in terms of general employment and gross economic development measures of increased GDP. Pressures to increase tourism come from many directions, without reference to the underlying costs to the nation, not least of which is the national demand to offset foreign debt with the rich foreign exchange resources international tourism brings. This fosters a mentality that countries must compete for the foreign exchange benefits that tourism can attract, especially from foreign direct investment.

By and large, tourism policy makers are held to account, not for conserving local resources or cultures as tourism assets for future generations but for increasing market demand and demonstrating strong economic growth statistics. Policy makers worldwide receive considerable pressure to spend more funds on tourism promotions as a means to benefit the tourism economy and help the national economy as a whole. Countries like India, which still have very high poverty rates, see tourism as an excellent strategy for pro-poor development. India seeks to improve its world competitiveness ranking and attract more foreign investment and has spent richly to achieve this with a total tourism economy valued at 6.6% of Indian GDP.[41] But competitiveness rankings, foreign direct investment figures, and even the percentage of tourism's value in the national GDP are poor indicators of how well the benefits of tourism are reaching local populations. And, worse, these figures may be deceiving, as they do not show whether the country is receiving a net benefit from high expenditures to foster growth. Empirical studies of how to link tourism to genuine economic development demonstrate that advance investment in energy and telecommunications infrastructure is required, along with greater attention to health and education facilities.[42]

In China, a country that also seeks to build tourism demand, there is a full awareness that some regions are developing rapidly, while other, highly rural areas have seen less dynamic growth. For instance, in the rural province of Sichuan, where 72% of its total land area is mountainous and poverty rates are high,[43] the regional government has made tourism one of the six industries for fostering new development in the next century. But rather than simply pumping up market demand, the local government seeks to ensure that investment in rural Sichuan is undertaken wisely. Chinese economists concluded that it would be best to improve their infrastructure for tourism and create a master plan before promoting their destination too much.[44] Such decisions show wisdom and an understanding of how to spread the economic benefits of tourism.

One of the barriers to achieving economic development from tourism is the need to succeed in attracting foreign investment, based on competitiveness metrics. Since 2006, the World Economic Forum (WEF) has included travel and tourism in its series of competitiveness reports. "The resulting Travel and Tourism Competitiveness Report provides a platform for multi-stakeholder dialogue to ensure the development of strong and sustainable travel and tourism industries capable of contributing effectively to international economic development."[45]

The tourism competiveness ranking is of special significance to developing countries that do not perform well in these rankings for reasons more likely to relate to high poverty rates, high dependence on agriculture, and highly diverse cultures. This lowers their competitiveness ratings, despite the fact that these countries may have high cultural and biodiversity.

In the 2013 WEF Travel & Tourism Competitiveness Report, the top 26 countries are developed world economies.[46] Barbados achieved the highest ranking of all developing nations, with the rank of 27. United Arab Emirates achieved a 28, Cyprus 29, and Malaysia 34 out of 140 countries rated. The WEF competitiveness ranking system places a high, one-size-fits-all value on globalized infrastructure and business standards, and a low value on social and environmental capital. This leaves highly regarded countries like Rwanda with poor ratings, despite the country's spectacular success at serving the bottom billion (the people at the bottom of the pyramid), while giving the neighboring nation of Kenya higher ratings despite a questionable record in preserving its environmental or social capital.[47]

A competitiveness ranking system may work well for other industrial sectors that do not depend on social and environmental capital to the same degree, but there is an underlying methodological issue with trying to fit this type of metric to measure the benefits of sustainable tourism. The more nations seek to become competitive based on the standards of first world countries, such as Switzerland, the more likely they may be willing to sacrifice the unique cultures and environmental capital they preserve. New visions for determining what represents competitiveness are now called sustainable competitiveness programs. The WEF has incorporated these ideas into its country competitiveness indicators since 2011, using data from outside resources. But, it may be highly difficult for global competitiveness indicators to appropriately reflect whether countries are sustainably managing their resources or responsibly investing in their local populations. New metrics are required that can measure the capital that needs to be preserved and livelihoods that can be improved through pro-poor investments.

Triple bottom line approaches to sustainable tourism development

Pro-poor revisionists have sought to foster a triple bottom line approach to investment in local economies that takes into account societal and ecosystem well-being, also known as livelihood needs. The goal is to foster long-term change and allow local people to access the capital required to do so. Such access depends on building equity while retaining societal well-being, which is often closely related to the conservation of natural environments and the preservation of social cohesion. The livelihood needs of people in developing countries are different from those in well developed economies. More people in developing countries depend on human skills, cash on hand, nature as a source of goods (such as timber) and food, cultural networks, and land upon which people can both reside and retreat. In such societies, there is generally much less dependence on the cash economy and much more reliance on local resources and the sharing of responsibilities to achieve a common good.

It has long been known that ecosystems (the natural capital upon which humans depend) are not factored into the frameworks of global domestic product accounting. As a result ecosystems continue to degrade without proper valuation.[48] Likewise social capital, which refers to the norms that bind communities together and help create well-being, is largely not considered when standard economic development accounting is used. As globalized norms take the place of traditional values, social cohesion is lost. The massive migrations away from agricultural societies in China and India, for example, have reduced the ties between generations – a "modernization cleavage" caused by the rapid pace of growth. Such dramatic transitions can unravel social fabrics and create a loss of common values. Some have suggested that a rapid loss of social capital between generations can cause social unrest, issues that must be resolved as part of economic development strategies.[49]

The social and natural forms of capital serve as psychologically profound touch points for humans, which can be linked to well-being and better human health. Leading environmental economists found that the preservation of the natural environment and social capital are statistically linked to measures of human development and well-being.[50] Tourism projects that succeed at preserving nature and social capital create balanced, happier societies. Conversely, the destruction of nature and community-based values leads to the breakup of societies, conflict, and an increasingly toxic social environment that is often marred by internal jealousies and a sense that there is no common goal for tourism development.

Tourism is a highly unusual economic development tool that is well suited to preserving both environmental and social capital, but it is extremely easy to distort its outcomes and create an environment where cooperation breaks down and every individual seeks to gain from each tourist. But tourism has special characteristics that can be fostered to allow it to remain more in balance with sustainability goals. Travelers around the world seek out nature and beauty to find respite from our increasingly urbanized world. And the ever growing phenomenon of seeking internal peace through yoga and other mindfulness practices indicates that humanity is in real need of retreats where they feel renewed, not only physically but within their busy minds. Travelers consistently report after their journey that the most refreshing experience is interaction with local people or the opportunity to be in a pristine environment. Extensive research on the physical and psychological benefits of being in forests strongly indicate that there are positive impacts on the immune system and perceived health and mood states are improved, but these studies still are still in their infancy.[51] It is likely that in future it will be possible to statistically link how travel and tourism maintains not only natural and social capital but human health and mental well-being in regions that will be treasured for their ability to create happiness, well-being, and inner peace.

Livelihoods analysis framework for balanced growth

As we have seen, while tourism grows rapidly in the developing world, mainstream economic success measures rarely reveal how the industry can leverage and protect social or environmental capital. There is every chance that tourism can be extremely effective at achieving more balanced, equitable economies if fostered with well thought-out methodologies. The United Nations Conference on Trade and Development (UNCTAD) affirms that the tourism value chain incorporates many sectors of the economy across the board from telecommunications and retail to health services. It builds the construction industry and fosters the growth of hospitals. These linkages remain underexploited, and most of the value is still captured by foreign investors, tour operators, and airlines.[52] Yet regulatory frameworks can stimulate private investment in local supply chains. Public–private strategies for investment are the key.

The pro-poor tourism research undertaken by ODI and the IIED in the UK advanced a global effort to measure the outcomes of tourism on poor societies in the 2000–2010 period. They used a system of assessing the livelihood needs of a community, which are much broader than simple economic measures and can be considered the building blocks of well-being. They are categorized as follows:

1 Financial capital – cash on hand or that can be borrowed
2 Human capital – the skills base of individuals or groups
3 Social capital – the social cohesion of a group and the strength of its networks[53]
4 Natural and cultural capital – the resources of the environment available, such as water resources, forest, land, and wildlife, and historic buildings and traditional lifestyles

Livelihood improvements can be categorized under each of these categories to ascertain whether a holistic approach is being taken. This framework, which will be now outlined, offers a methodology to review potential tourism projects with goals to measure financial, environmental, and societal benefits from tourism and their connection to well-being.[54]

1 Financial capital: has tourism contributed to the expansion of local business opportunities?

Economic well-being comes from direct participation in tourism, such as working as guides or transport operators. But these types of business opportunities are often fewer and less economically profitable than participating in key tourism supply chain sectors such as food, beverage, construction, furnishings, and services such as gardening, floristry, and laundry.[55] Pro-poor researchers have found that the poor may earn more from the tourism supply chain than from direct participation in tourism businesses and that they may also have greater opportunities to build and grow their businesses and incomes by working in the supply chain: "Poor people can operate as micro entrepreneurs, selling direct to hotels, restaurants and operators or they may have unskilled jobs in larger companies in those supply sectors."[56] This varies enormously by destination, depending in particular on the proportion of food that is sourced domestically, as food and beverage are a major component of tourism supply chains.[57]

Angkor Wat, Cambodia: case example

According to international hotel groups, in the Angkor Wat region of Cambodia in 2005, 95% of all foodstuffs were imported into Cambodia from Vietnam and Thailand. Locals thus lost important business opportunities. All businesses agreed that buying more local products and services would be an important contribution to the economic development of communities in the country.[58]

All hoteliers interviewed agreed that they would be willing to buy more local fresh fruits and vegetables in Cambodia, if a coordinated effort were put together to make Cambodian produce available in the quantity and quality they need. It was proposed by one hotel that a group of hotels in Siem Reap task their purchasing managers to set up a cartel for purchasing local produce in bulk. All interviewees agreed that coordinating on purchasing local produce could be a highly beneficial, cost-saving initiative for the hotels. One small hotel stated its commitment to building a small farm that would be established specifically to supply its own kitchen, and to meet the quality standards the hotel required not only for fruits and vegetables, but also for meats.

The United Nations Development Program (UNDP) has studied the potential for supporting the development of an agricultural supply chain in Cambodia that would meet the hotel industry's needs and has presented recommendations on how to proceed.[59] Several such initiatives have already failed. But the overwhelmingly positive interest in creating a more dynamic link between tourism and agriculture suggests that further investigations and pilot projects are worthy of priority consideration by the nation of Cambodia.[60]

Scholars have highlighted the importance of finding ways to increase the use of local food produce to maximize the benefits of tourism to local communities. A study on the constraints to achieving better linkages to the tourism economy in Siem Reap found that small plot farmers lacked adequate farm equipment, skilled labor, and storage to produce food in the quantities required by restaurants and hotels in the region. Leading farmer association were unable to compete with imports in terms of quantity, quality, and appearance. Farmers found that they lacked communication with hotel and restaurant buyers and could meet demand if given the opportunity. Hotel managers agreed but suggested farmers needed improved coordination through their associations. Financial capital to scale up was also lacking for the farmers, who also required technical assistance in the form of agricultural extension, which the government was not offering.[61] Such problems could be greatly assisted by donor investment, with support to both farmers and hotels. The GIZ development agency of Germany's researchers agree that support to food supply chains for hotels would "constitute a profitable market with no less than four winners: the hotels, the farmers, the local populace, and the environment."[62]

Empirical studies suggest that, at best, between one-fifth and one-third of total tourist expenditures in destinations are captured by the poor from direct earnings and supply chains. The impact on poverty depends on the employment, skill levels, pricing, and ownership, but most evidence points to the importance of fostering local tourism supply chains as a highly effective means of improving local benefits.[63] South Africa is one of the leaders in this field. In the 2015–2016 budget of South Africa Tourism's Department of Social Responsibility $17.8 million was allocated for skill development and community-based tourism initiatives at the municipal level.[64] In future, there needs to be far greater reinvestment of tourism revenues into the enabling environment for local businesses, especially in rural areas to ensure tourism can be a genuine tool for economic development.

2 Human capital: are tourism business and employment opportunities reaching new segments of the population?

Nearly all capital is held largely in the cities with the economic elite, and the marginalization and lack of attention to rural areas are a significant problem. As a result, the motivated, socially responsible private sector companies, based in cities around the world, have difficulty successfully investing in rural areas because the potential for yields are reduced, often due to poor infrastructure and the lack of capital availability for rural development projects. Investment in local, socially responsible companies that seek to develop in rural areas of developing countries is an excellent system for reaching new segments of the population.

I led an International Finance Corporation (IFC) study that found that traditional loans are often unavailable to companies that seek to work in remote rural areas, partly because the lag time between start-up and profitability can be longer than for businesses located in urban areas.[65] The resulting report recommended providing incentives, loans, and grants for successful ecolodges to undertake expansions, including such case examples as Sukau Rainforest Lodge in Malaysia, Chalalan in Bolivia, Kapawi in Ecuador, and Lapa Rios in Costa Rica.[66] It was found that these lodges had done exceedingly well at spreading the economic development benefits from their enterprises and had the potential to do even more.[67] No investment was ever made in this strategy, even though the required funds were small by IFC standards. The $80 million in IFC financing given to large-scale Malaysian business interests to build luxury hotels in Myanmar (Burma), reported on in the press in 2016, raises questions about how committed the IFC is to using tourism as a tool for alleviating poverty.[68] This is a shame because in fact such donor investments are truly needed to assist with the required enabling environment for tourism in poor rural areas and for loans to triple bottom line businesses to genuinely build human capital. Such methodologies for enhancing the economic development benefits of tourism are fully validated by IFC's own studies.

2A Human capital: has tourism provided greater opportunity for women?

The tourism industry offers genuine opportunities for the advancement of women. The industry's flexible work hours, low educational requirements, and many entry-level jobs open doors for women who have not yet joined the globalized economy or ranks of the educated workforce. According to a UN study, the primary occupations occupied by women in the tourism industry include catering, waitressing, maids, babysitters, cleaners, housekeeping, launderers, and dry cleaning. Ninety percent of these jobs are filled by women.[69] While these jobs are typically poorly paid, they are often an improvement over livelihoods in rural areas where subsistence agriculture, caring for children, and housekeeping are the only options for women who generally lack access to cash or capital of any kind. According to the World Bank, women are an estimated 70% of the world's poor.[70]

Once poor women enter the tourism economy, they do face problems moving upward in business dominated by men and do not have wage equity or the same benefits as men. Women in tourism generally earn about 80% of a male's wage.[71] Notwithstanding these disadvantages, there has been a broad increase in the participation of women in the tourism industry worldwide.[72] In Sub-Saharan Africa, for instance, women are 10% more likely to be the "boss" in food and beverage operations than in other industries.[73] A review of how tourism enterprises benefit the poor in Malaysia confirms that restaurants have over double the likelihood of benefitting the poor than hotels, retail establishments, or tours and excursions – and this may well be because women are more likely to run restaurants.[74]

In regions of the world where community-based tourism projects are being supported by donors, women are often likely to benefit because donors stress equal treatment and screen for projects that benefit women. In these cases, women can move up in the ranks, become equal partners in tourism development projects, gain more financial independence, and invest more in their children's education.[75] But even with the benefit of substantial investment from donors, women are not always beneficiaries.

Gunung National Park, Indonesia: case example

The island of Lombok, Indonesia, promoted itself as a destination, attracting visitors with the idea that it would be less touristic and more authentic than some of Indonesia's more popular destinations. Offering unique wildlife, birds, and trekking opportunities, the island's Gunung National park was seen as a worthy investment by New Zealand Aid (NZAID). Between 1999 and 2005, NZAID spent $3 million on ecotourism development in the Gunung region, with the goal of fostering community development on park boundaries by bringing benefits to rural women and men. After extensive consultation, the program focused on an ecotourism trekking program. However, this decision to develop a trekking project unintentionally weakened women's chances to benefit from the investment. Within short order, the project came under the control of a small number of men.

Guiding is traditionally seen in Indonesian culture as a male activity, which often excludes opportunities for women. The project trained women in "soft trekking" with the idea that they could provide tours of traditional houses and rice fields. But according to the project reviews, these tours were not adequately promoted by the men. Women also reported that the male tour operators did not give the same percentage of the tips due to them that they gave automatically to the men. In frustration, the female trek leaders discontinued guiding. NZAID gave a grant to women to make snacks for trekking groups in compensation, but the trekking agencies already purchased all their food in town and had no time to purchase near the trails where the women resided. The snack business failed, and the women returned to agriculture. They reported a feeling of prejudice against them, which included issues related to gender, class, and ethnic differences. Husbands became angered by the snack business, and they complained that the women were neglecting their child care duties.[76]

There are barriers to women joining the tourism work force in many parts of the world and also some risks for women and their families who may have little experience in the outside world. Religious and cultural norms may prevent women from leaving the home for work. Homestays may call upon women to juggle more duties in the home than men, at times without compensation. And young women are easily influenced and may even leave home or ignore their parent's wishes if they come under the influence of outside norms, particularly if men find them attractive and encourage them to do so. There is the danger that such young women can be misled and even be accosted under the worst of circumstances. In my work, I have had to carefully screen for the potential of unwanted sexual encounters in villages accepting visitors, a problem I see routinely overlooked or even ignored. Backpackers or foreign men may see innocent encounters and even romantic ones as all in good fun, but the local traditional families may not.

These barriers and risks to women's inclusion in the tourism workforce should not preclude participation. For example, home-based arts and crafts are a particularly fertile area to foster women's business skills and involvement. Although artisan work is being abandoned in societies around the world, travelers genuinely value locally made goods, more than ever around the world. There are numerous cases worldwide of women benefitting from this market. In Panama, the Kuna women living in the San Blas islands have been producing embroidered *molas* (beautiful panels of embroidered animals and plants using brightly colored fabrics) for generations, often earning more than men. I have observed how this raised women's status in Kuna society. In Peru, women produce world-famous textiles, which have given them economic security. Similarly, after the arrival of tourists in the Okovengo Delta increased demand for traditional baskets, the women of the Delta organized the production, sale, and marketing of the baskets by forming a collective. NGOs have provided them with training in bookkeeping, public relations, and English, as well as in sustainable management of natural resources that they use for basket making. With these types of simple interventions, women have become more self-sufficient, and basket making has become an important part of the local economy. Women use the funds they earn to buy food, school uniforms for children, and clothing for the families.[77] Women in Gunung National Park have similarly had success with weaving, which has "emancipated the women."[78] Women who never left the village now speak to outsiders and raise funds for their husbands' farming activities for seeds and fertilizers. Husbands who used to become angry if their wives did not make breakfast now care for themselves without complaint.[79]

Many cultural questions need to be asked, and cultural mores must be carefully considered as part of the process of working with women in traditional societies. The most important part of working with social capital and cultural change is to rely on local experts who have a deep understanding of local cultural mores and how they are evolving over time, as no culture is in a static state.

3 Social capital: have collective benefits to the communities been enhanced?

The study of collective benefits shifts the focus away from specific benefits to individual workers and rather offers a view of how much tourism development generates change at the community and societal levels. Understanding collective benefits helps us understand whether tourism is contributing to the overall well-being of local people. Social capital can be measured via opportunities for locals that derive from tourism such as:

- Access to information from visitors,
- Elevated status in the community and with policy makers,
- Market opportunities, livelihood options,
- Reinforcement of cultural pride,
- Reinforcement of community norms and values.

In many parts of the developing world, actors from the private sector have worked with nontraditional partners, including NGOs and communities, to successfully build profitable enterprises that generate funds for these local communities. Private investment in local social capital is particularly useful because once the private sector has invested in partnering with local communities, they have a long-term commitment to make this work.

In Peru, the award-winning company Rainforest Expeditions partnered with the Ese'Eja indigenous people of the Amazonian Tambopata region to create an enterprise that was 60% owned by the community and 40% by the company. This has proven to be a success for both parties, with $2 million in net revenues paid out to the community as of 2013, and a full transfer of ownership to the community destined to transpire in 2019. Research has shown that community ownership of tourism does reinforce social capital in Amazonia, where local people spend their newfound wealth in important social health and education programs, which are sorely lacking in the region.[80] Similar community-based tourism projects, with co-ownership strategies, have been especially common in Amazonia, particularly Ecuador, where dynamic indigenous federations have partnered with tourism businesses to develop large-scale tourism projects in remote regions. The process of developing these projects has enabled outside researchers to track the human and social capital investment this represents.

Achuar people and Kapawi Ecolodge: case example

The Kapawi ecolodge in Ecuador is an example of how tourism can be used to foster economic development while maintaining and even enhancing social capital. Built as a partnership between a large Ecuadorian travel company, Canodros, and the local Achuar indigenous federation, the lodge was constructed only after the development team conducted an extensive evaluation of the societal values of the Achuar. The Achuar live in a vast rain forest territory on the southern border of Ecuador, which is almost entirely roadless with little access to the outside world except by bush plane, and they have lived a traditional life based on hunting and gathering in the rain forest before ecotourism came, except for significant influence from missionaries. Social and environmental scientists evaluated the Achuar worldview, capturing their social, cultural, and environmental vision of the cosmos. I personally visited this site several times and was able to work with the original scientists to come to an understanding of the Kapawi socioeconomic and cultural world.

When visiting, we were introduced to the Achuar culture, which is unforgettably haunting. They recount their dreams every morning as part of their rituals and also share *chicha* liquor made by women who masticate yucca to give them strength – a source of resilience for activities throughout the day. Kapawi was built on respect and even honor of the Achuar's way of being in the rain forest even though there are vast geographic, social, and cultural differences between the entrepreneurs from Guayaquil who created the vision for the lodge and the structure of society in the Amazon rain forest. (See Figure 3.1.) A unique partnership was built that connected the most traditional rainforest society of 7,000 people in 64 communities to the bustling, mainstream city of Guayaquil where Conodros is located.[81] Researcher Arnaldo Rodriguez writes:

> When ecotourism projects are developed in consultation with host communities there are several mutual benefits. Tour operators gain access to local villages or remote areas. Local people derive income from hosting visitors while

FIGURE 3.1 Kapawi Achuar guide, Ecuador[82]

> elders pass on cultural knowledge to ecotourists. The benefits of ecotourism for rural or indigenous communities include preservation of cultural traditions, conservation of the natural environment, and maintenance of social, cultural, and spiritual values. In remote areas with limited development, ecotourism ventures can improve the quality of life, self-esteem, and well-being of local and indigenous communities.[83]

Work with indigenous people of the Amazon and across the world has fostered an awareness that exchanging worldviews with these disappearing cultures is valuable for both the host and guest. Psychology researchers point out that our reliance on small samples of Westernized subjects to understand the human psyche has left out crucial samples of mankind, particularly indigenous people who have evolved with highly different mores and values. These researchers concluded that Western psychology "lacks essential information about the evolutionary history of our species and the potential impact of (bio)diverse environments on our psychology."[84] The unique cultural consciousness of Amazonia is part of our evolutionary history and has created an adaptation pattern that is highly different from that of other humans. Exposure to this knowledge brings benefits to all who visit, builds global social capital, and changes personal perceptions about the role of nature in our lives.

3A Social capital: has tourism improved access to information and allowed for more participation within communities?

While the rights of communities to reject outside development is a critical golden rule that must be considered by any community, most locales in the world do chose economic integration with global commerce. One of the advantages of choosing integration is access to information, which helps local societies to become less marginalized. Information offers innumerable new opportunities to join community, municipal, and civil society processes. Tourism can be evaluated by the number of "strategic alliances" that the community has gained as a result of its involvement in tourism, either with the private sector, with NGOs, with other communities, or via civil society associations or indigenous federations, as has just been chronicled for the Oriente (East) of Ecuador.

Throughout the world rural, poor communities lack the social capital to directly influence their governmental leaders, but access to tourism development and interactions with visitors raise the level of empowerment that local communities experience. In my work as the executive director of the Planeterra Foundation from 2010 to 2012, I was able to observe this directly while preparing for a new project in the Sacred Valley of Peru.[85]

Huchuy Cosco, Peru: case example

As the leader of the design team, working in cooperation with the Multilateral Investment Fund (MIF) of the Inter-American Development Bank, I visited the village of Huchuy Cosco in Peru several times in 2011–2012 to determine whether a community-owned restaurant might be located there. The community leadership was deeply moved that we sought to locate our project in their modest, traditional village of 65 families, most of whom were farmers. They spoke of their remarkable journey. They had only 30 years before lived as serfs, farming for a share of produce for the large hacienda in the region. Now they had legal rights to manage their land due to land reform in Peru in the 1970s and had moved closer to the road to create the village we were visiting. At the time of the official launch of the partnership between Planeterra and the village, each elder spoke of their painful memories of what was virtual enslavement and of the joy they felt that they now had the right to create their own business and make their own business decisions.

The tourism economy flowed like a raging river of travelers heading straight toward Macchu Picchu, within meters of Huchuy Cosco, and yet left it isolated. Our team designed the project to be a "farm-to-table" program that would connect the supply chain of the large tour operator G Adventures to this community-owned restaurant. We knew from the beginning it was going to be an enormous success, given that the tour operator managed some 10,000 passengers yearly in the region and could decide annually how many might stop at the restaurant as part of their Sacred Valley tour. The community received MIF grants and MIF-financed training to build, equip, and run their own restaurant and even learned cutting-edge cuisine, based on native traditional ingredients. They were trained by a leading chef who moved to the village, whom I worked hard to recruit. The local farmers were given support and training to produce vegetables, quail eggs, quinoa, coca, and *kiwicha* (an herb) of the quality required by the restaurant. The restaurant turned a profit margin after six months of operation, and in 2014 was receiving 1,000–2,000 guests per month, with revenue in the first year of $200,000.[86]

These once landless peasants have become successful business owners with the empowerment they need to interact knowledgeably within their own social and governmental systems. This village is now part of an economy that is based on their culture and their cuisine, and they have received the information, training, and knowledge to operate on their own terms.

4 Natural capital: to what extent has tourism contributed to the cost of protecting and managing natural areas?

The tourism industry as a whole receives many more benefits from natural areas than the industry financially provides to preserve and maintain them.[87] Throughout this book, we will demonstrate that while various forms of natural capital, such as coral reefs, are essential assets that drive revenue generation in the tourism industry, very little is done by mainstream tourism to contribute to the protection of our earth's natural assets, especially when it comes to hard cash.

A small but dedicated industry niche, which calls itself either ecotourism or conservation tourism, contributes greatly to conservation. And when they do, the gain for conservation is substantial. Many case studies provide solid data that tourism has the potential to support more of the earth's wild lands, if a dedicated effort and appropriate changes are applied. A great deal has been written about this topic, with the best overview presently available in the excellent book, *Conservation Tourism* (2010) by Ralf Buckley. His research shows that the net contribution of tourism to conservation is still unknown.[88] But the case studies in his book alone continue to build a conclusive body of work that demonstrates net positive contributions from ecotourism to protected areas.

Though there are no international databases to record tourism's contribution to conservation or even of how much tourism contributes in fees to parks and protected areas annually, a recent Cambridge University study found that protected areas generate $600 billion in tourism revenue annually worldwide, based on an examination of the records of 556 parks in 51 countries from 1998–2007. The study's authors concluded that the money spent conserving our wild areas is "grossly insufficient," particularly when considering the many benefits that protected areas provide and the revenues they generate for nations. The study recommends substantially increasing investment in protected area maintenance and expansion, which could generate significant economic returns.[89] As of 2013, IUCN reports that 40% of protected areas worldwide have either basic management with major deficiencies or inadequate management.[90] It is without doubt that tourism is one of the few sources of revenue to bolster park budgets to protect natural resources and wildlife. But a global effort to boost these revenues is still not transpiring.

The two main tools used in conservation efforts include private reserves, such as the pioneering Londolozi Reserve in South Africa and parks and protected areas (PA). While private reserves are one of the most effective tools for achieving conservation and the protection of natural areas, and possibly the most important step that can be taken by lodges in natural areas that seek to support conservation, they are a good idea only if local owners or communal land users are not being disenfranchised from their traditional lands in the process.[91] Community partnerships to manage natural areas are an absolute must if people still reside on the land. For instance, in Namibia, Botswana, and South Africa, the contribution of private and community landowners to the expansion of conservation lands has been "phenomenal."[92] While public reserves represent approximately 6% of the land base of South Africa, private reserves now represent 18% of the land base, triple what the government preserves. These private reserves make significant contributions to the biodiversity conservation of the region, but this does raise questions about how privatized conservation lands affect local people.[93] Are private reserves causing the redistribution of land to well-off outsiders who seek not only to conserve land but to reserve it for the 1% wealthiest members of our global economy? This may be true in certain locales, but in Namibia, 5% of the total land base is in community conservancies. These conservancies generated 7% of the GDP of Namibia in 2004 from tourism revenues, all going to local villages, both from non-consumptive (wildlife viewing) and consumptive (hunting) tourism.[94]

4A Natural capital: are there initiatives to manage tourism impacts via land-use and environmental management regulation in buffer zones and in all areas outside protected areas?

While some attention is generally paid to visitor impacts on areas within the borders of protected areas, startlingly little attention has been paid to the management of visitor impacts outside the borders of protected areas. Land use by both private landowners and commercial developers outside protected areas can have a devastating effect, and this must be tracked in order to understand more broadly the impacts of tourism on natural areas. As discussed, tourism does not generate economic development benefits unless specific investments are made in required infrastructure for the poor. And there must be well laid-out plans to protect natural capital in buffer zones. The case of Tela in Honduras provides a cautionary tale of how poor planning can lead to a wide variety of debilitating economic outcomes for local communities.

Tela, Honduras: case example

In 2005, the government of Honduras presented a national strategy for tourism that gave high priority to tourism as an economic development engine with donor support of $174 million for its implementation between 2006 and 2016. The goal was to maximize the contribution of tourism to economic prosperity via securing reliable growth, maximizing visitor spending, and reducing leakages. The plans highlighted three growth areas: Copan and the Bay Islands where mainstream tourism was already in place, and Tela, a previously undiscovered coastal area for tourism where the Afro-indigenous Garifuna people have resided in relative isolation for 200 years. The Tela Bay Project, supported by national loans, was developed according to an official master plan to occupy more than 300 hectares along 3.2 kilometers of coastlines, with up to 4- to 5-star hotels, 168 residential villas, a mall, casino, tennis, and equestrian centers, and an 18-hole golf course. I arrived there in 2005 to perform an ecotourism evaluation along the coast.[95]

The planned Tela development was just 60 miles from the airport along an undeveloped coastal road that passed through an important protected area corridor with a coastal estuary and highland park with low-density populations and high biodiversity. Despite the high probability that this region would be affected by intense pressures for development along a major corridor between the resort and the airport, no land-use plan was in place to protect this locale. I was asked by my USAID contractors to evaluate the prospects for sustainable development in this area from ecotourism.

I reviewed the local businesses in the region but also reviewed the development plans for Tela, which would have long-term impacts on any small-scale business plan for tourism in the region. I confirmed there was "a limitation" in the budget for the national tourism program that would prevent any further investment in Tela buffer zones.[96] While I was there, a mild tropical storm caused flooding throughout the entire corridor, wiped out the low-lying airport road for days, caused the collapse of bridges, and generally demonstrated how this corridor was likely to be deeply affected by climate change. (See Figure 3.2.) In spite of these obvious

FIGURE 3.2 Airport road to Tela, Honduras, flooded in minor tropical storm[97]

problems, in 2007 the Honduran government announced that the country planned to build between 8,000 and 10,000 rooms in this vulnerable corridor and called it an ecobeach region because of its proximity to national parks.[98] The lack of protection of important corridors between locations where tourism is growing causes sprawl, skyrocketing land prices, destruction of vulnerable natural resources, and the displacement of local people. This Tela boondoggle could have been prevented via regional planning efforts. (See Chapter 8.)

Such efforts to develop tourism on a grand scale with public funds open the possibility of corruption and misuse of the public purse, a terrible loss of a genuine opportunity for local people.

Negative economic consequences of tourism

Economic Dependence

As tourism increasingly dominates emerging countries' foreign exports, there is imminent peril that entire countries will become too dependent on tourism. Throughout the world, tourism is now on the main stage of the global economy with 1 in 11 jobs, 1.4 trillion in exports, and over 1 billion international travelers expected annually in emerging economies by 2030.[99] It offers a leg up and a real chance for the advancement of poor countries. But there are dangers. Small island states are seeing their economies buffeted by global competition and diminishing natural resources. Fisheries, once the mainstay of island economies, are under severe threat worldwide. Tourism is often the only vibrant and growing source of hard cash coming across their borders, and they are increasingly desperate for the deals required to make more tourism companies take an interest. But once tourism becomes more than 25% of an export economy, island states are on the road to dependency.

Caribbean nations once exported their bananas and sugar successfully to Europe with trade protection that helped them maintain a strong foothold in these markets. In the mid-2000s, these benefits were discontinued due to strong sentiments that they were creating artificial incentives to grow uncompetitive crops.[100] This has forced Caribbean nations to rethink their economies and turn even more to tourism development. Countries such as Jamaica found themselves in serious debt, with 55% of all government spending going to paying the national debt by 2012. But despite this, they gave tax relief and concessions to tourism operations to stay competitive, undermining future prospects of generating revenue for the common good from tourism.[101] This is clearly undermining the well-being of their citizens.

What's more, island nations, such as Jamaica, are so dependent on natural resources, pristine coastal areas, and well preserved coral reefs that a lack of investment in conservation can be disastrous. The threat of climate change, warming of the seas, and possible frequent storms and hurricanes puts island tourism at risk in scenarios that will likely force many coastal properties to revamp or go "under" literally. Most Caribbean islands also suffer from rising costs of energy, which make them increasingly uncompetitive with coastal areas with easier access to fossil fuels, such as the coast of Mexico, which has long offered highly competitive pricing compared to the islands of the Caribbean. Embattled island states must push to lower prices undermining all prospects of achieving economic development from tourism. Such dependencies are caused by a lack of net proceeds from tourism, often called economic leakage.

Economic leakage

According to the United Nations Environment Programme (UNEP), of the $100 spent on a vacation tour by a tourist in a developed country, only $5 remains in a developing country economy. Leakage is the money spent on a vacation that is captured by international, not local markets. There are two types of leakage. Import leakage occurs when equipment, food, and products are purchased overseas to meet the requirements of the tourism trade. Export leakage occurs when visitors spend a percentage of their dollars on board a cruise ship, on a tour, or within a resort that is foreign owned and does not repatriate the funds within the destination.[102] To calculate leakage, gross revenues must be calculated and the total revenue that is diverted to global markets or leaves in the hands of foreign vendors can then be presented as a percentage of gross that does not remain in the local economy.

Leakage has been investigated for decades and has long been understood as one of the negative impacts of tourism. But some of the standard assumptions about what constitutes leakage are being questioned. For example, it is rare that local tourism offerings can survive without a global supply chain. Efforts to close out foreign business is self-defeating, leaving local businesses not only without international markets but without foreign direct investment. While donors for decades have mistakenly sought to invest 100% in local community tourism projects, UNCTAD and other agencies have made it clear that this is not the road to success. Tourism is a highly interconnected, global supply chain that requires international overhead costs to be built in and numerous global partners to be involved. What in fact needs to transpire is that donors and governments need to encourage linkages to global value chains and "backward linkages" into local economies by ensuring that there is adequate finance for local enterprises to connect to global enterprises.[103]

Leakage, as defined by UNEP, is a somewhat antiquated indicator of how much a local economy is losing benefits because of the costs of managing foreign supply chains. Instead, advocates for local benefits should be avidly attracting investment by foreign partners to make tourism viable in emerging economies but be certain it is taxed properly to ensure that a good share of the total revenue and local benefits are generated locally. GIZ's director of sustainable tourism suggested that All-Inclusive resorts in the Dominican Republic generates the equivalent of 80 small or medium-sized tourism enterprises, and salaries worth approximately $300,000–$500,000 per hectare of land used, much higher than other industries.[104] Even research supported by Tourism Concern, a long-time opponent of All-Inclusives, suggests that in fact All-Inclusives in the Caribbean offer a steadier source of employment than other types of hotels, which is a highly valued benefit in a tourism-dependent economy that often has long off-seasons with high unemployment.[105]

But the All-Inclusive hotel economy may be avoiding tax to a much greater degree than local businesses. Because All-Inclusive operations can shelter a majority of operating revenues or transfer it to tax havens, these businesses are likely to pay a much lower percentage of tax to local municipalities than local tourism businesses.[106] The investigation of how the tourism business is paying tax and to what extent it is being sheltered, to the detriment of local needs, is a highly overlooked question that needs much more investigation in future.

The most valuable way to move nations ahead without fostering a "blame game" is to study how to generate benefits equitably throughout the tourism supply chain. This is called value chain analysis and is increasingly chosen by development banks as the best means of understanding how to bring business and local destinations together to find approaches that help the local economy to prosper while giving international markets a fair return on their investment. Value chain analysis is typically performed in three stages:

1 Mapping industry processing chains graphically and quantitatively by disaggregating components along various segments of each chain and looking at each segment's performance separately, in terms of its economic performance
2 Establishing benchmark indicators for comparing domestic performance against international performance to understand how much the domestic performance is enabling the overall performance of the product vs. the international performance
3 Assessing the relative performance of the different links in the value chain and prioritizing binding constraints on garnering more domestic gains for local producers[107]

When stakeholders look at the graphic and quantitative maps of their supply chains, it can reduce polarization and bring parties together to discuss how to improve benefits and performance on both sides. For example, value chain analysis can discover inefficiencies in transport or problems with bureaucratic permitting procedures. The results given in an objective manner can foster commitment to improving transport systems or smoothing out government permitting procedures. Conversely, if it is found that permitting processing is fair and easy and transport efficient, international businesses may need to be prepared to review their own procedures and look at possible prejudices within their corporate structures against giving local government and business services a fair shake.

Large companies can achieve lower costs and higher value especially in tourism if they take advantage of local talent before assuming they need to "bring in their own systems and people." And governments need to consider how they can achieve greater cooperation from international business not only by giving tax incentives, which lowers government revenues, but by offering green incentives for public private cooperation to create efficient waste, water, sewage, and energy systems for tourism development. Such infrastructure is, according to all research, essential for achieving economic development results from tourism that is sustainable.

Conclusion

Tourism is an interconnected industry that survives via highly complex supply chains that offer local attractions through a wide variety of buyers, many of which are increasingly digital. This highly international service economy is so interdependent that there is no useful way to invest in the supplier at the local level without also engaging international buyers in the economic development strategy. Time and again, donors seek to finance the supply on the local level, such as community lodges or homestays in poor villages, instead of creating locally owned supply chains that link upward to gateway cities through triple bottom line businesses and link downward to local pro-poor vendors, who often are women and indigenous people who not only need the opportunity but who are often guardians of local culture and ecosystems. The improvement of local livelihoods through tourism must be done through the supply chain and not in isolation.

The common assumption that tourism development will automatically improve local livelihoods must be reevaluated. A new set of tools to evaluate net benefits or yields from international tourism to local coffers is urgently needed. Excellent research points the way to the essential infrastructure that is required, both in health and social services and in green development that must be factored in to ensure local economic development from tourism transpires. And the protection of natural capital cannot be left solely to large protected areas, leaving buffer zones to suffer the consequences of unmitigated sprawl and poor planning. While countries in the past have assumed that tourism is a glamour industry that needs few services to succeed, this is wrong and cannot be justified. Nor can donors fully justify providing large loans to upscale hotel projects being fostered by large business interests, rather than focusing their funds on loans to socially and environmentally sound businesses that can scale up. National policies that are the result of increasing dependence on tax incentives are a product of desperation and will not help nations to improve the economic development results of tourism.

New metrics that capture more than the ratio of economic GDP and employment that tourism delivers will help governments to place greater value on local services to tourists in their statistical systems. Livelihood analysis can become an important separate rating that provides much needed context to competitiveness rankings. Financial, human, natural, cultural, and social capital is all essential for sustainable tourism and cannot be undervalued in the game to win more foreign direct investment.

To achieve economic development from tourism that is focused on underserved populations, more funds must go from tax coffers directly into the local economy to build the capacity to offer a wide variety of local services, while ensuring there is synergy and buy-in from regional and international markets. Local indigenous people, for one thing, have great potential to offer more services for the tourism industry with their unique cultural knowledge, crafts, native crops, and the ecosystems they help protect being of vast import. Local farmers have the capacity to deliver healthy and organic foods on a vast scale yet to be explored. Women can take more responsibility for local micro, small, and medium-sized enterprises that serve tourists and at the same time be their own bosses. And local small business services of all kinds can be boosted in ways that improve the economic development benefits of tourism in a more equitable way that will allow local people to live with just access to the same benefits that tourists enjoy. New livelihood investments need to be prudently researched first in order to be certain that investments are being made in the areas where real gaps in local capacity and infrastructure exist. And with the appropriate measurement of the actual ground-up value of tourism, not the top-down impacts, there can be a genuine enhancement of tourism products, with more local value added, while ensuring that local commerce is genuinely connected to the international tourism trade.

Notes

1 The World Factbook (Central Intelligence Agency), Bangladesh, https://cia.gov/library/publications/the-world-factbook/geos/bg.html, Accessed September 13, 2016
2 EplerWood International, 2009, *Teknaf Peninsula Community-Based Ecotourism Strategy*, International Resources Group, USAID Bangladesh, https://academia.edu/23667453/Teknaf_Peninsula_Community-Based_Ecotourism_Strategy_for_Bangladesh, Accessed March 24, 2016
3 Epler Wood, Megan, January 17–February 8, 2009, *Field Trip Report, Integrated Protected Area Co-Management, Bangladesh*, Unpublished, EplerWood International for IRG, Washington, DC
4 Christie, Iain, Fernandes, Eneida, Messerli, Hannah, and Twining-Ward, Louise, 2013, *Tourism in Africa: Harnessing Tourism for Growth and Improved Livelihoods*, The World Bank, Washington, DC, page 1
5 Epler Wood, M. 2007, The Role of Sustainable Tourism in International Development: Prospects for Economic Growth, Alleviation of Poverty and Environmental Conservation, in *Critical Issues in Ecotourism*, ed. James Higham, Butterworth-Heinemann, Burlington, MA, page 159
6 UNWTO, 2014, *UNWTO Tourism Highlights*, Madrid, Spain
7 Christie et al., *Tourism in Africa*, pages 24–25
8 Ibid., page 17
9 UNWTO, 2015, *Tourism and the Sustainable Development Goals*, UNWTO, Madrid, Spain
10 OECD, 2016, *OECD Tourism Trends and Policies 2016*, OECD Publishing, Paris, Box 1.4, page 11
11 Towards Earth Summit 2, n.d., Sustainable tourism briefing paper, http://earthsummit2002.org/es/issues/tourism/tourism.pdf, Accessed June 13, 2013
12 UNEP and the International Council for Local Environmental Initiatives (ICLEI), 2003, *Tourism and Local Agenda 21, The Role of Local Authorities in Sustainable Tourism*, United Nations Environment Program, Paris
13 Capetown Declaration, August 2002, *Cape Town Conference on Responsible Tourism in Destinations*, Organized by the Responsible Tourism Partnership as a Side Event Preceding the World Summit on Sustainable Development in Johannesburg in 2002
14 Epler Wood, Megan 2008, An Ecotourism Project Analysis and Evaluation Framework for International Development Donors, in *Ecotourism and Conservation in the Americas*, eds. A. Stronza and W.H. Durham, CAB International, Cambridge, MA, pages 207–233
15 Goodwin, Harold and Rosa Santilli, 2009, *Community-Based Tourism: A Success?* ICRT Occasional Paper 11, from Santilli R, 2008, *Community-Based Tourism; an Assessment of the Factors for Success*, Master's Thesis, University of Greenwich, Unpublished Manuscript http://haroldgoodwin.info/uploads/CBTaSuccessPubpdf.pdf, Accessed September 29, 2016
16 Epler Wood, Megan, 2008, An Ecotourism Project Analysis and Evaluation Framework for International Development Donors, in *Ecotourism and Conservation in the America*, eds. Amanda Stronza and William H. Durham, CABI, Oxfordshire
17 http://planeta.com/planeta/04/0411donors.html, Accessed March 17, 2015, site is no longer available
18 Epler Wood, M., 1998, *Meeting the Global Challenge of Community Participation in Ecotourism, Case Studies from Ecuador*, Nature Conservancy, Arlington, VA
19 Pulido-Fernández, Juan Ignacio, Pablo Juan Cárdenas-Garcia, and Marcelino Sánchez Rivero, 2014, Tourism as a tool for economic development in poor countries, *Tourism*, Vol. 62, No. 3, 309–322
20 Ibid., 318
21 Cárdenas-Garcia, Pablo Juan and Marcelino Sánchez-Rivero, 2015, Tourism and economic development: analysis of geographic features and infrastructure provision, *Current Issues in Tourism*, Vol. 18, No. 7, 609–632

22 ProPoor Tourism, Library, http://propoortourism.info/Library.html, Accessed September 14, 2016

23 Ashley, Caroline and Dilys Roe, April 2001, *Pro-Poor Tourism Strategies: Making Tourism Work for the Poor, a Review of Experience*, Pro-Poor Tourism Report No. 1, Overseas Development Institute, Nottingham, page 2

24 Ibid.

25 ProPoor Tourism, Library, http://propoortourism.info/Library.html, Accessed September 14, 2016

26 Spenceley, Anna, ed., 2008, *Responsible Tourism, Critical Issues for Conservation and Development*, Earthscan, London, Sterling, VA

27 Spenceley, Anna, 2008, Impacts of Wildlife Tourism on Rural Livelihoods in Southern Africa, in *Responsible Tourism, Critical Issues for Conservation and Development*, ed. Anna Spenceley, Earthscan, London, Sterling, VA, page 168

28 Snyman, Susan Lynne, April 12, The role of tourism employment in poverty reduction and community perceptions of conservation and tourism in southern Africa, *Journal of Sustainable Tourism*, Vol. 20, No. 3, 395–416

29 Okavango Delta Explorations, Tourism, http://okavangodelta.com/general-information/tourism/, Accessed September 14, 2016

30 Mbaiwa, Joseph E. and Amanda L. Stronza, June 2010, The effects of tourism development on rural livelihoods in the Okavango Delta, Botswana, *Journal of Sustainable Tourism*, Vol. 18, No. 5, 635–656

31 Ibid., page 643

32 Ibid., page 646

33 Ibid., page 652

34 Mbaiwa, Joseph, 2008, The Realities of Ecotourism Development in Botswana, in *Responsible Tourism, Critical Issues for Conservation and Development*, ed. Anna Spenceley, Earthscan, London, Sterling, VA, page 208

35 Ibid.

36 Frechtling, Douglas C., 2010, The tourism satellite account, a primer, *Annals of Tourism Research*, Vol. 37, 136–153

37 Ibid.

38 Ambrosie, Linda M., 2015, *Sun & Sea Tourism; Fantasy and Finance of the All-Inclusive Industry*, Cambridge Scholars Publishing, Newcastle upon Tyne, page 168

39 OECD, *OECD Tourism Trends and Policies 2016*, page 11

40 Ibid., page 11

41 Kochi, 2013, Tourism to contribute $26 billion to India's forex by 2015, says Assocham, *Business Standard*, September 29, http://business-standard.com/article/economy-policy/tourism-to-contribute-26-billion-to-india-s-forex-by-2015-says-assocham-113092900745_1.html, Accessed September 14, 2016

42 Cárdenas- Garcia and Sánchez-Rivero, Tourism and economic development, 609–632

43 Shen, Mao Ying, n.d., Poverty and anti-poverty in China mountain region, http://fao.org/docrep/ARTICLE/WFC/XII/0510-A3.HTM, Accessed September 14, 2016

44 He, Li-hua and Xun-gang Zheng, March 2011, Empirical analysis on the relationship between tourism development and economic growth in Sichuan, *Journal of Agricultural Science*, Vol. 3, No. 1, 212–217

45 Brende, Borge and Robert Greenhill, 2013, *The Travel and Tourism Competitiveness Report 2013, Reducing Barriers to Economic Growth and Job Creation*, World Economic Forum Insight Report, Geneva, Switzerland, Preface

46 Ibid.

47 Terrill, Steve, 2012, Economic growth pulls Rwandans out of poverty, *PRI*, April 1, http://globalpost.com/dispatch/news/regions/africa/120328/rwanda-economic-growth-pulling-rwandans-out-poverty, Accessed September 14, 2016

48 Natural Capital Accounting, The World Bank, http://worldbank.org/en/topic/environment/brief/environmental-economics-natural-capital-accounting, Accessed September 14, 2016

49 Ostwald, Kai, 2008, *India and China: Trust, Social Capital and Development*, Master's Thesis, Department of Political Science, National University of Singapore, Singapore, pages 110–111
50 Vemuri, Amanda W. and Robert Costanza, 2006, The role of human, social, build and natural capital in explaining life satisfaction at the country level: toward a national well-being index, *Ecological Economics*, Vol. 58, 119–133
51 Meyer, K. and R. Burger-Arndt, 2014, How forests foster human health – present state of research-based knowledge (in the field of forests and human health), *International Forestry Review*, Vol. 16, No. 4, 421–446
52 United Nations Conference on Trade and Development Secretariat, March 2013, *Sustainable Tourism: Contribution to Economic Growth and Sustainable Development, Trade and Development Board*, Expert Meeting on Tourism's Contribution to Sustainable Development, Geneva, Switzerland
53 Jamieson, W., Harold Goodwin and Christopher Edumunds., 2004, *Contribution of Tourism to Poverty Alleviation, Pro-Poor Tourism and the Challenge of Measuring Impacts, for Transport Policy and Tourism Section*, UN Economic and Social Commission for Asia and the Pacific
54 Developing Markets for a Sustainable World, EplerWood International, http://eplerwood.com/projects.php, Accessed September 14, 2016
55 Ashley, C. and Gareth Haysom, 2008, The Development Impacts of Tourism Supply Chains: Increasing Impact on Poverty and Decreasing Our Ignorance, in *Responsible Tourism, Critical Issues for Conservation and Development*, ed. Anna Spenceley, Earthscan, London, Sterling, VA, page 130
56 Ibid.
57 Ibid., pages 130–131
58 Epler Wood, Megan, 2005, *Cambodia – Corporate Responsibility and the Tourism Sector*, Investment Climate Department, The World Bank, Washington, DC, page 14
59 UNDP recommendations on developing backward linkages from hotels to the agricultural sector, *Unleashing Entrepreneurship in Cambodia – UNDP's Proposed Activities in Private Sector Development*, Draft for Discussion
60 Epler Wood, *Cambodia*, page 14
61 Mao, Nara, T. DeLacy, H. Grunfeld, and D. Chandler, May 2014, Agriculture and tourism linkage constraints in the Siem Reap-Angkor region of Cambodia, *Tourism Geographies*, Vol. 16, No. 4, 669–686
62 Ibid., page 50
63 OECD, *OECD Tourism Trends and Policies 2016*
64 Ibid., Box 1.4, page 36
65 International Finance Corporation, 2004, *Ecolodges: Exploring Opportunities for Sustainable Business*, The World Bank Group, Washington, DC.
66 Ibid.
67 Ibid.
68 Provost, Clair and Matt Kennard, 2016, The World Bank is supposed to help the poor. So why is it bankrolling oligarchs? *Mother Jones*, January–February
69 United Nations World Tourism Organization and UN Entity for Gender Equality and the Empowerment of Women, 2011, *Global Report on Women in Tourism 2010*, Madrid, Spain, page 51
70 Christie et al., *Tourism in Africa*, page 27
71 Gender & Tourism: Women's Employment and Participation in Tourism, Summary of UNED-UK's Project Report, Toolkit for Women, http://earthsummit2002.org/toolkits/women/current/gendertourismrep.html, Accessed September 14, 2016
72 Ibid.
73 Telfer, D.J. and G. Wall, 1996, Linkages between tourism and food production, *Annals of Tourism Research*, Vol. 23, No. 3, 635–653
74 UNEP and UNWTO, 2011, *Tourism, Investing in Energy and Resource Efficiency*, United Nations Environment Programme, Paris, France

75 Gender & Tourism: Women's Employment and Participation in Tourism, Summary of UNED-UK's Project Report, Toolkit for Women, http://earthsummit2002.org/toolkits/women/current/gendertourismrep.html, Accessed September 14, 2016
76 Scheyvens, Regina, 2007, Ecotourism and Gender Issues, in *Critical Issues in Ecotourism*, ed. James Higham,, Butterworth-Heinemann, Burlington, MA, pages 196–201
77 United Nations World Tourism Organization and UN Entity for Gender Equality and the Empowerment of Women, *Global Report on Women in Tourism 2010*, page 51
78 Ibid., page 202
79 Ibid.
80 Epler Wood, The Role of Sustainable Tourism, pages 167–173
81 Ibid.
82 Kapawi: a model of sustainable development in Ecuadorian Amazonia, n.d., Cultural Survival, http://culturalsurvival.org/publications/cultural-survival-quarterly/ecuador/kapawi-model-sustainable-development-ecuadorean-ama, Accessed February 12, 2015
83 Epler Wood, Megan, photo taken at Kapawi, Ecuador, 1996
84 Henrich, Joseph, Steven J. Heine, and Ara Norenzayan, 2010, The weirdest people in the world? *Behavioral and Brain Sciences*, Vol. 33, No. 61, 135, 64
85 Planeterra, "Parwa" Sacred Valley Community Restaurant, http://planeterra.org/sacred-valley-community-restaurant--mif--projects-79.php, Accessed September 14, 2016
86 Galaski, Kelly, March 2015, personal communication with Megan Epler Wood
87 Buckley, Ralf, 2010, *Conservation Tourism*, CABI, Oxfordshire, page 177
88 Ibid.
89 Gaukel, Candice, 2015, Protected areas bring in $600 billion, so why cut funds? *Good Nature Travel*, http://goodnature.nathab.com/protected-areas-bring-in-600-billion-so-why-cut-funds/, Accessed September 14, 2016
90 Annual Report for the IUCN Framework Partners, 2013 IUCN, May, http://cmsdata.iucn.org/downloads/annual_report_for_the_iucn_framework_partners_2013_2.pdf, Accessed September 14, 2016
91 Ibid.
92 Castley, J. Guy, 2010, Southern and East Africa, in *Conservation Tourism*, ed. Ralf Buckely, CAB International, Oxfordshire, 146–175
93 Buckley, *Conservation Tourism*
94 Spenceley, Anna, 2008, Impacts of Wildlife Tourism on Rural Livelihoods in Southern Africa, in *Responsible Tourism*, ed. Anna Spenceley, Earthscan Press, London, pages 159–186
95 Epler Wood, M., 2008, An Ecotourism Project Analysis and Evaluation Framework for International Development Donors, in *Ecotourism and Conservation in the America*, eds. A. Stronza and William H. Durham, CABI, Oxfordshire, page 173
96 Epler Wood, The Role of Sustainable Tourism, pages 167–173
97 History of Tela Bay development has had many ups and downs. A good resource is this website: http://en.centralamericadata.com/en/search?q1=content_en_le%3A%22Bah%C3%ADa+de+Tela%22, Accessed September 30, 2016. The original article by Mark Chesnutt, 2007, *Travel Weekly*, is no longer available
98 Epler Wood, Megan, photo taken on road to Tela, Honduras, 2005
99 UNWTO, 2014, *UNWTO Tourism Highlights*, Madrid, Spain
100 Gillson, Ian, Adrian Hewitt, and Sheila Page, 2004, *Forthcoming Changes in the EU Banana/Sugar Markets: A Menu of Options for an Effective EU Transitional Package*, Report to DFID, ODI, London
101 Ambrosie, *Sun & Sea Tourism*, page 200
102 Negative Economic Impacts of Tourism, n.d., United Nations Environment Programme, http://unep.org/resourceefficiency/Business/SectoralActivities/Tourism/FactsandFiguresaboutTourism/ImpactsofTourism/EconomicImpactsofTourism/NegativeEconomicImpactsofTourism/tabid/78784/Default.aspx, Accessed September 14, 2016

103 United Nations Conference on Trade and Development, *Sustainable Tourism*
104 Lengefeld, Klaus, *Community-Based and Small Enterprises or Mass Tourism: Which Type of Tourism Do Developing Countries Need?* Unpublished manuscript provided to author
105 International Union of Food, Agricultural, Hotel, Restaurant, Catering, Tobacco and Allied Workers' Associations (IUF), 2013, *The Impacts of All-Inclusive Hotels on Working Conditions and Labour Rights in Barbados, Kenya and Tenerife*, Tourism Concern research report, Tourism Concern, London
106 Ambrosie, *Sun & Sea Tourism*, page 220
107 Christie et al., *Tourism in Africa*. Adapted from Box 2.1 on page 30

4

HOTELS

The backbone of the tourism industry

Introduction

Zanzibar is a magical destination, off the coast of Tanzania in East Africa, with outstanding beaches, an ancient city that bustles with traders from every walk of life, and a host of fascinating hotels, small and large. Tourism growth rates hit double digits before the downturn of the global economy in 2009. Since 2012, Zanzibar's tourism numbers have resumed their upward momentum with 6.5% growth in 2013.[1]

Zanzibar has a panoply of hotel types, ranging from ultra luxury hotels that can cost up to $2,000 USD per night, 5-star hotels up to $500 per night, 4- and 3-star up to $200, and budget hotels below this level. It is particularly known for a wide variety of small guesthouses attractive to more bohemian travelers and youth, who assume they travel in tune with local culture and in harmony with local mores. Before 2009, journalists bemoaned the fact that Zanzibar's once tranquil setting might be transformed by mass tourism, predicting that larger hotels would destroy the hip ambience.[2] But behind all the chatter, other threats went unnoticed and unreported by the mainstream travel press.

Fresh water on Zanzibar is becoming increasingly scarce with every year of tourism growth. Rainfall on the east side of the islands is the lowest, where tourism is concentrated along the coast of the main island of Unguja. In the village of Nungwi on Unguja Island, the water supply is being diminished by the "wall to wall hotels and guesthouses."[3] Stark inequities between tourists and locals are widening as villagers are faced with increasing water scarcity. Over-extraction of groundwater by the tourism industry is causing the salinization of local wells. With every new tourism facility, the extraction of fresh water is likely to cause sea water to rush in, destroying local wells. As the situation worsens, residents have felt so threatened they have begun to cut the water lines, prompting hotels to police their wells.[4] Scientists have found that tourism on Zanzibar is highest when rainfall is lowest, a problem that could also be exacerbated by climate change, which may make pressures on water supplies even worse due to higher temperatures and rising sea water.[5]

Research by Tourism Concern, in the study *Water Equity in Tourism – a Human Right and Global Responsibility*,[6] lays bare the growing conflict between tourists and residents over access to scarce fresh water, with tensions between villagers and tourists clearly outlined as a growing problem for the destination. Recent research documents conclude that average tourist consumption per room on Zanzibar is 14 times higher than daily household use for local villagers, and in 5-star hotels it is 30 times higher.[7]

Few hotels surveyed by the Tourism Concern authors use water conservation measures, and only one hotel treated its gray water for garden use. A minority used sewage treatment plants, and most use unlined soak pits allowing sewage to leach into the water table. This is not necessarily because the property owners are unwilling to do more but rather because the urgency of working to reduce water consumption may not be clear enough to small local owners, who tend to be absorbed with day-to-day management.

Environmental impacts of hotels: key issues

The footprint of hotels worldwide is expanding rapidly, and there are key issues that each hotel corporation must address in order to be certain they are not overburdening local resources. The industry is consolidating rapidly, and as it does, an increasing number of sub-brands will become the responsibility of name brand umbrella corporations.[8] These multinationals include Hilton, Accor Hotels, InterContinental Group, Marriott, and Wyndham, all of which are positioned to become the dominant global players in hotel development and management.[9] This chapter discusses how these large brands will need to oversee a global effort to protect declining renewable resources, conserve and generate alternative energy, reduce carbon impacts, and bring their waste products under strict control, including food, solid waste, and waste water. With the green economy becoming increasingly affordable, many innovative technologies are now available that not only reduce corporate footprints but help to achieve the larger goal of setting common goals with local destinations, such as becoming a carbon neutral destination.[10] Such public–private efforts will help corporations to ensure they are not competing for vital resources with local users or pushing the costs for management of waste, waste treatment, and energy generation on those who can least afford it.

This chapter lays out the barriers the hotel community faces to achieving further sustainability outcomes. First the global hotel brand community is using franchising and management contracts on an unprecedented scale. This business model is allowing the dominant brands to consolidate and manage their companies with less liability or responsibility for the management of each property. But this evolution away from hands-on ownership to arm's-length agreements makes it incumbent on these rapidly consolidating umbrella corporations to place sustainability and environmental management under brand agreements and not leave local owners and managers with 100% of the bill. The gaps in sustainability systems are obvious to consumers who can easily see how frequently recycling and towel reuse programs are not carried out. This undermines credibility and reveals the lack of oversight across hotel brands and properties.

The second barrier is the dominance of small, independent hotel properties that are often not highly profitable and have owners with little time or expertise to manage questions of measuring their impacts. A wide variety of owners show little predilection to adjust their personal lifestyles to adopt more rigorous sustainability management approaches. While the management of energy, waste, waste water, and questions of sustainability may seem remote to these individuals, their operating costs will rise rapidly due to the growing costs of water, energy, and waste management – a problem they can little afford that can be alleviated with more attention to the efficiencies and cost savings that are available to them.

Systems for reducing resource use and monitoring operations are essential. While the hotel industry has been measuring, benchmarking and improving performance in energy, waste, and water for decades, there are many inconsistencies in measures because of the diverse design and operational requirements of hotels. It does not help that most tools were designed for proprietary use without the intention of sharing the measures beyond the hotel group using them.[11] Carbon, energy, and water use in hotels must be measured, and these metrics must be designed to allow for comparisons between properties. There is no lack of tools to assist with this process, but cross-platform measurements will allow regional measures of hotel impact a goal that cannot be ignored.

The essential reduction of carbon and energy use should include investment in efficiencies first and in renewable energy where possible, as well as simple design fixes such as passive solar and good fresh air circulation. Large facilities can adopt combined heat and power (CHP) systems. And new buildings can achieve Leadership in Energy and Environmental Design (LEED) certification to meet green building standards and lower total energy use. Hotels in emerging countries, which demand a high percentage of commercial energy from grids, will need to be educated to contribute to achieving COP 21 goals for lowering the earth's total carbon emissions by 2030. The hotel community has great potential to help finance efforts for bundled energy projects, which are required to create economy of scale for nations seeking to introduce higher percentages of renewable energy to their grids. These projects are poised to lower dependence on fossil fuels, level out energy costs, and achieve savings for all users.

Solid waste problems from tourism come in a variety of forms, all of which require reducing, reusing, and recycling; establishing better metrics for lowering waste; and creating more sophisticated and well executed solid waste management programs. Construction waste generated by the boom of new hotels worldwide, particularly in Asia, is only rarely documented by local authorities, but Harvard Extension student work presented here indicates that while the problem may be under the radar, it is serious. Experts working on lowering food waste have convincing metrics to show this is the hotel world's number one waste problem. In emerging destinations around the world, small properties are not managing solid waste and litter, a dangerous problem that causes toxic, unmanaged plastics, batteries, and other consumer goods to foul wells, pollute local ecosystems, and attract innocent children and residents to pick through the waste.

Competition for precious water resources and a serious lack of sewage treatment worldwide are two of the most perplexing problems for the growing hotel community. Net Zero water use and the rechanneling of gray water for safe reuse will provide a gold mine of new water resources that must be utilized. The danger of sewage outflows from untreated systems around the world requires immediate solutions. The implementation of alternative waste water treatment systems that can serve single facilities or have the ability to serve regional needs on a cost-effective basis is a critical next step for a wide variety of hotels and lodges worldwide.

Hotel business structure challenges

The unregulated expansion of hotels presents numerous challenges for local authorities who must balance both the needs of the tourism industry and local citizens. Hotel impacts are accelerating worldwide, despite efforts to improve results. While an ever growing number of initiatives have been laid out by the industry to bring hotels into better compliance with environmental global and local concerns, the industry is structured in ways that make the problem difficult to address.

There are two primary models for the management of hotels: the branded franchised model and the small independent hotel management community. The leading brands are making rapid strides to put performance measurements in place, as will be presented here. But the growth of global hotel development, particularly in developing country destinations, takes place with little or no environmental monitoring, at either the property or the municipal level. The larger hotel brands have influence over only about 10% of the hotel market and cannot take responsibility for much of the rapidly escalating growth in many destinations.[12] A "gold rush" environment makes this situation worse. Many properties are built over a short period of time, and local authorities are very unlikely to put appropriate permitting or resource protection monitoring in place, even when very scarce resources, such as water, might need to be protected.

Hotels generally take responsibility only for their buildings, not for the resources upon which their buildings' depend. In general, hotels do not have a strong track record for managing the environmental impacts of their properties worldwide. A 2012 study found that only 20% of all hotel companies were making corporate commitments to environmental management, taken from a list of the leading 300. Just 13% were reporting on their environmental impacts based on measurements allowing them to track performance from year to year.[13]

Global brands

Marriott International wants to be the number one lodging company in the world and touts its business model of managing and franchising hotels rather than owning them as pivotal to their success. The benefits of this model are clearly articulated in their annual report. Franchising reduces financial leverage and risk and provides more stable earnings during economic downturns. At the end of 2011, only 2% of the hotels branded as Marriott were owned by the company.[14] This business model has very important implications for the future of environmental management in the hotel world.

The business relationships between owners in the hotel world is far from obvious and can be extremely puzzling to the outsider. A witty article in *The Economist* offers a humorous view of the issue:

> You book a room on the website of a famous international hotel chain. But the hotel is owned by someone else – often an individual or an investment fund – who has taken out a franchise on the brand. The owner may also be delegating the running of the hotel, either to the company that owns the brand or to another management firm altogether. The bricks and mortar may be leased from a property firm. In some cases, yet another company may be supplying most of the staff, and an outside caterer may run the restaurants.[15]

The fact that the hotel industry rarely owns and operates its own buildings is one of the primary impediments to a greener, more responsible industry. The growth of the lease, management contracts, and franchise operations puts the brand manager in a position where they rarely oversee the environmental management of the hotels they are branding.[16]

In the last ten years, international hotel chains, under pressure from shareholders to return capital, are franchising at the highest rates in history.[17] This trend is influenced by recent economic downturns, which have pushed hotel companies resistant to the model into the franchising system.[18] Franchising protected the large brands from economic shocks by removing the company from day-to-day operations. Hotels franchised to major brands are more profitable in recessions, and their profit rates are more stable, making them a safer investment.[19] Although the economic crash of 2009 pushed hotel revenues down, roughly 6.5% for InterContinental, Marriott, and Wyndham and 12.5% for Starwood, the brands themselves effectively protected themselves from the worst of these recessionary impacts via franchise and management contracts. Hotel owners in 2009 experienced bankruptcies at a high rate, but the banks continued to finance the operation of the hotels until new ownership took over, including the payments of franchise fees.[20] After the experience of 2009, Starwood plans to double its franchised hotel percentage from 40% to 80%.[21]

For this reason, franchise agreements are increasingly becoming a worldwide model. Even in China, which once was dominated by state-owned hotels, franchise agreements are becoming a popular new business model to build and develop more hotels in order to meet international expectations and develop more consistent brand performance for travelers moving around the country.[22]

Franchise agreements define the brand's responsibilities to their property owners. And, conversely, the arrangement lays out precisely what the property owner must do to meet the requirements of the brand. They include initial fees and continuing fees. Initial fees are roughly based on the size of hotels, with 100-room property fees at $17,000–$66,000, 200 rooms from $35,000 to $71,000, and 300 rooms from $70,000 to $150,000. The continuing fees are as follows:

- Royalty fees for the use of the brand name
- Advertising or marketing contribution fee
- Reservation fee for the cost of the central reservation system
- Frequent Traveler Program fee for the incentive programs
- Miscellaneous fees for additional services, such as development planning and construction services or interior design to ensure brand approvals are smoothly obtained

The total cost of franchise fees to the owner are calculated as a percentage of room revenue. For all classes of hotel properties from 100 to 300 rooms, the franchise relationship costs the owner about 10% of room revenues.[23]

The commitment of owners to sustainability is related to their ability to manage the cost implications of investing in their hotel. To achieve sustainability performance metrics, the owner must invest in the additional upfront costs. The brand is not responsible for this. These costs may appear to be unattractive to a new owner. The upfront costs of their branding and continual fees could influence an owner's decision whether or not to absorb additional investments in sustainable design, alternative energy solutions, energy efficiency, in-house air pollution, or the management of their waste streams.

If the brand does not specify that sustainability is part of the agreement, property owners are not obligated to work toward the benchmarks for sustainability being set by the brand. There are statements within annual and sustainability reports that must outline how many properties are included in the sustainability strategy, statistics that are crucial to review for all students of the industry.

For example, the Hilton Worldwide annual report provides environmental data on 2,380 properties out of the 4,202 properties in their corporate portfolio in 2013, or 57% of the total. Hilton has successfully created a benchmarking system that is tied to their legal franchising agreement. This important legal arrangement has allowed them to create a robust global platform for gathering data that will only improve with time, bringing more properties into the system with the ability to accommodate global expansion. The ease of adopting the data platform should give most of their franchisees a strong incentive to become involved. Hilton "gives all their properties the tools" to track their carbon footprint and offers metrics that allow hotels to compare themselves with comparable properties in similar climate zones. They have leveraged their Lightstay tool, which is their global environmental reporting data system, to become one of the most powerful tools in the industry.[24] Marriott operated 4,000 hotels in 2013. By contrast, their sustainability report for the same year covers the sustainability performance of 1,101 company-operated hotels, or just 27% of their properties.[25] For the industry at large, it is critical to understand how many hotels are reporting and with what comparative performance. Such statistics are only beginning to emerge as of the writing of this book.

For example, the International Hotel Group (IHG), which includes the Inter-Continental Luxury brand and the more modest Holiday Inns, among others, has aggressively sought to include as many hotels as possible in its Green Engage program, with over 2,500 participating in 2013, or roughly 50% of their total properties, allowing the company to set standards and actively benchmark a growing number of hotels in its portfolio. The IHG Green Engage website is extremely clear and a model of transparency, using graphics and charts to specify how many of their hotels are being measured, and whether the hotels being measured are franchised or owned and managed. This sets an excellent standard for all hotel brand corporate sustainability reports.[26]

While all of these hotel groups are making substantial contributions, the franchising and management contract business model is impeding their progress and making it doubly challenging for them to create a global platform for gathering corporate data on sustainability, no less manage their portfolios of properties with consistent metrics.

Small independent hotels

Most travelers are fully aware that they can choose between the branded hotel and the many small proprieties found in every destination, designated as guesthouses, bed and breakfasts, and lodges. In fact, the small hotel or lodge model dominates the industry. For example, in 2009, 75% of all European hotels were categorized by the European Union as small businesses (defined as employing under 50 people). And in countries with lower GDPs, such as the Czech Republic, Poland, and Greece, 90% of all hotels were small businesses.[27] The dominance of small-scale hotels in the hotel world is one of the most important impediments to achieving a more coordinated environmental reporting strategy on a worldwide basis.

Small hotels are more often than not run by less than experienced owners who frequently lack management experience and often do not have the expertise or profit margins to develop a consistent approach to environmental management. And to make matters even more challenging, most guest houses and small-scale properties have little awareness of their properties' environmental impacts. Even if these small property hotel owners are aware, their management style is very often "the opposite of strategically planned behavior and can be defined as ad hoc,"[28] making them averse to any rigorous efforts to measure their impacts.

Small owners focus almost entirely on keeping their properties booked, cleaned, and operating. As a result, they are not managing their water and waste or seeking to even economize through basic efficiencies on energy. Most small hotels do not see environmental management as a source of competitive advantage, making it difficult to get them to pay attention to the opportunities to save money through efficiencies and improve their bottom line.[29]

In Greece, in the study done by Harvard Extension student Sofia Fotiadou, few small property owners on the island of Santorini in the Cyclades were seeking to conserve water. While water was coming at a growing cost to owners, they did not choose to try to economize on their use of water. Rather, their response was to charge the tourists more for water, clearly not a good business model over time. With some good metrics and a look at the potential for water-saving devices, Fotiadou found that small-scale property owners could have a substantial payback if they made the move to conserve. In one case study of a hotel on Santorini, she found the following:

> The only water conservation practice that is incorporated in the operational routines in this hotel is the low flow basin taps, leaving room for ample other improvements. There are opportunities to cost effectively reduce water consumption without compromising the quality of services provided to customers. Practices proposed are retrofits with low to medium initial cost. Average payback period is 8.9 years, though if the icemaker is not considered (with minimal difference in water savings); it can be as low as 3.3 years. Annual savings are $5,400 and 36% in water consumption.[30]

The structural challenge of unifying small and large property owners around the goal of lowering tourism impacts is a challenge that needs to be fully understood and acknowledged. Twenty-first century hotel managers will have to deploy management systems that can be effective for highly different operational styles. While large-scale brands are moving toward increasingly sophisticated systems of tracking the use of resources, their own legal frameworks to reduce risk and lower liability must addressed, via new environmental management provisions in legal franchise and management contracts. For small property owners who eschew rigorous management procedures, more assertive efforts to demonstrate the value of environmental efficiencies targeted at next-generation owners may be the most effective means of getting an educated response.

Hotel environmental management issues and reporting

Every hotel seeks to maintain good revenue per available room (RevPAR) ratios, one of the most important ratios used in the hotel industry to measure success.[31] Lowering the cost of water and energy per client/room is an effective and smart means for hotels to improve their revenue per room, while at the same time reducing their environmental impact. All hotels have this data; they simply need to apply these cost savings per room to their RevPAR ratios to see how well efficiencies improve their RevPAR numbers. As energy and water prices continue to increase across the world, hotels will be under increasing pressure to lower these costs to prevent erosion of revenue per room.

There is no question that hotels can benefit financially from efficiency measures, which are almost always cost-effective when utilized. In their book, *Responsible Hospitality*, Rebecca Hawkins and Paulina Bohdanowicz argue that all hotels do have the potential to manage their facilities with care and ensure that they do not irresponsibly consume local resources.[32] It is a question of both training and higher awareness, particularly in middle management, but every manager and small business owner can benefit from measuring the environmental impacts of a facility. The next generation of owners or managers may become more motivated as the costs of their operations come under greater scrutiny, as the price of water and energy rises, and the demand to manage the carbon impacts of the travel industry escalate.

The hotel industry has a tendency to focus exclusively on the guest room when evaluating hotel environmental performance, but public use areas such as conference areas and spas – often used by clients not staying at the hotel – must be scrutinized as well.[33] As such, it is helpful to think of every hotel as an ecosystem of services all managed by a different departments with different objectives and procedures. Hotel environmental performance must be broken down by the metrics of their restaurants, kitchens, conference rooms, laundry services, spas, and resort activities such as golf courses and swimming pools, all with very different resource use patterns that cannot be measured as part of a basic room night. Efforts to properly benchmark these diverse uses cannot be easily put into one monitoring system with one simple metric for performance. For example, the energy and water usage of a hotel depend on the size and number of their restaurant facilities. Likewise, the presence of laundry facilities on the property will greatly influence energy and water metrics. Comparing hotels that have numerous restaurants and host in-house laundries with those that do not is not a fair comparison. Hotels, even in the same brand and star category, all have unique characteristics. Every monitoring system utilizes a different set of reporting "boundaries." If these boundaries are not clear and consistent within the hotel industry, all of the other efforts to make reporting useful are ineffective.

Although global reporting systems, such as the Global Reporting Initiative (GRI) and the Carbon Disclosure Project (CDP), have been adopted by the hotel community, the lack of standardized reporting presents a challenge. In an evaluation of the GRI reports filed by 9 global hotel companies in 2009, it was found that the lack of standardized reporting did not allow for comparison of performance of the different brands. For example, specific documentation was lacking on what resources were used to calculate Greenhouse Gas Emissions.[34] Similar problems were found with certification programs for hotels. Consistent methodologies were not available, despite the fact that the Global Sustainable Tourism Council's (GSTC) criteria were designed to harmonize tourism sustainability platforms. The GSTC do not specify within what "boundaries" hotels must measure their metrics (which leaves them free to measure only a portion of their properties).[35] The methodologies for all of the top certification systems had inconsistent criteria and measurement systems.[36] The widely quoted Energy Star ratings by the U.S. Environmental Protection Agency (EPA) are not tailored to specific regions or class of service, from budget to luxury. Another reporting challenge is the quantification methods used to deliver each metric for performance. To create a set of metrics that are comparable between brands or properties, all facilities must use the same "denominator" – the figure used to divide the gross amount of energy, water, or waste to demonstrate the average use of resources.

The hotel community already measures its performance according to revenue per room and has long managed its energy bills by measuring cost of operations per square foot or square meter. But, as we have investigated already, the guest room is actually a small portion of hotel operations. Additionally, the total square footage of a hotel facility does not properly weigh all the types of uses within the building. One solution may be that carbon, energy, and water use in hotels will have to be divided by square footage and consumer use using the metric of Per Room Sold (PRS), which would properly value the number of users consuming resources as the best possible "common denominator."[37] Corrections for weather and subcategories of service levels will also be needed because present hotel categories do not properly acknowledge the wide differences in facilities within each category.[38] Sub-meters that can analyze the energy usage of specific systems within the larger utility bill will also make hotel reporting a much more straightforward process than in future.[39]

Weighted measures for the facility as a whole, with reports for each facility division, must be supported by comprehensive IT and reporting tools. A spa manager needs to track the spa's performance in order to be effective, just as restaurants and laundry units will need to do, in order to appreciate what they are achieving and what percentage their division is responsible for within the total performance of the hotel. Leading certification programs that offer benchmarking tools are growing and are a user-friendly way for hotels to move forward, such as Earthcheck and Green Globe. While sustainability reporting is still evolving, the use of varying boundaries for reporting, inconsistent metrics and denominators that vary make comparisons year on year impossible. In general, hotel sustainability reports at the corporate level need much greater clarity and consistency.[40] Experts in hotel sustainability systems agree that a standardized set of methodologies is needed.[41] The major hotel brands have created a wide variety of internal benchmarking systems, including the Hilton LightStay Program and InterContinental Group's Green Engage, which have successfully measured impacts to the degree they are applied within the brand system.[42] Efforts to create more uniformity and consistency in reporting via certification are hindered by a lack of specificity in such systems as the GSTC, which was established to harmonize tourism sustainability platforms.[43] Achieving uniformity of metrics has without doubt been more challenging than hotel experts might have imagined originally:

> Benchmarking in the hotel industry has tended to lean toward reliance on single tell-all numbers such as RevPAR. However, when analyzing energy

> and carbon for example, it is important to recognize that a wide range of complexities that affects performance, such as the source of energy itself – which can vary from dirty coal fired power plants, to a local wind farm. There is little chance that a single number will be all-telling for whether a property is managing its energy and carbon footprint well.[44]

Certain basic agreements could be made within the hotel industry on how to report and measure. The Hotel Carbon Measurement Initiative (HCMI) is one such agreement. Published in 2013, the HCMI unites hotel industry efforts to both measure and report on their carbon impacts using one standardized methodology. This reporting protocol is a landmark effort to average out normal operations of hotels over a 1-year period. It uses "occupied rooms" as the denominator to ensure that the metrics are not based on the lower-energy emissions caused by low occupancy rates. It also makes certain that laundry operations are calculated into the emissions metrics, even if outsourced. And emissions are apportioned to two main services: guestrooms and public meeting spaces, to account for these very different types of services.[45]

While not perfect, the HCMI is an important advance in the collection of global data on hotel carbon emissions. This initiative has allowed researchers to conclude that some specific attributes of hotels, called energy drivers, cause higher energy use. For example, the luxury sector on average has a higher energy use per square foot than other full-service hotels. And the main causes of higher energy use are the laundry services, room size per client, and air conditioning usage. Lowering these energy drivers within hotels will be a means to lower global hotel footprints.[46]

Hotel water use: issues and solutions

On a global scale, tourism represents less than 5% of total water use according to one of the world's leading experts on the topic, Stefan Gössling and his co-authors.[47] While 5% may seem to be rather insignificant, Gössling anticipates that conflict between locals and tourists is bound to increase due to the growing costs of providing fresh water to the expanding population of local users, the relative scarcity of water in many areas where tourism is a major industry, and the lack of waste water treatment across the globe, which will spoil healthy water sources.[48] Access to water will become more controversial if the hotel community does not respond or if governments do not coordinate a response. With climate change multiplying the potential risks, the hotel community will not be able to passively procure water for its clients while ignoring local needs but must rather actively review all means to both conserve water and ensure that local water infrastructure for local people is adequately in place. Some may argue this is not the private sector's role, but while efficiencies are certainly the first priority, active participation in planning will also be essential in future.

Water scarcity is particularly severe during the dry seasons, exacerbating the conflicts. Some countries have started to import fresh water on tanker ships, including the Bahamas, Antigua, Mallorca, the Greek Islands, Fiji, and Tonga.[49] Gössling and his coauthors suggest that water use by hotels may threaten local resources in tourism-dependent regions, where demand is often concentrated in dry seasons, and will be acerbated by climate change.[50] Most governments in developing countries offer less adequate government oversight and frequently do not finance universal access to public drinkable water.

According to Harvard Extension student Sofia Fotiadou's research on the Greek Islands:

> Water consumption may increase up to or even more than 300% during tourism's high season and accommodation is most likely the top contributor. At the same time as of 2013 authorities have failed to integrate externalities (additional costs) from tourism's increased water consumption in the tourism product, mainly because tourism is seen as one of the vehicles for the country's financial crisis exit.[51]

The use of water varies according to the type of hotel, typically rated by the number of stars. The more stars a hotel has, the greater the use of water by its clients. Water consumption ranges from 80 to 2,000 liters per tourist per day on average, with 5-star clients documented above 3,000 liters per day in hot tropical destinations such as Thailand and Zanzibar.[52] Higher temperatures increase water demand, irrigation of grounds increases, and swimming pools need more frequent replenishment of fresh water, demands that will increase with global climate change.[53] Limited research indicates that continuous irrigation of gardens may cause 50% of water usage on resort properties that have extensive grounds in tropical environments, with significant water consumption also spiked by swimming pools, spas, and golf courses. In these types of resorts, restaurants represented only 15% of water use, laundry 10%, and cleaning 5%.[54] Low-star guest houses without significant grounds, have a much better record, closer to 150 liters per tourist per day, but their cumulative impacts cannot be ignored.

The Zanzibar case is likely to be typical in future as emerging country regions face the rapid development of tourism. A dangerous cocktail of rapid growth of small-scale properties and a growing number of medium to upscale resort properties on tropical islands is a trend. And though travelers are often perceived as being a potential part of the solution, it is difficult to determine whether they can really be of assistance. Despite growing environmental concerns expressed at home, travelers frequently are unaware of these problems when they are on vacation overseas. In Zanzibar, the tourists visiting considered themselves to be committed to environmental sustainability, but less than half indicated to Tourism Concern that they were aware of water issues on the island.[55]

The increasing importance of tourism as a generator of foreign exchange for developing countries, or cash-strapped economies such as Greece, has allowed governments to remain silent about the industry's resource hungry ways. Small traditional hotel businesses are the least likely to seek environmental solutions, as we shall see, and the creation of efficient management systems for larger hotels and resorts is only just beginning.

Hotel energy use: issues and solutions

Hotels are major energy users, ranking in the top 5 in the commercial/service building sector.[56] They use energy for heating, air conditioning, hot water, lighting, and miscellaneous other services. Additionally, the average hotel emits between 150 and 200 kilograms of CO_2 per square meter of room per year.[57] As the cost of energy rises, responsiveness to energy efficiency in the hotel community will increase. As one major hotel analysis firm conclusively reports, the potential for the reduction of operating costs provides a compelling incentive for hoteliers to invest in environmental technologies and more efficient operation procedures.[58]

For example, Serena Hotels, an East African–based hotel chain with many awards, reports that their profits declined by 15% in 2011 due to increased energy and food costs. Their response was to install solar hot water heating systems for 5 of their resort properties, together with inverter systems that create quieter, efficient production of energy for 7 of their safari camps within National Parks, which depend on generators. This led to significant savings of approximately $1.3 million USD, $1.1 million liters of fuel, and approximately 4 years of 24-hour generator use.[59]

Hotel chains differ markedly in efficiency even within the same chains because of the differences between owners within franchises. While budget hotels are generally more efficient than upscale hotels, which use about 25% more energy per room night, efficiencies could be easily obtained at all star levels. Modern hotels use significantly more energy per room night than older hotels, even with the growing interest in LEED and Energy Star management systems for commercial buildings surging in the U.S. This can be explained by the increasing energy demands of the average client in hotels worldwide. U.S. hotels consume approximately 4% of all commercial building energy and produce nearly 35 million metric tons of CO_2 annually.[60]

Large U.S. hotels with over 300 rooms constitute about 20% of the total hotel community, and it is these properties that have the highest energy needs not only by size but also per room.[61] A study of larger hotel energy needs concluded that large resort hotels, particularly casinos, are the most energy intensive facilities in the U.S., with convention hotels the second most energy intensive."[62] This study, which had access to the EPA Energy Star Program data for 1,222 hotels, was able to document with certainty that the energy cost per room increased with price and star classification, correcting for all climate differences. Resort hotels had energy costs of over $2,000 per room, nearly 50% more than the next most intensive user. Convention hotels stood at $1,500. The average energy cost per room was $1,254 per available room for all types of properties. With the figures available in 2012, the hotel sector in the U.S. likely pays $6.2 billion in energy bills.

Energy savings of 10–20% can be expected if any hotel goes through a technical review of equipment and operations, and energy savings can be achieved through conservation measures. The most recent analysis suggests that average heating and cooling costs even for upscale and luxury hotels is roughly 18% of total energy bills in a sample of hotels across the U.S.[63] Heating in northern climates accounts for approximately one-third of energy consumption costs for hotels and is routinely the focus of energy savings efforts in these climates. Among the many efficiency approaches to be considered are good pipe insulation, reduction of flue gas losses, and temperature control systems that lower the temperature in unoccupied rooms.[64] Both the front of the house and the back of the house must be considered. For new hotels, site selection alone can make a tremendous difference and enable passive solar warming.

Air conditioning can be a major expense for hotels operating in more tropical climates. In a study on energy use of hotels in varying climates, the number one energy expense was decisively air conditioning in locations between the Tropic of Cancer and Tropic of Capricorn, expending 30–42% of a hotel's total energy bill.[65] Savings of 20–30% in energy used for space conditioning can be achieved by zoning or autonomous temperature control systems in individual rooms.[66] Energy-saving measures include:

- Installing frequency controllers on the fans (VSD systems that vary according to use),
- Recovering heat from the extraction air (CHP systems that use waste heat to warm water),
- Optimizing running hours by using timers and other systems to automatically raise temperatures for rooms not in use, and
- Optimizing the temperature and humidity.[67]

Few travelers in the world have not experienced how often hotels and other commercial spaces are cold, even when the temperature outside is warm or hot. Questions about the necessity of this are rare not only among consumers but even among professionals who may suffer from being cold on business trip after business trip. Overly air-cooled temperatures have been shown to possibly lower productivity and can cause illness.[68] Yet little mention is made of the simple idea of raising temperatures to more productive, healthful, and energy-efficient temperature ranges in the literature on hotel energy efficiency.

Apart from simply raising the temperature a few degrees in hotels worldwide, there are many other important methods to avoid thermal temperature losses, be they in hot or cold climates. Good thermal insulation and efficient glazing coating on glass, which prevents heat from escaping, can reduce losses and allow for more efficient space conditioning and allow passive cooling and heating.[69]

Design for energy efficiency

For years, the field of ecotourism has fostered the idea of creating proper ventilation and energy efficiency using low-tech and indigenous architectural concepts that do not require high investment. Appropriate, climate-wise design can eliminate many problems with energy efficiency in tropical climates and offer more comfortable surroundings for both work and play. Natural cooling design elements should stress ventilation, shade, and insulation. Porches, courtyards, high thatched roofs, and natural vents that allow cooled air to circulate through shaded areas make visitors feel comfortable and have the benefits of not containing clients in box-like environments without ventilation, which require air conditioning.

Ecolodge design guidelines were published in 2002 by The International Ecotourism Society. They help guide hotel planning and foster deep respect for natural features and offer many helpful and commonsense ideas for designing resorts. They consider how to blend the design with natural features to maintain tree cover, take advantage of solar energy for bathing water, and use the natural cooling of water features to create comfortable temperatures. To prevent the need for excessive energy to cool lodge rooms and public spaces, ecolodges in tropical climates are sited to take advantage of cool breezes. Ecolodges build venting systems that maintain natural cross ventilation and use ceiling fans to eliminate or reduce the need for air conditioning.[70]

In the hotel industry, ecolodges still set the standard for energy-efficient design. But their architectural principles have been widely adopted by luxury brands such as Banyan Tree and Six Senses. These new environmentally designed upscale resort chains are delivering beautifully designed spaces that offer traditionally inspired architecture in spectacular natural settings, often for extravagant prices in the $500 and above per night category. They command a higher price because they offer both luxury and real sensitivity to the natural resources of their destination.

New concepts in ecolodge design continue to break new ground by using what once may have been considered to be less interesting agricultural landscapes. For example, in Sri Lanka, Jetwing's Vil Uyana lodge is situated among paddy fields, which were once destroyed by slash-and-burn agriculture. Re-established without agrochemicals, the rice paddies are part of a new, larger constructed wetland habitat that is planted with native rushes and reeds and that now is home to native crocodiles and otters. Each of the 30 dwellings is integrated into the new habitat, reducing energy needs by virtue of maintaining a naturally cooled environment for a price of approximately $260 per night.[71]

Renewable energy

Lowering a hotel's carbon footprint is an imperative for any hotel seeking to manage its environmental impacts responsibly. Increasingly, renewable energy solutions are an important part of this mix, such as solar thermal energy for heating water, which is cost-effective and offers an excellent return on investment. Opportunities for solar electric energy are becoming increasingly affordable for commercial properties when they can obtain financing that includes incentives and paybacks from governments. Wind power, biomass fuel, and micro hydro are all options for developers that are still not being commonly implemented, and heat pump plants are a very cost-effective solution that have enormous benefits for larger properties.

While renewable energy is becoming increasingly affordable, it continues to be too expensive for most mainstream hotel properties. Not all renewable energy solutions are out of reach, such as solar thermal energy for heating water, which is also cost effective and offers an excellent return on investment. Whitbread PLC, the UK's largest hotel and restaurant company operating budget hotels, has laid out firm policies for meeting energy efficiency goals, including high-quality metering systems for tracking, low-energy lighting, and energy-efficient water heaters. Whitbread is motivated and has made a commitment to reduce its carbon emissions by 26% by 2020 and has been carefully measuring its consumption of energy since 2008.[72] It has managed to double its renewable energy supply to 10%[73] and opened a Premier Inn in London's Covent Garden in 2014 that is using 100% renewable certified energy supplied from the grid.[74] Like many companies, Whitbread is more able to take advantage of renewable energy if it is supplied from the grid.

While there are a growing number of outstanding cases, the hotel industry is not matching the energy efficiency and carbon reduction achievements of other industrial sectors. The barrier to achievement may be due to the long period required for return on investment for renewable energy technologies if the hotel company cannot procure the energy from the grid or find government incentives. Few hotel investors are prepared to bear the additional cost burden that these technologies imply without price breaks. With the franchise and leasing system, owners typically have 15 years or less to show a return on investment for the hotel. As it is, they are forced to weather the cost of operating in economic downturns, while still paying 10% to the brand, leaving their margins vulnerable and thin. Under these circumstances, it should not be too surprising to learn that hotel owners for the most part are wont to invest in renewable energy technologies when they cannot recoup the value for themselves.[75]

Government's role in hotel energy efficiency

Governments are important actors in the development of renewable and efficient energy generation options for commercial buildings of all kinds. Governments can:

- Help finance research for the best energy-efficiency techniques,
- Provide training on energy technologies and efficiency methodologies that are the most appropriate for the size of the business,
- Offer smart metering to help customers monitor and benchmark energy use,
- Offer help with financing tools and low interest loans
- Provide tax incentives to help subsidize the cost of renewable energy installations,
- Develop net metering policies to enable private business to earn funds from renewable energy generated from their own privately financed installations, and

- Play a major role in offering an increasing percentage of renewable energy to their customers in their own energy portfolios.

A detailed study on the benefits of combined heat and power (CHP) systems prepared for the U.S. EPA in 2005 with outstanding statistics on the cost savings benefits of these systems began the process of helping the largest hotels in the world to recognize the benefits of managing their energy more efficiently. The study found that casinos of over 1,000 rooms operated more like small college campuses than hotels because of their constant need for power at all hours of the day and night. Most of these huge facilities are found in Las Vegas and California and occupy millions of square feet. Their energy needs were found to be much higher than typical business hotels.[76] The demand for year-round, 24-hour high-capacity service at casinos justifies thermal activated technologies that run off the waste heat of on-site generators. The EPA-funded study concluded that casinos and mega resorts could earn substantial savings and lower carbon impacts with the installation of CHP systems.

With the data EPA provided, a whole new generation of casinos and hotels were put into the pipeline with more energy-efficient systems. In October 2013, MGM Resorts gave a tour to the prestigious Pew Foundation's Clean Energy team to show off their brand-new CHP system at the ARIA Resort and Casino in City Center, Las Vegas. The Pew team toured the unprecedented, 18-million-square-foot mixed-use resort on the Las Vegas strip. At the heart of the project is a state-of-the-art 8.2-million-megawatt CHP system that uses natural gas to generate electricity, which in turn captures waste heat and warms domestic water. This brings a 37% improvement in energy efficiency compared with comparable resorts, and all heated air emissions from the system are avoided. The decision to embrace energy efficiency and CHP systems has been moving across the U.S., with large corporations and native gaming corporations adopting the technology.[77]

Even Dubai has been moving toward adopting a more energy-efficient approach to its tourism development. Since its 2014 State of Energy Report, the emirate has been discussing how to manage its energy consumption via public policy. Key to this strategy is investing in more efficient energy systems, in particular, by its oil-rich financing partner, Abu Dhabi.[78] The report optimistically states that the rapid growth in the market's interest in green initiatives and the evolution of alternative and innovative finance instruments are opening up the space for large-scale financing for a green economic transformation. This transformation is transpiring at a speed never seen before, particularly in emerging economies.[79] But more careful reading of the report indicates that the financing of renewable energy is still not at all easy; it requires a very long-term view and partners that have deep pockets, as UAE and other governments found primarily in the Persian Gulf region.

Green building standards are to be required for all new buildings in Dubai under the Integrated Energy Plan, but the "vast majority of buildings that will be occupied for the next 25 years have already been constructed."[80] These have been constructed with almost no regard for energy efficiency and operated with poor operation and maintenance practices. In addition, utility bills are not high enough to motivate investment in improvements.[81] This and continued rapid growth explains why Dubai's energy-efficiency profile will not change dramatically.

As the tourism industry becomes the leading economic development category and source of foreign exchange in country after country, nations need to take leadership before unsuitable infrastructure is built. Rather than following the flamboyant model of Dubai, which is often admired for its bold buildings and architectural wonders, nations need to use the early flush of development dollars to pay for and build energy efficiency and renewable energy systems right into the cost of doing business. The next Dubai must astound the world not with its astoundingly inefficient use of energy but with its remarkable energy efficiencies. Such leadership needs to come from both governmental leaders and private sector who recognize that profligacy is not a business model for the long term, nor is it beneficial to most nations and their citizens.

Examples in India and Europe demonstrate how public–private partnerships can lead to innovative solutions for reducing energy consumption. In India, where only 43% of villages have a permanent power supply, hotels cannot depend on government, and reliable energy sources are not fully available, according to a case study published by Harvard Business Review.[82] To cope with the problem, the ITC Ltd. hotel company built its own wind farms in Tamil Nadu, with additional wind turbines in Maharashtra and Andhra Pradesh all connected to regional power grids. By 2011, ITC had nearly 31% of its energy from renewable wind sources. Working with the government of India, ITC found a public–private solution to secure its own source of renewable energy while offering renewable energy it did not use to the nation of India.[83] In Europe, public–private solutions are offered on a small-scale with the Nearly Zero Energy Hotel project, which is co-financed by the European Commission to reinforce businesses operating in the hospitality sector to meet the challenges of adopting green energy and accomplish large-scale renovations. Spain, Greece, Italy, Romania, Croatia, and France have agreed to support small and medium-sized (SME) hotels, which represent 90% of the European market base for hospitality, to help advise on the most appropriate techniques and technologies for retrofitting SME hotels into Net Zero Energy facilities. The program offers information on financing opportunities for the retrofits, networking, e-tools and pilot projects as a testing ground.[84]

Nearly Zero Energy Hotels is a beginning and an important form of outreach to the underserved SME hotel community, but as tourism grows in destinations around the world, more dramatic projects like the ITC Hotel wind power project are needed. Nations that are barely covering the needs of their citizens must capture developers when they are ready to invest and prepare attractive public private programs that can co-finance renewable energy programs on a large scale, offer tax incentives and net metering benefits, and develop financing packages at attractive rates.

Hotel waste generation: issues and solutions

The Accor Hotels is a French hotel conglomerate that owns a broad portfolio of hotel brands, including Sofitel and Ibis. One of the world's leading hotel chains, with hotel properties in 90 countries and over 4,000 rooms, its leadership has made a substantial commitment to managing its global environmental footprint by reviewing the company's global impacts using Life Cycle Analysis (LCA). Their Planet 21 report (2011) was the first environmental footprint study by a major hotel group and provides insights into the underlying causes of the hotel industry's environmental impacts.[85] According to the report, hotels have the most impact on energy and water. Hotel waste is less significant, and the majority of hotels' impacts on global waste streams come from construction, which constitutes 70% of their total waste impact. While Accor did not diminish the importance of carefully managing hotel waste, they found that only 5% of their total waste stream was generated by hotel operations, including roughly 40% from food waste, 10% from inorganic plastics, 23% from recyclable paper products, and 14% from glass.[86] Reducing, reusing, recycling, and disposing of waste properly are the most effective means of achieving waste management.

Harvard Extension student Angelina Jao reported on the growth of the casino destination Macao in China. Macao is a semiautonomous special administrative region of the Peoples Republic of China located in the Pearl River

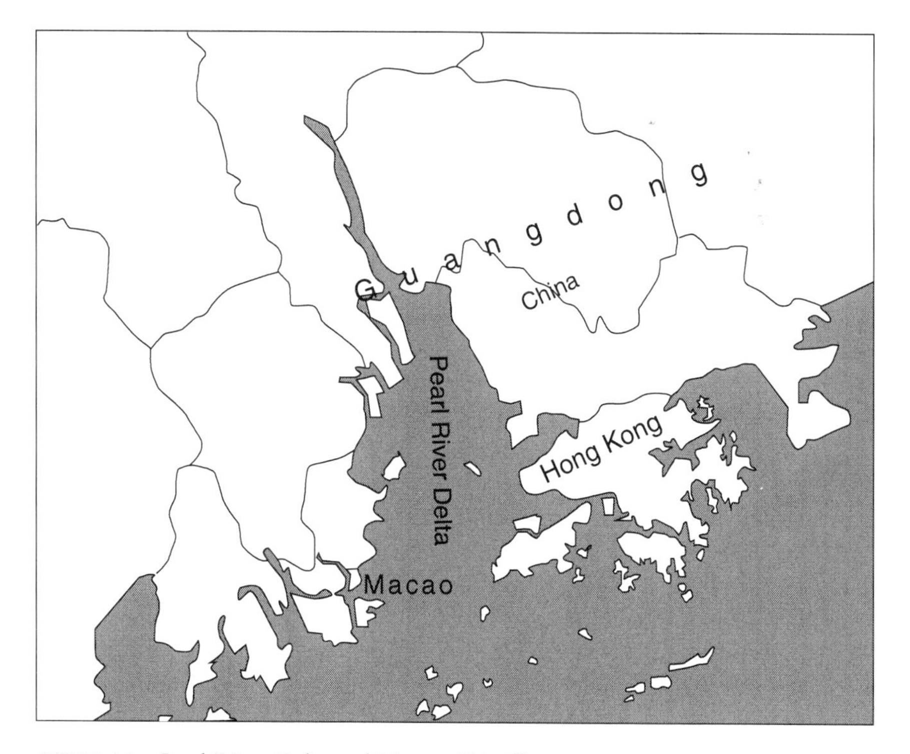

FIGURE 4.1 Pearl River Delta and Macao, China[88]

Delta near Hong Kong in southern China. Macao is no stranger to rapid economic growth and has become the largest gaming destination in the world, hosting 28 million tourists in 2012. The Delta is one of the most densely urbanized regions of the world and a major hub for Chinese economic development. Integrated resort complexes that include casinos comprise 97% of Macao's GDP. (See Figure 4.1.)

The Las Vegas Sands (LVS) corporation funded reclaimed land between 2 islands in Macao in 2005 and built the Venetian Macao there in 2007, the largest resort complex in the world. While the environmental management challenges of Macao are numerous, this study reviewed solid waste. Jao focuses on construction waste, which normally reaches 25% of landfills in other countries, but is unaccounted for in China. She investigated whether there were any systems for recycling or reuse or programs to handle toxics and found none. While it cannot be known how extensive the solid waste problem is in Macao, the enormous size of facilities being constructed and the restricted space on the island for landfills leads to a dangerous cocktail. Where is all this waste going? To make matters worse, the seventh most endangered coral reef in the world is just off the coast of Macao. This reef is not protected by any marine protected area system.[87]

Food waste

Food waste is the number one waste management problem for hotels around the world. Research from the UK cites that some 920,000 tons of food is wasted by the hotel and food service sector in the UK, worth approximately $4.1 billion USD each year. While 46% of wasted food is recycled in the UK by either composting or anaerobic digestion, nonetheless some 75% of the food wasted could have been eaten. Preventing this amount of food waste could reduce the hotel sector's greenhouse gas emissions considerably, avoiding 2.7 million tons of greenhouse gas emissions. The UK hospitality sector is working with the NGO WRAP and the UK municipal governments to reduce food waste by 5%. In the baseline study on food waste of the hotel and food service sector, hotels were losing the most money, some $6,500 USD per ton of food wasted, due to the higher value of the food types being wasted, such as fruit, meat, and fish versus potatoes, bread, and rice.[89]

A consultant in Bangkok who undertakes Food Excess Audits was able to demonstrate that one hotel wasted 1,300 kilograms of edible food in 7 days, or 70 tons per year. After implementing their recommendations, the hotel was able to save 8% of the food wasted for a savings of 2.29% per month on monthly food costs.[90] There is a generous return on investment when food is managed as a valuable commodity that should not be wasted and when careful menu and portion size planning is implemented. Measuring and monitoring the amount of plate waste over time removes the guesswork of determining portion sizes. Precise estimates of food resources needed for menus can replace the more intuitive styles chefs have used in the past when ordering from their suppliers. While suppliers are happy to deliver fresh foods in large quantities, smaller, more frequent deliveries lower the chance of waste and improve freshness. Finally, developing more sophisticated estimates of fluctuating consumer demand can help the hotel and food service community to tailor their purchasing to seasonal demand.[91]

The social question of food waste is also of considerable importance, given that 12% of the global population remains undernourished. Reducing food waste reduces pressures on the globe's food resources and the water that is required to produce food. In the Accor Hotels' Planet 21 report on their groundbreaking Life Cycle Analysis, food and beverage operations were discovered to contribute 86% of the hotel group's total water footprint, and hotel restaurants can make environmentally responsible choices. For instance, feeding and bathing a cow to produce beef takes 15,000 times more water per kilo than apples and about 4 times as much water as chicken.[92] In order to reduce their impacts, Accor Hotels has sought to create more environmentally friendly menus.

Solid waste in emerging economies

As the hotel community improves its waste management performance, its overall growth trajectory will put increasing pressures on waste facilities worldwide. Commercial waste disposal systems are weak in emerging economy municipalities worldwide and often are not covered adequately by taxes.

In one evocative case study, on the island of Langkawi in Malaysia, hotel waste was overwhelming an informal system where local residents had previously generated only organic waste. Malaysian researchers discovered that tourists generated two times more solid waste per person than local people, much of which was not organic. An increasing amount of this plastic waste was being burned near roads, fouling the air. The rapid "urbanization" of Langkawi Island finally led to a problem of illegal dumping, causing environmental health issues for local residents. The study's authors requested more action from the municipality to protect local citizens. They concluded that the current system of solid waste could not meet the needs and that a legal and institutional framework for municipal solid waste management was required. They also suggested that the resident hotels, resorts, and companies on the island contribute to the cost of waste collection and to the establishment of a recycling facility, which did not exist on the island.[93]

Experts and travelers around the world are well aware of the problem of trash. In my work at a major tour operator, the professional guides frequently wanted to focus on trash cleanup as projects for the company foundation. They wanted to "teach" local residents not to litter. It was hard to explain to them that, in fact, residents were only the victims of the introduction of the throwaway society made possible by easy access to consumer goods, brought in part by the travel and tourism trade. One sustainable tourism consultant posted a social media message in 2013 from Morocco saying he was working with over 1,000 properties that had no access to legal, regulated solid waste treatment facilities. The litter problem, he added, was shocking to European tourists.[94]

A complete change in the system for the management of waste is needed in locales around the world, but few are ready to tackle this. As the growth of tourism escalates, there will be a point where the environmental consequences of growth will despoil the very qualities visitors are coming to appreciate. For instance, Angkor Wat Temple, which attracted 90% of the 4.2 million tourists to Cambodia in 2013, is located next to the city of Siem Reap, which has not yet instituted proper waste or sewage management systems. The waste is deposited in a rubbish dump where poor local children gather daily to feed themselves. (See Figure 4.2.) At present, growth is outstripping all efforts to create waste treatment systems, and local residents are the unwitting victims.

In emerging destinations around the world, the hotel community's calculations of cost when determining how to manage their hotels does not include participation in the creation of the solution, and local government has not even considered the problem in most cases. In my work in the village at the base of Macchu Picchu, a locale without real governmental authority to manage waste, I discussed the problem of trash burning with local environmental advocates. Every year, in the dry season, fires were set by a growing population, there to work the tourism trade, who lacked services and burned trash. These fires became so hazardous that they were threatening the vulnerable plant life in the Macchu Picchu reserve and blackening the region. One hotel and its intrepid Peruvian team was seeking to

FIGURE 4.2 Siem Reap dump, Cambodia[95]

create a coalition of hotels and residents who could work together to recommend a better system for managing solid waste. The ideas were good, but support for the effort was limited. The fires at the base of this world heritage monument continue to threaten the biodiversity in the region.

In Peru, there have been concerted efforts to create solutions through master planning and national governmental oversight. But at the municipal level, there is rarely capacity or budget to manage the problems of instituting vastly more sophisticated waste management services. This problem cannot be left to local governments alone; hotels, along with national and local governments, must embrace the problem together. While the sustainability reports of global brands show that impressive reductions in waste are being achieved that can be touted in annual reports, this still does not address the larger question of ensuring the waste that remains in major destinations such as Angkor Wat or Macchu Picchu is managed properly. In each nation, financing for proper waste treatment cannot sensibly be left to local advocates. Financing, training, and municipal governance strategies are needed across the board worldwide. Larger solutions to these problems need to be researched and discussed on a global level as soon as possible.

Hotel waste water generation: issues and solutions

Pristine beaches and crystal clear waters are the most iconic selling point for hotels in the leisure tourism business. Who has not dreamed of diving into the aqua waters of the Italian Amalfi coast or floating in the shimmering waters off the coast of Turkey? Who has not gazed at the spectacular covers of travel magazines with lovely whitewashed hotels perched on the fabled coasts of the Aegean Sea in Greece? The Mediterranean Sea and its residents have offered their beautiful coastlines to travelers for centuries in return for handsome payments. But they have not always invested this wealth in the protection of their precious aquamarine resource. It is difficult to know how much waste water runs off tourism facilities around the world. Tourists are estimated to use an average of 300 liters of water per person per day, with highest usage measured up to 2,000 liters per person per day, with about half returned to the sewer system.[96] Waste water also runs off the impermeable surfaces of hotel buildings and grounds, causing overflowing storm water runoff problems particularly in rainy seasons. Golf courses and hotel gardens contribute to the pollution of both fresh and sea water thanks to the use of organic and inorganic fertilizers that can reach water bodies and result in clogged, smelly, algae-burdened canals, lakes, and ponds. Agriculture is generally the primary cause of this problem, and hotels cannot be considered a primary driver of runoff from fertilizer use, but they are a contributor.[97]

Even the best research available cannot always pinpoint the source of marine pollution, due to the high variety of outfalls into the sea, which can originate from point (from piping) or nonpoint (from runoff) sources. In 2000, it was estimated by the Worldwide Fund for Nature study that tourism contributed to 7% of all pollution in the Mediterranean, including sewage, crude oil, detergents, mercury, phosphates, and fertilizers. Approximately 20% of all sewage was still untreated in the Mediterranean as of 2003.[98] On a global level, 80% of waste water is discharged into rivers, lakes, and oceans, while in emerging economies just 7% of all sewage is treated according to the agency UN Water.[99] Tourism is an important source of the problem, as it continues to be a dominant form of development in coastal areas, and the industry largely depends on local waste water treatment facilities to manage the problem, if they exist. Given the low rate of sewage treatment globally, especially in emerging economies, there is every reason to believe that tourism's contribution to the problem is growing and will aggravate the degradation of local ecosystems, particularly coral reefs. Higher levels of nutrients from sewage are linked also to the growth of algae and the destruction of coral reefs.[100]

It is both ironic and frustrating that global sustainability advocates tend to look away from the problem of sewage. While plan upon plan is presented on creating a sustainable tourism environment, there seems to be a lack of real discussion about what infrastructure will be needed at what cost to prevent the most obvious and unsanitary environmental problem of them all, sewage.

There is little doubt that hotels often dump sewage into the sea. The following cases provide some good examples:

- In Mombasa in 2013, tourists complain that untreated sewage is being discharged onto their beaches. The government states that a modern sewage plant is a top priority, but they await help from the World Bank to build it. Of Mombasa's 1 million residents only one-third have sewage treatment. Residents have been demanding a modern sewage plant for 20 years. An environmental watchdog group inspected 19 hotels and found only a handful have treatment plants.[101]
- In Honolulu, Hawaii, on the island of Oahu, in 2012 residents complained that raw sewage is still flowing into local waterways despite legal agreements to prevent it, due to Oahu's aging sewage system. In 2006, 48 million gallons of untreated sewage was dumped into the Ala Wai canal in downtown Honolulu and flowed into the sea during a heavy rain that had overwhelmed the waste water system. Honolulu signed an agreement with the U.S. EPA to upgrade its system but failed to fully comply. Completion of the upgrades is expected by 2020.[102]
- Along the beautiful Pacific coastline of Costa Rica, where scores of visitors have flocked to visit or buy second homes and enjoy Tica hospitality, there are troubling trends rarely seen in the media. An in-depth study by CREST, the advocacy organization that investigates how tourism is affecting destinations around the world, shows that 77.5% of Costa Rica's Pacific beaches are threatened by direct and indirect sources of pollution, including fecal contamination. Between 2007 and 2009, important Pacific coast beaches were found to have polluted waters with fecal matter and waste water contaminants and were subsequently closed temporarily. A large tourism complex was found to be illegally dumping untreated waste into the ocean, river, and an estuary in

> 2007–2008. Even the famed coastal park, Manuel Antonio, was found to have high coliform levels, due to poorly managed bathrooms in the park. These specific situations have been rectified, but according to CREST, the limited capacity of many development's wastewater plants and the lack of public sewage treatment systems in most Pacific coast towns and cities will cause continuing persistent problems in the region.[103]

Hotels are the point source for sewage, but the responsibility for the waste water that comes out of hotel pipes is not necessarily seen as the responsibility of the hotel. I have heard it argued in a UN forum on sustainable tourism that in fact the tourism industry cannot take responsibility for sewage treatment because this is a governmental responsibility. As tourism continues to grow in emerging destinations, which in nearly all cases lack the sewage treatment services required, can the hotel community really transfer the total cost of sewage treatment to the community?

The market demand for coastal hotels continues to escalate, and the extensive infrastructure solutions required will reach billions of dollars. In the case of Hawaii, there is a $14.3 billion price tag projected for the new sewage treatment system for Oahu Island that will be paid by higher fees for water and waste water. Bond financing materials indicate that local citizens saw a 175% increase in fees between 2006 and 2011, reaching 2.1% of household incomes, and will see double-digit increases between 2010 and 2015 and 4–5% annual increases thereafter until the system is completed.[104] While the commercial sector, primarily hotels, also pays fees, no increases are mentioned in the same documents. In less prosperous locales, such as Mombasa, the tax burden for centralized waste water treatment will be impossible to shoulder for local residents. Clearly, public–private solutions that include the hotel sector are urgently needed worldwide.

Solutions for waste water treatment

Governments are caught in an infrastructure gap for the management of water and waste water across the world. Taxes are not adequate to build treatment plants for residents and commercial uses, and services are simply not meeting demand.

One solution to spreading the cost burden is to move the treatment process to "on-site" water systems. The idea of Net Zero water projects is being promoted by environmental engineers, and these systems have real applicability for hotels in urban and rural areas. They aim to ensure that 100% of storm water and building water discharge is managed on-site. Net Zero systems rely on investments in rainwater harvesting, water-conserving fixtures, dual plumbing for water reuse, and on-site treatment systems. The Net Zero approach also incorporates the reduction of water use from toilets, not only using low-flow toilets and urinals but also installing composting toilets. Pathogens caused by microorganisms can be effectively treated via membrane bioreactors, sand filters, or biofilters, and/or use the natural biological systems of constructed wetlands to treat waste without odor.[105]

Such investments elevate the cost for the building owner but solve the long-term problem of inadequate municipal sewage treatment facilities. Robust financial incentives for hotels to create Net Zero water projects are a crucial part of the solution. New York City's Comprehensive Water Reuse Incentive Program provides project owners a 25% discount on water services for reducing demand on the city's infrastructure.[106]

Best practices for water management are required as a starting point for such Net Zero water solutions. First, there should be proper systems for recycling gray water, which is roughly 50–80% of the total amount of water that is generated for treatment by municipal systems. To recycle gray water on property, it must be separated and piped for distribution in areas that are safe, preferably underground for irrigation purposes. Gravity feed systems are effective and inexpensive, and they eliminate the need for pumps and energy use. While gray water does have contaminants, it is much safer than sewage for reuse. It biodegrades quickly and can be a good source of nutrients for landscaping. Municipalities are waiving permits in the U.S. for small-scale systems to be installed in water-scarce areas, as the cost savings are high and the risk is very low. Gray water recycling can remove 35–40% of the water that normally reaches municipal treatment facilities, therefore greatly lessening the load and expense to the community at large. Hotels around the world generate higher percentages of gray water than most buildings because of commercial kitchens and laundry operations. Gray water recycling is a relatively inexpensive step that can recharge aquifers and save funds on maintaining hotel landscaping.

Alternative treatment systems

Constructed wetland systems are effective and offer comprehensive treatment. They work in most climates and can be designed to operate in areas with space limitations. They are durable, relatively affordable with good payback on financing, and no longer an exotic solution to on-site waste treatment.

Ecolodges have been applying this type of solution for decades. For instance, when I visited in the ecolodge Alandaluz on the coast of Ecuador in 1995, it had a functioning wetland system right in the center of the property for treating all visitor waste. Some hotels around the Masai Mara Game Reserve in Kenya are building alternative waste water treatment systems to protect the Mara River, which was becoming contaminated with fecal waste.[107] Similarly, the Jean Michel Cousteau Fiji Islands resort built a constructed wetland in the mid-1990s to protect the pristine waters and neighboring reefs of Fiji.[108] All waste water is treated with biofilters and is circulated through the wetlands built by residents on the property in simple plastic containers, which have the advantage of keeping the treatment system to a relatively small footprint on the resort property. The system prevents any nutrients from reaching the reef, is tested regularly, and maintained properly. All gray water is used to irrigate and fertilize the resorts' edible landscaping.[109]

In my work as a Senior Fellow at the Institute at Golden Gate in 2010, I had the opportunity to meet and work briefly with the designer of the Jean Michel

Cousteau resort constructed wetland system, Peter Haase. He joined our working group, and together we studied the urgent need for sewage treatment at affordable prices worldwide. The Institute encouraged him to take the Fiji design to scale and offer it to resorts around the world. He took our advice and formed the business, now called Acqualogic, which offers low-cost, easily understood waste treatment called Advanced Biological Treatment (ABT). ABT is energy efficient, has a small footprint, meets strict discharge regulations, is easy to install and expand, and offers real-time monitoring off-site by AquaLogic's team to ensure its systems are properly maintained.[110]

Mainstream property owners are often hesitant to even consider a sewage treatment system for their own guests when local hookups are not available. But in fact these systems are increasingly economical. Harvard Extension student Giles Jackson compared the financial incentives for a 350-room hotel to install a constructed wetland as a means to manage sewage at the property level and found the following outcomes:

> Constructed wetlands (CW) are an effective means to meet sanitation standards with proven results. CW systems cost 50% as much to build and operate as conventional systems, while reducing water demand by 50% or more. They may also be scaled up to serve small communities.
>
> Adoption of CW technologies in developing countries has been slow. While many of these obstacles are now being addressed, incentives are lacking for industry (including tourism) to adopt CW technology – a situation perpetuated by lack of knowledge among public and private and public decision-makers. 'Business-As-Usual' tends to externalize the social and environmental costs of development.
>
> The findings of our scenario analysis, which explored the viability of installing a CW system in a 350-room resort in three different sanitation regimes, confirmed the critical role of public policy in determining the economic viability of CW technology, and thus its likelihood of adoption.
>
> - The US project broke even in Year 3 due to the avoidance of sewer and water fees (almost $650,000/year savings).
> - The Mexican project broke even in Year 2 due to the avoidance of higher water fees (over $800,000/year water/sewer savings).
> - In the Lesser Developed Country scenario, where the cost of water was lower and sewage is unregulated, the project never broke even. CW seems least viable in the places it's most needed.

Jackson was able to easily connect the importance of managing sewage to the protection of coral reef ecosystems worldwide and recommended policies that would provide incentives to hotels to adopt this solution, via grants, pricing incentives, and tax breaks. Regional strategies that serve both communities and hotels are also a very promising option that can spread the cost and benefit local rural communities, and yet such options are still rarely explored.[111]

Advanced alternative technologies like CW and ABT for waste water treatment are becoming increasingly accessible and affordable for small resorts. Not enough attention is being paid to the importance of using these innovative waste water systems as a solution for hotels worldwide. There are good manuals for assisting with the process of decision making, such as the UNEP GTZ Manual for Water and Waste Management: What the Tourism Industry Can Do to Improve Its Performance.[112] This manual gives a helpful overview of the standard technologies for primary, secondary and tertiary waste water treatment. But the next generation of waste water systems needs to become much better known.

The demand for new, cost-effective solutions to waste water management is now acute. The example set by ecolodges is not just for those interested in the green economy or ecological design. They set a standard that can be scaled up in a cost-effective manner in an increasingly water-scarce world. More work remains to be done by hotels on biologically sound management techniques for the treatment of waste water worldwide.

Future economics of hotel development

Asia Pacific region

The hotel development community is being driven by market demand in the Asia Pacific region. International arrivals in Asia have grown over 7% annually since 2012, double the rate of arrivals worldwide.[113] Hotel construction has boomed, with 13% growth in hotel capacity projected in Asia from 2014 to 2016.[114] Domestic tourism demand in China in 2013 was five times greater than it was a decade before. Almost 1-billion-room visitor arrivals within China were logged, the equivalent of 1.2 room nights per capital, up from 0.3 ten years previously.[115] Among Asian nations, the Chinese are the most eager to travel. They are the biggest spenders overseas as well, thanks to their growing middle class and well documented interest in luxury lifestyles. There is little doubt that the Chinese economy will be the dominant driver of new hotel development in the next decade.

In a study I supervised for Wyndham Worldwide,[116] Harvard Extension student Linda Tomasso found that China is under great pressure to reduce energy use, particularly in the building sector, which is using 45% of all of the country's energy. Most of this energy is still generated by coal-based power plants, which have become renowned for generating high rates of CO_2 and pollution that are affecting many of the great cities of China and the health of their citizens. New hotel buildings, in China or elsewhere, are the most damaging to the environment because of the tremendous consumption of energy-intensive materials required in the construction process.

The prospects for managing the rapid growth of the hotel sector are hard to evaluate. On the one hand, tremendous financial resources are in place to use green construction techniques, renewable energy, and water-saving devices. And the government is beginning to apply green building standards. On the other hand, the growth of tourism in China is likely far outpacing any effort to reduce energy and water consumption. The reduction of wasteful practices is very much in the planning stages. According to Tomasso, the present cost of power undercuts all savings due to state subsidies for dirty energy. And the Chinese construction sector is plagued with inefficiencies and is certainly affected by corruption, which could easily affect efforts to regulate and green the sector.

But it is easy to underestimate the capacity of China to put broad mandates into effect. The outcomes will likely be mixed. The 12th Five Year Plan (2011–2015) addresses energy savings in buildings with a goal of reducing energy use by 20%, with aggressive targets to improve insulation, recycling, and material reuse. New preconstruction acceptance codes have been authorized to encourage compliance with efficiency goals at the design stage to encourage the acquisition of appropriate materials for green construction from the beginning. Nonetheless, according to all sources, compliance with any building codes in China has been "abysmal" in the past, with 50% of new buildings failing to meet national energy conservation standards in 2005.[117] Time will tell.

Similarly, tourism in India is projected to grow at nearly 8% per year from 2013 to 2023, and the hotel industry has been growing at 14%. The government seeks to deliver rapid growth and speedy implementation of new tourism projects throughout the nation. While tourism was once viewed as a luxury industry, travel and tourism is now understood by the government of India to be an important economic driver.[118] But questions of how local infrastructure can meet the demands for speedy growth are surfacing. Water-stressed regions of India are struggling to deliver water to local populations, and solid waste disposal and waste management are serious concerns. Most landfills and dumping sites pose serious health threats for people living in their vicinity, and there are problems of contaminated soil and water.[119] The hotel community will either need to create their own systems or work with governments to solve these issues. The management of waste, water, and energy will be critical concerns for the hotel community in India for the foreseeable future.

All-Inclusive hotels

The hotel industry struggled in the difficult years of the great recession of 2009–2011 and is coming back with excellent growth prospects. But lessons were learned. Franchising jumped, and traditional chains rapidly offloaded fully owned properties into arm's-length branded relationships. Since then, another business model is also gaining momentum – All-Inclusive hotels, which package all costs at the point of sale into one competitive price.

Born in the Mediterranean with Club Med in the late 1950s, All-Inclusives have evolved from their origins as party resorts for "swingers" and are now marketed as safe, affordable, family and honeymoon friendly Caribbean vacations in destinations like Jamaica. Sandals, known as the King of All-Inclusive Resorts, with 24 properties in 7 Caribbean countries, pioneered the upscaling of All-Inclusives and continues to be a pioneer in the delivery of an experience that includes absolutely every cost: gourmet food, premium brand drinks, airport transfers, taxes, and all land and water sports activities.[120] All-Inclusives have long been the bête noir of the responsible tourism community. The advocacy organization Tourism Concern has watchdogged this industry over the last decade, with documented concerns about labor rights and the lack of local benefits to regional economies.[121]

Since the 2008–2010 recession, hotels in leisure destinations have lost increasing market share to cruise lines, which are in essence, mobile All-Inclusives. With their megaships now rolling out annually, cruises are able to offer glamorous, upscale holidays at highly competitive costs, making them the fastest growing, most profitable leisure industry in the world. In order to compete, the hotel industry must consider other options to maintain their market. All-Inclusives are becoming a more attractive business to offer affordable pricing to the middle-class family trade and to create a one-stop shop alternative for hotel customers. Industry sources confirm that the All-Inclusive model is now taking hold and being explored by all the major brands. For the hotel trade, this is a no-brainer. There are higher margins, with occupancy rates that are much less seasonal. As hotels struggle to contain costs, they must avoid waste and inefficiencies, and this relies on having a more steady market flow that can allow companies to better project what to procure when and how to avoid waste. It is also key to managing a labor force, which in highly seasonal destinations like the Caribbean is a real challenge. Another important key to the All-Inclusive trend is that the owner of the hotel is more likely to control the booking of the hotel, thus lowering commissions and removing the entire model from the fierce pricing models and constant discounting found online. While tour operators in the U.S. have seen lower returns for their once steady market for Florida and California, there is now a move toward providing the same secure and reliable set of services by offering the entire trip through the All-Inclusive booking engine, which is owned by the *hotel company*.

With the investment of the German development corporation (GIZ), sustainable tourism leader Klaus Lengefeld has been able to look at the question of economic leakage from All-Inclusives. In the first analysis anywhere in the world, GIZ looked at four All-Inclusive resorts in the Dominican Republic, a prime All-Inclusive destination. GIZ found substantial salary benefits to the poorest sectors of society and 40% of all agricultural produce coming from small and medium-sized farms and cooperatives. And in a similar study in Nicaragua, he found that waiters and chambermaids working for All-Inclusives in Nicaragua were paid double what was paid in nearby small hotels.[122] Lengefeld argues that the All-Inclusive trade is much preferable to most other forms of economic development, such as coffee. He notes that 20% of a $2,500 All-Inclusive trip to Sandals in Jamaica remains in country and is redistributed via wages, supplies, and services.

That same $500 from one client would take years to produce by buying fair trade coffee because only 10% of every cup stays in Jamaica, and the value per cup is so much less, about 1 penny![123]

In Europe, recent research shows that the market share of the All-Inclusive hotel offer is growing by double digits.[124] Here, vertically integrated tour operators now seek to control the entire travel experience and are making their All-Inclusive hotels exclusive to their clients. TUI Travel, one of the largest tour operators in the world, has moved quickly to create "unique and differentiated" hotels that they own from top to bottom, to battle the online trade. Their strategy is giving the company greater operating profits, up 20% in 2013.[125] TUI is proudly touting unique holiday visits at hotels that are exclusive to their clients. The predictability of costs and profit margins may be helping the company to achieve higher standards. They now offer 1,200 hotels with sustainability certifications. Their airway, TUIfly, is delivering the lowest carbon airline experience in the world, and they make increasingly strong commitments not only to lowering environmental impacts but to the delivery of economic benefits to local communities. (See Chapter 6.)

The response to this trend is tepid among environmentalists and sustainable tourism advocates, which have long criticized All-Inclusives for exporting the majority of their economic benefits. Dr. Linda Ambrosie, in her groundbreaking research on tourism taxation, reveals that tax incentives for All-Inclusives leads to deficits in local municipalities and that these businesses are more likely to maintain their operational profits offshore,[126] a problem of policy that will be investigated in greater depth in Chapter 8. If municipalities are losing money due to offshoring and tax incentives, this can lead to the industry costing more than it generates in tax revenues, leaving regions such as Cancun in a deficit without the resources to cover the costs of replacing its aging infrastructure.

The question of how beneficial the All-Inclusive trend can be for local economies is not clear-cut. While hotel corporations expand into All-Inclusive terrain, they may be using the system to pay lower taxes, which will undercut the destination. But the business model may be inevitable as a means for the hotel industry to gain and keep loyalty and to provide exclusive offerings affordable for middle-class families. The All-Inclusive trend helps travelers to enjoy their time traveling and relieves them of planning. While the trend may lead to lower tax revenues at the local level, more attentive public policy could help to alleviate this problem.

Hotel business models in transition

Young Asian and Indian travelers are transforming the hotel industry with tastes and preferences that demand new offerings. The new Asian travel market will account for one-third of all travel spending by 2020. And they are no longer traveling alone strictly for business. They are taking their business and family life with them, enjoying an increasing number of holidays, and, like the Japanese of yesteryear, they often travel in groups.[127]

Airlines, airports, and hotels found in travel hubs define the new travel economy, for both transport and the provision of digital communication zones. Connecting to mobile devices is surely by far the number one demand from travelers when they deplane, even perhaps above the need to visit a restroom. Travelers want hotel lobbies, restaurants, and airport concourses that provide good wireless, cellular, and Internet service. Smart gateways, such as the Dubai airport, offer wireless for free, an easy and effective means of attracting the global travel marketplace. The electronic, digital environment quickly connects travelers to restaurants or local sightseeing options. TripAdvisor reviews and Google maps quickly guide us to preferred attractions, restaurants, and hotels. The enabled digital traveler knows that jumping on OpenTable, Yelp, or TripAdvisor gives instant access to a universe of options that are reviewed objectively.

Nonetheless, travelers of the future need personal guidance to connect to the real culture of the places they are staying, and hotels still have an essential role here, if they personalize services and avoid overcommercialized approaches. Hotel restaurants that are capturing this trend now seek to attract local guests and help visitors mix with locals to create genuine authenticity. Local arts, culture, and conversation awaken the visitor to local ambience. And this will be increasingly important, as an expanding universe of options makes it increasingly easy to break out of the hotel bubble. Travelers can now easily stay with families in local homes or even on couches in small apartments thanks to Airbnb, which has become a multibillion-dollar business by connecting local home owners to the travel economy through a simple and effective booking engine. And this may well just be the tip of the iceberg of a whole new, less formal "hotel" industry. The digital market for hotels is expanding rapidly, making Airbnb the manager of more rooms than any other hotel company in the world as of 2015 with over 2 million rooms for rent.[128]

Hotels and sustainability in the green economy

Digital tools are facilitating a shared economy in hospitality, which is booming, and is without question the most energy-efficient way to travel. In a study released in July 2014, Airbnb announced that Airbnb guests use 63% less energy than hotel guests, based on a survey of 8,000 hosts and guests worldwide.[129] It does not take a survey to prove that resource sharing within existing buildings is much more efficient than building new hotels. The digital economy has created a market for helping to arrest the rapid growth of hotels worldwide, and this trend is of great significance for slowing the growth of new hotels even if that is not fully recognized as a benefit to society, yet.

As the bulge in middle-class travelers moves upward on the bell curve, resources will become increasingly scarce and in demand, and hotels will more often be in competition for water and energy with local users. The travel and tourism economy may be increasingly essential to trade, but it cannot justify placing the burden for energy, water, waste, and waste water treatment 100% on local societies that cannot afford to improve their infrastructure. This will cause conflict, and with conflict there is business risk.

ITC India hotel: case study

The case study of ITC in India is an important one because this company is one of the few that is speaking out about the role of the hotel community in solving the larger problems of society as part of their investment in development. Their proposal is to achieve proper incentives for greening the travel industry via cooperation between government and the powerful and efficient private sector. Chairman of the Board of the ITC Corporation Shri Y.C. Deveshwar speaks about social imbalances and the global challenges of food energy, water security, and sustainable livelihood creation, calling for India to enlarge its Green GDP and ensure sustainable growth. He claims that there is "immense transformational capacity of business in innovating business models that can synergistically deliver economic and social value simultaneously."[130] As we explored in the early part of this chapter, water will become one of the most contentious issues as hotels continue to use over 10 times more water per person than local residents. ITC has already set policies to recycle gray water and use it for toilets and drip irrigation, saving 40% in water consumption.

And while the hotel community begins to explore their role in society at large, they will certainly recognize that hotels can play a major role in transforming local agricultural models, which throughout the developing world have depended upon exports, with a low-value proposition for the local producer. ITC is already supplying 30% of its food and beverage operations from providers within 100 miles. As a conglomerate, the firm has been assisting farmers through an initiative called e-Choupal, which connects farmers to regional markets via an Internet system on mobile phones that allows them to bypass middlemen and capture a larger portion of the profits.

The economics of hotel development must be looked at anew, and the management of growth will require significant new investment to cover the long-term costs for necessary water, solid waste, and sewage treatment infrastructure. If the hotel community invests in renewable energy sources for their destination, the reward could be a total of 40% reduction in carbon impacts. Given that the growth of tourism will outweigh all efficiency gains, it is incumbent on the hotel community to consider investment in alternative energy to pay for a relevant portion of clean development infrastructure at a time when nations must reduce their footprint and help destinations to work toward carbon neutrality.[131] Such a strategy is likely to be impeded by the global franchising system, which places cost burdens on local owners and encourages short ownership life spans with frequent turnover. This business model is likely the reason the hotel world has not taken a stronger stand on the creation of renewable energy grids, which will lower carbon footprints but require a certain investment in energy infrastructure.

Dominican Republic's hotel energy use: case example

The COP 21 Agreement in Paris will affect hotel infrastructure worldwide, and umbrella brands and small hotel owners will need to lower the total carbon emissions on a global basis, with each nation establishing specific goals. The case of the Dominican Republic illustrates the importance of the hotel community in this process. And it helps students of future national tourism management to begin to picture how a country, such as the Dominican Republic, will have to work with the hotel community to meet the goals of the carbon-cutting agreement set out in Paris.

The Dominican Republic (DR) has ambitions to lower its carbon footprint by 25% by 2030, and the country has set out to secure climate financing to make itself attractive to sources of international financing for new renewable energy projects in order to meet its carbon reduction goals. Worldwatch Institute created a road map for transition to renewable energy for the Dominican Republic,[132] an island known for its extensive hotel infrastructure. And, in fact, hotels account for 43% of all commercial sector energy use on the island according to the report. Energy accounts for 40% of expenditures for hotels with fewer than 300 rooms, and efficiency measures for hotels could lower costs and total carbon impacts by 24%. The Dominican Republic is home to 25% of all of the hotel rooms in the Caribbean region; consequently, the decision to move towards renewables would have an impact on the entire region. But the long-term question for the nation is how the private sector and government will raise the capital to convert a good portion of their grid to renewable energy.

The DR has excellent renewable energy potential. And with projected growth of demand, renewables could fill the gap, making the DR increasingly less dependent on imported fossil fuels. The conversion is also technically feasible from an installation and distribution perspective. The cost to convert infrastructure and add new capacity is estimated to be $78 billion, but this investment could save the country more than $25 billion versus the business-as-usual projection by 2030, or nearly 40% per kilowatt-hour, based on detailed calculations and scenarios depending on energy efficiencies achieved and the growth of the economy. If the investment is not made, the annual cost of electricity generation will increase sharply, and the DR could face a near doubling of energy costs.

The country is at a crossroad and must make a decision. Financing options are tricky, but one sector has reliable access to finance: the hotel and tourism industry. The hotel community is a clear candidate for what is known as bundling, which reduces financing and capital costs for the development of new grid projects. Of course, hotels cannot take full responsibility, but the report states that the barrier to involvement is largely "the need for education, outreach, capacity building regarding the benefits and opportunities of climate finance, as well as the will to implement energy upgrades."[133]

This vision of the hotel as a central driver of local economies is new. Travel is connecting powerful sectors of the economy and bringing much higher value to local people than is usually acknowledged. Hotels can purchase renewable energy and food and develop alternative systems for water and waste to bring down their own operating costs while significantly improving local livelihoods. These ideas demonstrate that large-scale tourism can go beyond efficiencies to opportunity. It is truly time for development agencies, financiers, and governments to partner with the hotel economy to finance the energy, waste, and water treatment systems required to ensure the new booming travel economy meets its true potential as a sustainable development tool.

Conclusion

The expansion of hotels across the world will continue to be driven by a global economy that is bringing hundreds of millions of new travelers into the marketplace in the next several decades, one-third of whom will be from Asia. Most will stay in hotels, a large proportion of which will be small hotels or shared accommodations purchased through such companies as Airbnb, primarily welcomed by nonprofessional hosts. The number of small property owners without formal management experience has impeded progress on the environmental management of the industry even as the larger brands seek to advance – though often slowed by their own legal franchise and management contract arrangements. All hotels need to respond to 21st-century challenges of managing their footprint, and a much larger proportion of the industry must begin to measure and manage their water, waste water, solid waste, and energy systems in a systematic way. Governments in growing destinations can no longer afford to underwrite the cost of these sustainability systems for the private sector, under the guise of providing tax incentives for economic growth. Such economic growth is coming at the expense of local citizens, and new public–private systems to manage tourism growth must be the goal for achieving more sustainability in future.

New agreements to manage and measure the hotel community's carbon impacts through the Hotel Carbon Measurement Initiative will help advance global reporting and make it more transparent for all parties to work with. This should be the beginning of a new era where more information from the hotel community is not only compatible but publically reported in a consistent fashion to help nations to better understand how to collaborate with the private sector on lowering impacts.

Not all hotels will be able to manage these sophisticated reporting systems, given that most are small businesses. But all hotels can benefit from making their systems more efficient, with an average of 10–20% savings on energy alone. Water is becoming an asset that cannot be overused by visitors without any awareness of how it is affecting local villagers, as the case in Zanzibar demonstrates. It is far too precious, and case studies clearly show that even small hotels can afford to install water-saving devices that will not only pay them back in short order for the trouble but reduce the potential of regional conflict with residents.

The landslide of solid waste generated by travelers is very often unanticipated by innocent locals, who begin to burn the plastics or feed themselves from waste dumps, often without any knowledge of the danger of the toxic substances they are exposing themselves to. This is a growing emergency in beautiful destinations around the world, and while hotels may be seeking to lower their waste, studies show that there must be local cooperation and efforts to work with communities to help them manage the new responsibilities that they have for managing waste. Hotels must become active in conserving and managing not only their solid waste but also food waste, which can be cut dramatically with substantial savings for hotel restaurants in the category of 2–3% monthly.

A significant percentage of hotels worldwide operate without waste water treatment. While figures differ, this chapter demonstrates that hotels can now easily take advantage of cost-effective alternative waste water treatment systems that would preserve the vital coral reef ecosystems and beautiful transparent waters where many hotels operate at affordable prices per room. It is impossible to imagine that hotels can justify allowing their sewage to flow into the sea. And yet it is common practice to simply not pay for waste water treatment at all, therefore leaving local residents to pay for the problem or allow local waters to become polluted.

The greatest challenge ahead for the hotel community is to become part of the financing formula for achieving a lower-carbon economy by 2030. Hotels in tourism-driven economies, such as the Dominican Republic, compose nearly 50% of commercial sector GHG emissions. The conversion to renewable energy at the grid level will require financing and the bundling of energy needs to create projects at scale. Hotels have a great deal to gain not only from lowering their footprint but from contributing to the transition of global grids, and they must join in the effort to attract capital, donor funds, loans, and long-term financing to lower their own costs and create a lower-carbon economy wherever they are located worldwide.

Surprisingly, travelers seem to accept that hotels may not be responsible for the many environmental problems they observe around them during their visits to emerging economies, but global information systems are continuing to improve. The time has arrived for the global hotel community to invest in sustainable development on a proactive basis. There is much that the next generation of managers can achieve to lower the hotel industry's footprint, and this will help them to maintain the confidence of their market as it becomes aware of the issues of sustainable tourism and lower their operational costs as resources become increasingly scarce and more expensive.

Notes

1 United Nations Development Programme (UNDP), n.d., *Tourism Lab, Zanzibar Development Vision 2020, Results for Prosperity*, UNDP. New York, page 10
2 Blair, David, 2010, Zanzibar's island idyll under threat, *Telegraph Travel*, September 23, http://telegraph.co.uk/travel/travelnews/8019912/Zanzibars-island-idyll-under-threat.html, Accessed September 14, 2016
3 Ibid.

4 Tourism Concern, 2012, *Water Equity in Tourism, a Human Right – A Global Responsibility*, Tourism Concern, London, page 10
5 Global Climate Partnership, May 2012, *The Economics of Climate Change in Zanzibar, Vulnerability, Impacts and Adaptation*, Technical Report, UKAID, for Department for International Development
6 Tourism Concern, *Water Equity in Tourism*
7 Global Climate Partnership, *The Economics of Climate Change*, page 41
8 Trejos, Nancy, 2016, Hotel CEOs talk mergers, Airbnb and robots, *USAToday*, March, 22, http://usatoday.com/story/travel/roadwarriorvoices/2016/03/20/hotel-ceos-talk-mergers-airbnb-and-robots/81423222/, Accessed September 14, 2016
9 Business Wire, 2014, Global Top 10 Hotel Operators: Company Guide- Peer Analysis & Company Profiles, Research and Markets, February, http://researchandmarkets.com/research/5ptq7f/global_top_10, Accessed September 14, 2016
10 Gössling, Stefan and Kim Philip Schumacher, April 2010, Implementing carbon neutral destination policies: issues from the Seychelles, *Journal of Sustainable Tourism*, Vol. 18, No. 3, 377–391
11 Ricaurte, Eric, July 2011, Developing a sustainability measurement framework for hotels: toward an industry-wide reporting structure, *Cornell Hospitality Report*, Vol. 11, No. 13, 2011
12 Hawkins, Rebecca and Paulina Bohdanowicz, 2012, *Responsible Hospitality: Theory and Practice*, Goodfellow Publishers Ltd., Oxford, page 60
13 Grosbois, D., 2012, Corporate social responsibility reporting by the global hotel industry, commitment, initiatives and performance, *International Journal of Hospitality Management*, Vol. 31, 896–905
14 Marriott, 2011–2012, *Sustainability Report*, Revised September 19, 2012, page 7
15 Gulliver Business Travel, 2009, Why hotel chains don't own many hotels, *The Economist*, February 19, http://economist.com/blogs/gulliver/2009/02/why_hotel_chains_dont_own_many, Accessed September 14, 2016
16 Ceballos-Lascurain, Hector and Hitesh Mehta, 2002, Architectural Design, in *International Ecolodge Guidelines*, eds. Ana L Baez, Paul O'Loughlin, and Hitesh Mehta, The International Ecotourism Society, Burlington, VT
17 *The Economist*, February 19, 2009, Outsourcing as you sleep
18 Ibid.
19 O'Neill, John W. and Mats Carlback, 2011, Do brands matter? A comparison of branded and independent hotel's performance during a full economic cycle, *International Journal of Hospitality Management*, Vol. 30, 515–521
20 *The Economist*, February 19, 2009, Outsourcing as you sleep
21 Ibid.
22 O'Neill and Carlback, Do brands matter?, pages 515–521
23 Rushmore, Steve, Jaime I. Choi, Teresa Y. Lee, Jeff S. Mayer., January 2013, *HVS Hotel Franchise Fee Guide*, Mineola, NY
24 Hilton Worldwide, 2014, *Travel with Purpose 2013–2014 Corporate Responsibility Report*, Hilton Worldwide, McClean, VA
25 Marriott, 2014, *2014 Sustainability Report*, Marriott International, Bethesda, MD
26 Our Global Presence, InterContinental Hotels Group, http://ihgplc.com/index.asp?pageid=749, Accessed September 20, 2016
27 Ecorys SCS Group, September 2009, *Study on the Competitiveness of the EU Tourism Industry*, Rotterdam, The Netherlands, pages 44–45
28 Bonilla-Priego, Maria Jesus, J.J. Najera, and X. Font, 2011, Environmental management decision-making in certified hotels, *Journal of Sustainable Tourism*, Vol. 19, No. 3, 361–381
29 Ibid.
30 Fotiadou, Sofia, 2013, *Potential Implications of Improved Water Management in Tourism Accommodation Facilities in Semi-Arid Destinations*, Unpublished Manuscript, Environmental Management of International Tourism Development class paper, Sustainability

and Environmental Management program, Harvard Extension School, Cambridge, MA, page 20
31 Revenue per Available Room (RevPAR), Investing Answers, http://investinganswers.com/financial-dictionary/ratio-analysis/revenue-available-room-revpar-807, Accessed September 14, 2016
32 Hawkins and Bohdanowicz, *Responsible Hospitality*, pages 60–61
33 Bodhanawicz, Paulina, 2006, *Responsible Resource Management in Hotels, Attitudes, Indicators, Tools and Strategies*, Doctoral Thesis, School of Industrial Engineering and Management, Department of Energy Technology, Royal Institute of Technology, University of Stockholm
34 Ricaurte, *Developing a Sustainability Measurement Framework*, page 9
35 Ibid.
36 Ibid.
37 Ibid., page 28
38 Chong, Howard G. and Eric E. Ricaurte, May 2014, Hotel sustainability benchmarking, *Cornell Hospitality Report*, Vol. 14, No. 11, Cornell University, Ithaca, NY, page 14
39 Ibid., page 15
40 Hawkins and Bohdanowicz, *Responsible Hospitality*, page 60
41 Ricaurte and Bohdanowicz comments in class, 2013, *Environmental Management of International Tourism Development*, Harvard University Extension, Cambridge, MA
42 Ricaurte, *Developing a Sustainability Measurement Framework*, page 9
43 Ibid, page 11
44 Chong and Ricaurte, Hotel sustainability benchmarking, page 21
45 World Travel and Tourism Council and International Travel Partnership, June 2013, *Hotel Carbon Measurement Initiative, vol. 1.1, Methodology*, World Travel and Tourism Council and International Travel Partnership, London
46 Chong and Ricaurte, Hotel sustainability benchmarking, page 21
47 Gössling, Stephan, Paul Peeters, C. Michael Hall, Jean-Paul Ceron, Ghislain Dubois, La Vergne Lehmann and Daniel Scott, 2012, Tourism and water use: supply, demand, and security: an international review, *Tourism Management*, Vol. 33, No. 1, 1–15
48 Ibid., page 4
49 Ibid.
50 Ibid., page 18
51 Fotiadou, *Potential Implications of Improved Water Management*
52 Ibid., The Zanzibar figure from the Tourism Concern report equals the amount documented in Thailand by Gössling
53 Global Climate Partnership, *The Economics of Climate Change*, page 41
54 Ibid., page 7
55 Ibid., page 10
56 Bohdanowicz, *Responsible Resource Management*
57 Legrand, Willy, Philip Sloan, and Joseph S. Chen, 2013, *Sustainability in the Hospitality Industry*, Second Edition, Routledge, Oxon
58 Goldstein, Kevin and Ritu V. Primlani, February 2012, *Current Trends and Opportunities in Hotel Sustainability*, HVS Sustainability Services, Mineola, NY and Gurgaon, India
59 TPS Eastern Africa Limited, 2012, *2012 Annual Report and Accounts, Serena Hotels, Safari Lodges and Camps Hotels*, Nairobi, Kenya, page 17
60 Brighter Planet, 2012, *Hotel Energy and Carbon Efficiency*, Vermont and San Francisco, page 8
61 2013 Lodging Industry Profile, American Hotel & Lodging Association, http://ahla.com/content.aspx?id=35603, Accessed September 14, 2016
62 Energy and Environmental Analysis, 2005, *CHP in the Hotel and Casino Market Sectors*, Prepared for US EPA CHP Partnership, page 3, epa.gov/chp, Accessed September 14, 2016
63 Chong and Ricaurte, Hotel sustainability benchmarking, page 15
64 Hendrix, Niki, 2008, *Power Quality and Utilization Guide, Hotels, Leonard Energy*, Leonardo-energy.org, Accessed September 14, 2016

65 Bohdanowicz, Paulina, Angeloa Churie-Kallhauge, and Ivo Martinac., 2001, *Energy-Efficiency and Conservation in Hotels – Towards Sustainable Tourism*, 4th International Symposium on Asia Pacific Architecture, Hawaii, Table 4, page 6

66 Ibid., page 8

67 Hendrikx, *Power Quality and Utilization Guide*, page 9

68 Zipkin, Amy, 2005, Some like it hot, *The New York Times*, January 23, http://nytimes.com/2005/01/23/jobs/23COOL.html?pagewanted=all&position=&_r=0, Accessed September 20, 2016

69 Bohdanowicz et al., *Energy-Efficiency and Conservation in Hotels*, page 8

70 Ceballos-Lascurain, Hector and Hitesh Mehta, 2002, Architectural Design, pages 55–92, in *International Ecolodge Guidelines*, eds. Hitesh Mehta, Ana L Baez, and Paul O'Laughlin, The International Ecotourism Society, Burlington, VT

71 Jetwing Eternal Earth Program, Vil Uyana – A Man Made Nature Reserve, http://jetwingeternalearthprogramme.com/jetwing-vil-uyana.html, Sirgiriya, Sri Lanka, Accessed September 30, 2016

72 Hawkins and Bohdanowicz, 2012, *Responsible Hospitality*, page 78

73 United Nations Environment Programme and World Tourism Organization, 2012, *Tourism in the Green Economy-Background Report*, Madrid, Spain, page 74

74 Whitbread PLC, Corporate Responsibility Report 2014/15: Another Good Year for Good Together, https://whitbread.co.uk/content/dam/whitbread/pdfs/corporate-responsibility/reports-presentations/CSR_2014_15_v7.pdf, Accessed May 1, 2016

75 Hawkins and Bohdanowicz, *Responsible Hospitality*, page 80

76 Energy and Environmental Analysis, *CHP in the Hotel and Casino Market Sectors*

77 CHP in the Hotel and Casino Market Sectors, https://epa.gov/chp/chp-hotel-and-casino-market-sectors, Accessed September 30, 2016. Site provides updates on the EPA CHP program and contact info.

78 Abu Dhabi bails out Dubai with $10B, 2016, CBS News, September 14, http://cbsnews.com/news/abu-dhabi-bails-out-dubai-with-10b/, Accessed September 20, 2016

79 Supreme Council of Energy, 2014, State of Energy Report Dubai 2014

80 Ibid., Section 5 Green Buildings, Existing Buildings

81 Ibid.

82 Boone, T., Nalin Kant Srivastava, and Arohani Narain, June 2013, *ITC Hotels: Designing Responsible Luxury*, Indian School of Business, HBR published ISB016 Business Case

83 Ibid.

84 Tsoutsos, T. Stavroula Tournaki, Carmen Avellaner de Santos, and Roberto Vercellotti, 2013, Nearly zero energy buildings application in Mediterranean hotels, *Energy Procedia*, Vol. 42, 230–238

85 Accor Hotels, Planet 21 Research, Open platform of shared knowledge about sustainable development in the hotel industry, includes all reports between 2011–2016 with new results in 2016, http://accorhotels-group.com/en/sustainable-development/planet-21-research.html, Accessed September 30, 2016

86 Hawkins and Bohdanowicz, *Responsible Hospitality*, page 86

87 Jao, Angelina Hsiaoping, December 13, 2012, *Las Vegas Sands Corporation's Environmental Responsibility Managing Its Waste Impacts*, Unpublished Manuscript, Environmental Management of International Tourism Development class paper, Department of Sustainability and Environmental Management, Harvard Extension School, Cambridge, MA

88 Pearl River Delta Area, Wikipedia Commons, https://commons.wikimedia.org/wiki/File:Pearl_River_Delta_Area.png, Accessed September 20, 2016

89 The Composition of Waste Disposed of by the UK Hospitality Industry, 2011, WRAP, http://wrap.org.uk/sites/files/wrap/The_Composition_of_Waste_Disposed_of_by_the_UK_Hospitality_Industry_FINAL_JULY_2011_GP_EDIT.54efe0c9.11675.pdf, Accessed September 20, 2016

90 Lephilibert, Benjamin, 2015, *Food Excess: The Neglected Issue in the Hospitality Industry*, Light Blue Consulting, Bangkok, Thailand

91 Ibid.

92 Accor Hotels, Planet 21 Research, Open platform of shared knowledge about sustainable development in the hotel industry, includes all reports between 2011–2016, http://accorhotels-group.com/en/sustainable-development/planet-21-research.html, Accessed September 30, 2016; Accor Hotels Environmental Footprint, 2016, impact of producing 1 Kg of different foods chart, page 35
93 Shamshirty, Elmira, Behzad Nadi, Mazlin Bin Mokhtar, Ibrahim Komoo, Halimaton Saadiah Hashim, and Nadzri Yahaya, 2011, Integrated models for solid waste management in tourism regions: Langkawi Island, Malaysia, *Journal of Environmental and Public Health*, Vol. 2011, article number 709549, doi:10.1155/2011/709549
94 MacGregor, James, in Linked In dialog, *Volunteer Tourism*, February 2014
95 Perawongmetha, Athit, 2015, Reuters from Cambodia: Children scavenge at Anlong Pi rubbish dump near Angkor Wat in Siem Repa by David Sim, March 20, 2015, *International Business Times*
96 Gössling, S., P. Peeters, C.M. Hall, G. Dubois, J.P. Ceron, L. Lehmann, and D. Scott, D., Tourism and water use, page 18 and UNEP, 2011, *Tourism Investing in Energy and Resource Efficiency*, United Nations Environment Programme, Paris, page 473
97 Ibid.
98 Scoullos, M.J., 2003. *Impact of Anthropogenic Activities in the Coastal Region of the Mediterranean Sea*, International Conference on the Sustainable Development of the Mediterranean and Black Sea Environment, May, Thessaloniki, Greece
99 UN Water, 2014, *Analytical Brief, Wastewater Management*, UN Water at the World Meteorological Organization in Geneva, Switzerland
100 Sutherland, K.P. et al., 2010, Human sewage identified as likely source of white pox disease of the threatened Caribbean elkhorn coral, acropora palmata, *Environmental Microbiology*, Vol. 12, No. 5, 1122–1131, and Reopanichkul, P., R. Carter, S. Worachananant, and C.J. Crossland, 2010, Wastewater discharge degrades coastal waters and reef communities in southern Thailand, *Marine Environmental Research*, Vol. 69, No. 5, 287–296
101 Orengo, Peter, 2013, Beach hotels poisoning our seas with raw sewage, *Standard Digital*, October 3, http://standardmedia.co.ke/lifestyle/article/2000094825/beach-hotels-poisoning-our-seas-with-raw-sewage?pageNo=3, Accessed September 20, 2016
102 Cocke, Sophie, 2012, Raw sewage still flowing into local waterways despite legal agreement, *Honolulu Civil Beat*, March 29, http://civilbeat.com/articles/2012/04/03/15367-raw-sewage-still-flowing-into-local-waterways-despite-legal-agreement/, Accessed September 20, 2016
103 Honey, Martha, Eric. Vargas, and W.H. Durham, April 2010, *Impact of Tourism Related Development on the Pacific Coast of Costa Rica*, Summary Report, Center for Responsible Travel, Stanford University, Washington, DC, pages 82–83
104 Masterson, Kathy and Douglas Scott, October 25, 2010, *City and County of Honolulu, Hawaii, Wastewater System*, Fitch Ratings, New York
105 Cascadia Green Building Council, March 2011, *Toward NetZero Water: Best Management Practices for Decentralized Sourcing and Treatment*, Program of International Living Future Institute, Portland Oregon, page 94
106 Ibid.
107 Some hotels in the Masai Mara Game reserve begin treating their wastewater. 2016, Video Report, Kenya NTV, https://youtube.com/watch?v=D3phKRaHZ-Q, Accessed September 30, 2016
108 Jean-Michel Cousteau Fiji Islands Resort, Ocean Futures Society, http://oceanfutures.org/about/collaboration/jean-michel-cousteau-fiji-islands-resort, Accessed September 20, 2016
109 Ibid., Accessed August 7, 2014
110 AcquaLogic Advanced Biological Treatment System, AcquaLogic, http://acqualogic.com/wastewater/, Accessed September 20, 2016

111 Jackson, Giles, December 2015, *Ecological Sanitation Strategies for Coral Reef Conservation: The Opportunity for Hotels and Resorts*, Environmental Management of International Tourism Development class paper, Department of Sustainability and Environmental Management, Harvard Extension School, Cambridge, MA
112 UNEP, 2003, *A Manual for Water and Waste Management: What the Tourism Industry Can Do to Improve Performance*, Paris
113 IPK International, *World Travel Trends Reports, 2013/2014*, Messe Berlin GmbH, Berlin, Germany, page 6
114 Ibid.
115 Oxford Economics for InterContinental Hotels Group (IHG), n.d., *The Future of Chinese Travel, The Global Chinese Travel Market*, Page 15
116 EplerWood International and Linda Powers Tomasso, September 12, 2011, Environmental Management of Tourism in China, White Paper prepared for Wyndham Hotel Group, EplerWood International, Burlington Vermont, Unpublished Manuscript
117 Ibid, page 32. According to Tomasso,

> Half of all new buildings fail to meet national energy conservation standards. Compliance records vary throughout China, with southern and rural areas lagging behind rigorous, multi-step inspections in larger cities. Laxness over compliance arises from inadequate training, too few qualified inspectors, and weak political muscle. The lack of transparency makes it hard to control standard construction basics, never mind sustainability criteria in China's frontier areas. Corruption, skimming, graft and the use of inferior materials by which to skirt building regulations are widespread. Cost-cutting often occurs between design submission and actual construction as with the use of cheaper insulation.
>
> Outside Beijing, building regulators are getting tougher as the central government threatens local officials responsible for failure to meet the new energy efficiency mandates of Five Year planning documents. The MEP has halted a number of projects for failure to meet environmental regulations. Steep penalties ensue – in theory – for violators. Builders must modify deficiencies or retrofit non-compliant buildings. Certifiers can lose their license. Even officials who fail to curb high-energy consuming or similar non-compliant projects can face criminal charges.

118 Tourism & Hospitality Industry in India, IBEF (India Brand Equity Foundation), http://ibef.org/industry/tourism-hospitality-india.aspx, Accessed September 20, 2016
119 Boone et al., *ITC Hotels*
120 Sandals Montego Bay, http://sandals.com/about/, Accessed September 20, 2016
121 Action for Ethical Tourism, Tourism Concern, https://tourismconcern.org.uk/all-inclusives/, Accessed September 20, 2016
122 Lengefeld, Klaus, n.d., *Community-Based and Small Enterprises or Mass Tourism: Which Type of Tourism Do Developing Countries Need?* no publisher, provided by author
123 Ibid.
124 Smalley, Sarah and David Trunkfield, 2010, *Master Class Travel Trends*, The Travel Convention in Association with Telegraphmediagroup, http://thetravelconvention.com/archived/2010/documents/travel_trends.pdf, Accessed October 1, 2016
125 Ashton, Jane, January 2014, Lecture for Harvard Extension School Class TUI Travel and Sustainable Development, PowerPoint, Cambridge, MA
126 Ambrosie, *Sun & Sea Tourism*
127 The Futures Company, 2013, *The New Kinship Economy, from Travel Experiences to Travel Relationships*, InterContinental Hotel Group, http://library.the-group.net/ihg/client_upload/file/The_new_kinship_economy.pdf, Accessed September 20, 2016
128 Chafkin, Max, 2016, *Can Airbnb Unite the World?* Fast Company http://fastcompany.com/3054873/can-airbnb-unite-the-world, Accessed September 20, 2016

129 New study reveals a greener way to travel: Airbnb community shows environmental benefits of home sharing, 2014, Airbnb, July 13, https://airbnb.com/press/news/new-study-reveals-a-greener-way-to-travel-airbnb-community-shows-environmental-benefits-of-home-sharing, Accessed August 2014
130 Chairman speaks – 2011, Address by Chairman, Shri Y C Deveshwar, ITC, http://itcportal.com/about-itc/ChairmanSpeakContent.aspx?id=1111&type=B&news=chairman-2011, Accessed September 20, 2016
131 Gössling and Schumacher, Implementing carbon neutral destination policies, 377–391
132 Konold, Mark et al., 2015, *Roadmap to a Sustainable Energy System Harnessing the Dominican Republic's Sustainable Energy Resources*, Worldwatch Institute, Washington, DC
133 Ibid., page 140

5

I'LL FLY AWAY

Airlines, airports, and the global circulation of travelers

Introduction

Long-haul flights over 1,000 miles cause over 80% of the greenhouse gas impacts of air travel. While environmentally aware individuals around the world agree that climate change is a proven high-risk challenge for our planet's environment, few have changed their travel behaviors. There are many ways to discuss how air travel impacts the environment. The German organization Atmosfair maintains that each individual should emit no more than 5,000 pounds of CO_2 per year to avoid overloading the planet with emissions that would change average global temperatures more than 2 degrees Celsius from measures taken in 2013.[1] While this personal measure of responsibility toward our planet's climatic stability is certainly just one way of looking at how we as human beings can respond to the global carbon budget, it is a benchmark to consider. One long haul flight of 4,000 miles can quickly exceed a single person's annual climate CO_2 budget. According to Atmosfair's measurements, a short-haul flight of under 500 miles would generate between 462 and 793 pounds of carbon, while an inefficient long-haul round-trip flight from Munich to New York can generate over 5,700 pounds of carbon.[2]

Even in Europe, where climate change awareness is quite high, individuals are loath to sacrifice their long-haul travel, and many remain unaware that travel is a high-climate-impact activity. In studies on consumer climate awareness in the UK and Norway, some subjects did not know there was any difference in carbon impacts between taking a train, using a car, or flying. Many did know the difference, but traveler guilt over carbon emissions caused by air travel was barely registering. Generally, well-educated travelers were not willing to cut back on their travel, particularly long-haul travel.[3]

The importance of travel to the subjects of this study was clear. They wanted to broaden the horizons of their families and considered world travel an important part of their personal identities. None of the subjects planned to postpone their ambitions to travel across the planet or make their travel behavior consistent with their efforts to cut their carbon impacts at home.[4]

One Norwegian interviewee commented,"I know it is problematic and I should be concerned more, but my conscience is not bad because of taking flights at all."[5] A certain carbon-conscious elite is more likely to condemn frequent short-haul travel, perhaps because trains and buses are available in Europe that are high speed, efficient and a much better alternative for those seeking to avoid carbon impacts. But this same elite is not willing to sacrifice their long-haul trips to the U.S. or places even much farther, such as New Zealand, despite the extraordinary carbon impacts. New Zealand is over 11,000 miles from Europe, making it one of one of the longest trips a European traveler can take. Researchers wanted to know how a sample of Europeans viewed the carbon impacts of such a trip. Most were not willing to stop flying long distances because they are not willing to forgo these more remote travel opportunities, even while they condemned frequent short-haul travel in Europe with their "carbon consciences."[6]

This less than rational decision making about travel is probably not surprising to any reader. There is almost no social ethic emerging that equates long-distance travel with poor environmental citizenship. Environmentally conscious citizens of the world continue to do all they can at home to reduce their footprint but are really not willing to put air travel into the equation. Outside of Europe, discussions about the carbon impacts of air travel are still confined to a very rare minority, although the media is finally starting to incorporate this discussion into articles about travel going green.[7]

For the researchers involved in the study on New Zealand travel, there is an urgent need to understand why because "there is hardly any other human activity that contributes to such substantial amounts of greenhouse gas emissions in a comparably short period of time."[8] Yet long-haul travel is considered to be acceptable by most individuals because it is infrequent and should not be sacrificed because of its enormous personal value.

One UK traveler gives his thoughts on the topic:"I do see the impact. [sic] I'm not happy about the fact that it's not particularly good for the environment. But not unhappy enough yet, that's the truth."[9]

Some of the subjects in the UK were beginning to calculate their air travel emissions using a carbon calculator. This was clearly bringing home the substantial impacts of air travel. Creating a global travel ethic that responds to the need to consider climate impacts would take considerable effort and heavily financed campaigns not dissimilar to antismoking campaigns. But there are few indications that the industry or even the general public would embrace this agenda. The consumer's responsibility to lower their travel carbon impacts is an issue that is largely avoided by the industry. Like so many issues in the world of travel, the industry is not prone to stressing the environmental impacts of travel with their customers. It interferes with the beauty and attractiveness of what they are selling. And, as a result, few consumers are highly aware that their long-haul travel is one of the most carbon-intensive activities in their lives.

Environmental impacts of air travel: key issues

The environmental impacts of air travel can be divided into two main categories: greenhouse gas emissions due to air travel and issues specific to airports, which include noise, air quality, waste management, and storm water runoff. In order to better understand the impact and potential mitigation of greenhouse gas emissions, we need to first understand the airline industry and the way it is managed and governed. Airport-specific environmental management issues have increasing significance in a global economy where airports grow without planning and frequently without input from local residents. All of these areas of concern will be discussed in the latter half of the chapter.

Pressing decisions are required to improve the aviation sector's carbon reduction performance. The blizzard of facts and figures can be daunting. This chapter will help the reader weigh the feasibility of current efforts to mitigate and reduce impacts, analyze the technologies that are being cued up to lower the industry's growing footprint, and come to an understanding of the most promising options to improve the airline industry's carbon performance in the first half of the 21st century.

Expert authors such as Stefan Gössling offer in-depth discussion of the industry's metrics and solutions that go beyond airline efficiencies and new technologies.[10] Gössling rightly points out if travel behaviors do not change, a reduction of only 36% in total emissions from the 2005 baseline can be expected by 2035, with no viable path to absolute reductions. To date, all of the scenarios show that carbon emissions for aviation will require more than what technology or best practice can offer.[11]

These statistics may vary, but the path to neutrality must be the ultimate goal. Every reader needs to understand how the different mitigation options can be managed and improved in future. Work on biofuels, better aircraft technology, and improved navigation will all yield better results with more investment and professional engagement. The process of reaching the best possible results for the industry have important implications and require full engagement from the global economic, business, NGO, policy, environment, and political leadership and their constituents worldwide. For all students of the industry, each one of these areas of engagement is required knowledge that can lead toward a wide range of new professional opportunity and options for contributing to lowering the industry's impacts.

The status of the airline industry and managing greenhouse gas emissions

Air travel is the essential ingredient of the new travel economy. It fuels globalization and trade and has become the lynchpin of global business activities. Aviation as an industry sector has created millions of new jobs and produced trillions of dollars of new GDP worldwide. It has opened up entire economies to the world and brought greater prosperity to the remotest parts of the planet. While once a luxury or a business necessity for a privileged percentage, it is now an essential part of the global economy. But swelling passenger numbers flying across the globe have set off alarm bells in the corridors of the Intergovernmental Panel on Climate Change (IPCC) and other organizations that review the most significant sources of greenhouse gases in our global economy. Until recently, aviation was believed to be a source of greenhouse gases that could be brought under control relatively quickly, but solutions have been slower to emerge than expected, and agreements to address these problems are being pushed into the future just as the industry's growth rate is skyrocketing.

At present, 2 billion passengers travel annually. But in the decade to come approximately 6 billion travelers are expected to fly every year, an astounding 200% increase.[12] And to make matters worse, travel distances are doubling.[13] Air travel is generally known to be the source of 2.5% of all carbon emissions from all human activity worldwide. While this may sound acceptably small, the acceleration of the aviation economy will cause carbon emissions to outstrip almost all other sources of carbon pollution if business as usual reigns.[14] The Intergovernmental Panel on Climate Change (IPCC) estimates that aviation will climb to 5% of the earth's total emissions by 2050, with worst-case scenarios as high as 15%. And the climate risks from air travel are not limited to CO_2. Nitrous oxide, nitrogen dioxide, sulfur oxides, soot, and water vapor emitted during air flights all impact the atmosphere where it is most vulnerable, the stratosphere. This makes aviation an increasingly dangerous climate actor and one that must act to reduce its use of fossil fuels in order to offset its increasingly ambitious global growth agenda.[15]

Responsibility for both aviation and shipping greenhouse gas emissions lies with the International Civil Aviation Organization (ICAO). The airline industry and ICAO have worked hard to address the problem. In 2010, ICAO set out optimistic goals for aviation to become carbon neutral from 2020 forward. But a flurry of analyses revealed that even with the most aggressive implementation of solutions, total greenhouse gas emissions could be expected to be reduced by only 6.4% by 2050,[16] far below the original expectations of a 10% reduction of total emissions by 2020.[17]

New solutions are being researched, and the industry is now committed to a global market-based mechanism that will set a cap on global emissions from aviation for the first time, to be implemented by 2020. But the aviation industry is running late due to overly optimistic projections that must now be laid to rest. While aviation enthusiasts continue to tout the amazing ability of the industry to innovate and respond, they must do so quickly. As the 2014 IPCC report lays out, delaying mitigation actions will reduce global options and lead to increasing risks. The longer the global community waits, the greater the chance is that warming will cause "severe, pervasive and irreversible impacts."[18]

Institutional efforts to manage aviation impacts

Even though public dialog on reducing and diminishing travel habits is barely registering worldwide, the airline industry is nonetheless under considerable pressure to lower their emissions. International emissions of CO_2 from aviation have not been seriously considered until very recently and were not part of the Kyoto Protocol, the landmark climate agreement that came into force in 2005 calling for industrialized countries to reduce collective emissions of greenhouse gases by 5.2%, compared to 1990 levels. And while the protocol applies only to nations who became signatories, famously leaving out both the U.S. and China, it nonetheless set the stage for all other international climate agreements, without any consideration of the crucial role of aviation.[19]

In October 2013, ICAO met to review all of the options for managing aviation's growing climate impacts. The debate was lively. But hopes had already been dashed that the industry could meet its aspirational goals set in 2010 to become carbon neutral by 2020. The dialog was thorough, with a great deal of technical expertise at the table. The main approaches identified by ICAO[20] for reaching carbon neutrality are:

- Market-based mechanisms, otherwise known as emissions trading,
- Aircraft efficiency,
- Air traffic control efficiency, and
- Alternative fuels.

As pressures to lower aviation's impacts have increased, the debate on how well each of these factors can contribute to carbon neutrality is intense. ICAO sets goals that are "aspirational," largely because there is no proven route to achieving neutrality. Projections are based on best available knowledge in each category, and the assumptions being used are constantly being updated. In 2010, leading aviation thinkers believed that aircraft efficiency, together with biofuels and improved air traffic control systems, could bring the industry into alignment much more quickly than projections ultimately showed. Now, due to influential research performed at Manchester Metropolitan University, it is now assumed that that only market-based mechanisms can save the day.

Market-based mechanisms

Dr. David Lee and his team from Manchester Metropolitan University (MMU) published three extraordinarily influential articles after the 2010 ICAO meeting that demonstrated with extensive modeling and detail that that carbon neutrality would not be possible in 2020 and would not even be likely by 2050, without the use of market-based mechanisms (MBMs).[21] Other studies by industry analysts have confirmed these results.[22] As the industry continues to tout progress on biofuels, there have been impediments to taking these fuels to scale, and aircraft efficiencies, while impressive, are not able to decrease jet fuel use adequately, even with the most efficient aircraft in history rolling out. One of the primary barriers to progress is the fact that global fleet replacement takes 25 years from research and development to service.[23]

Market-based mechanisms have attracted skepticism over the years because they allow corporations to invest in cheap carbon offsets and avoid investment in more costly long-term projects to produce carbon efficiency. But because the aviation industry escaped the Kyoto Protocol process, there was no cap set on aviation's carbon emissions and as a result no price set on excess carbon, as well as no mechanism to pull the industry into a global system of carbon trading. This led to inaction and a legal battlefield in Europe over their right to establish a system in their airspace to help meet global goals for lowering emissions.[24]

Originally, ICAO did not think a global system was necessary, and in 2004 delegated decisions for market-based mechanisms to regional authorities. The European Commission was ready and waiting, as they already had a plan called the European Union Emissions Trading System (EU ETS), which has become the largest international trading system for greenhouse gas emission allowances ever implemented.[25] This system puts a value on carbon and asks corporations within the EU to receive or buy carbon allowances in order to meet carbon reduction goals. In 2012, aviation was included for the first time in the system. The EU suggested that all flights worldwide, either landing or departing from EU airports, would have to participate. This idea that the scope of emissions to be regulated went beyond European borders attracted a great deal of controversy. Normally, regional trading schemes manage carbon emissions only within their region. But the EU ETS sought to lower the entire aviation industry's carbon emissions by regulating airlines from takeoff to landing. They set a cap for carbon emissions at 97% of the baseline measured in 2006–2007 for all airlines landing in Europe and began the process of putting a price on all additional carbon emissions over this baseline and requiring emissions trading.

The EU ETS system immediately faced a number of serious challenges. First, the price of carbon in the EU was burdened by a surplus of roughly 2 billion credits even before the airlines entered the system. The price was at $5 USD a ton in 2012, as opposed to $41 a ton in 2005, making carbon so cheap that the trading system was not providing an incentive to industry to improve. Second, 26 nations, including America, China, and Russia, refused to participate in the aviation component, declaring it a breach of sovereign rights to regulate. In the U.S., the Senate proceeded to prohibit participation in the EU ETS, and President Obama signed the legislation into law in January 2012 right after his 2011 reelection.[26] Due to these pressures, the EU ETS system for airlines was suspended in February 2013 until the important October 2013 ICAO meeting of all parties.

The influential CEO of IAG, the parent company of British Airways and Iberia, Willie Walsh, spoke of the many challenges ahead at an event not long after the February EU ETS suspension. He stated that regional systems could distort the market, causing pricing to be higher where compliance was required. Mr. Walsh urged patience, suggesting that time is needed to be certain there was a level playing field globally.[27]

But time is precisely what was lacking. Dr. Lee's Manchester team went into high gear addressing the issue of timing for action in their August 2013 publication right before the ICAO October meeting. They presented the fact that postponement of the implementation of market-based mechanisms would have serious consequences. Looking at every scenario, Lee and his team concluded that the EU ETS could produce the single most important improvement in climate impacts, a reduction of about 15%. The postponement of the EU ETS or of the effort to create a global system would not only slow down the process of addressing escalating carbon impacts, it would undermine the ability of the industry to respond before reaching a climate tipping point.[28] From a climate science perspective, patience was the wrong response.

But this advice was not heeded. The final decision was made in October 2013 in Montreal by the full ICAO Assembly. It was agreed that a global market-based mechanism for all aviation would be finalized in 2016 and implemented in 2020, without any binding commitment. Europe was allowed to regulate and implement the EU ETS only for the emissions from flights occurring over its own airspace until 2020, when the new ICAO system would theoretically replace it. The European Commission is putting this reduced system in place, which may still be the most effective effort to reduce aviation emissions in the world. Even though the system is now only within Europe, the response has remained unanimously negative from other countries, including the U.S.[29] Though the EU ETS did not survive as a global carbon trading market for airlines, the aviation sector has embraced the idea of a global market-based mechanism that could be an effective means to achieve a fair emissions trading system for aviation. What is of great concern is that ICAO will not make a global trading system a reality. Many observers note that ICAO may not have the vision or the leadership capacity to create a system that can work worldwide. Some experts in carbon trading argue that "given the proximity of ICAO to the industry it is meant to regulate, and its long track record of avoiding climate action, it is highly unlikely that it will adequately regulate greenhouse gas emissions from flying."[30] And, worse, timely solutions are sorely lacking in the next 6 years precisely when they are needed. Bill Hemmings of the Brussels-based NGO Transport and Environment (T&E) states that ICAO has halted the only real and effective system (the EU ETS) while they agree to agree on something in 2016. The Assembly's resolution looks like Swiss cheese, full of holes.[31] But most mainstream organizations and experts on climate negotiations agree the ICAO agreement is essential. It was passed on October 6, 2016 with 83% of international flight operations prepared to launch voluntary mechanisms in 2021.[32] According to the Environmental Defense Fund, there is an immediate need to develop implementation standards to ensure there is proper accounting and training packages for local authorities in developing countries to help them take advantage of global expertise on lowering carbon emissions impact from their burgeoning aviation sectors.[33]

Aircraft efficiency

The aviation industry is committed to improving its technology substantially to reduce its carbon impacts. This strategy relies largely on the manufacturers of aircraft. Boeing and Airbus are the most prominent players in the industry, and both have award-winning design programs that stress lowering environmental impacts.

Aircraft manufacturers have sought to improve fuel efficiency by streamlining aerodynamics of their newest aircraft bodies. They continue to enhance laminar flow over the aircraft bodies to reduce drag and use super lightweight laminate materials to construct planes that are more buoyant than ever but also safe and durable.[34]

Boeing, a $60 billion plus Seattle-based company, operates in 70 countries with 22,000 suppliers. Their corporate goal is to become the number one aerospace company in the world. Their most popular commercial aircraft are the Boeing 737, 747, 767, and 777, which earn 70% of their annual income. Boeing has an edge over Airbus and other competitors in using composite lightweight material in the development of their aircraft.[35]

Boeing's touts its ecoDemonstrator program, which, the firm states, accelerates technology that will improve environmental performance and sustainability. This is a public–private program in cooperation with the U.S. Federal Aviation Administration's CLEEN (Continuous Lower Energy Emissions and Noise) program. Some of the technologies being tested are:

- Wing-adaptive trailing edges to improve fuel efficiency at takeoff, climb, and cruising;
- Hydrogen fuel cells to power cabins when at terminals;
- Improved flight planning technology; and
- Exhaust nozzles made of ceramic that can withstand hotter high-efficiency exhaust gases and make engines lighter and more efficient.[36]

At the center of Boeing's efforts has been the introduction of the 787 Dreamliner, which offers double-digit carbon footprint reduction. The Dreamliner is made of a composite material made of carbon fiber, aluminum, and titanium, which is so lightweight it allows for more long-haul options and fewer fuel-inefficient stops at intermediate hubs. Boeing states the planes consume 20% less fuel, making it 20% more carbon efficient.[37] Many airline companies are purchasing the Dreamliner to meet sustainability standards and goals.

The Airbus Group Innovation-EOS eco-assessment program has been in direct competition with its new A380, a very large airplane that carries 42% more passengers than its nearest competitor.[38] The A380 was designed using life cycle analysis and ISO 14040 systems that seek to deliver a cradle-to-cradle cycle of production, recovery, and remanufacture of all materials. But academic research makes it clear that the manufacturing of the aircraft has a very small impact compared to its heavy fuel use during operations.[39] The aircraft's size ensures that its operating costs will be 15–20% lower per seat. Its massive body can hold as many as 800 passengers, which is forcing airports to create new facilities to accommodate it. Though Airbus can produce impressive efficiencies per passenger, its approach to creating a more efficient plane by supersizing it, raises questions. While the plane itself is constructed in a cradle-to-cradle system of perpetual recycling of all materials, it is surely contributing to the never ending cycle of new construction at airports that, in many parts of the world, is not governed by ISO 14040 design standards. This will lead to a great many new carbon-intensive buildings that could greatly undermine the lowered carbon results per passenger of the new aircraft, if the new construction was taken into account. This is a problem shared with the cruise line industry, which also is driving massive construction of new ports but does not take responsibility for these carbon emissions, when reporting on per-passenger carbon accounting.

Informed analysts who watch aviation design suggest this is a competition between two titanic firms with ambitious visions for delivering the most sustainable flying machine. The A380's massive size and the B787's carbon fiber composites are both aimed at making their aircraft the most fuel-efficient, technologically empowered aircraft in the 21st century.[40] But it remains to be seen if pre-passenger efficiencies will lower total carbon emissions or simply spur more growth. If the number of long-haul flights doubles, and manufacturers build aircraft to carry them at the lowest possible operating cost, this translates into bigger aircraft, more new airport construction, and higher and higher carbon emissions from aviation, even with lower emissions per passenger.

Harvard Extension student Tania Fauchon argues that the industry should put more investment in hybrid aircraft technology in the 2015–2025 time frame, which could provide at least 85% emissions reduction over entire flights – a figure that is well beyond other investigated options. The technologies she evaluates in her paper are:

- Green taxiing, known at EGTS (electric green taxiing system), which will lower emissions in airport suburbs but have little impact on the carbon emissions of flight;
- Distributed Electrical Aerospace Propulsion (DEAP), developed by Airbus and Rolls-Royce, which can reduce emissions by 50% during cruising and taxiing and to 0% during descent; and
- Boeing (subsonic ultra green research) SUGAR, which builds on improved aircraft aerodynamics and usage of electrical power from batteries during flight, offering zero emissions during the cruising period of flight.

Table 5.1 offers a good overview of the advantages of each technology. Fauchon suggests that it is urgent for the industry to put a much higher investment in hybrid technology than the projected $2 billion in order to make these advances available to the next generation of aircraft models to be developed and built between 2025 and 2030. If this investment is not accelerated, by either Airbus or Boeing with

TABLE 5.1 Emissions by technology as a proportion of typical emissions per aircraft mission segment (%)[41]

Technology	*Taxi*	*Takeoff*	*Climb*	*Cruise*	*Descent*	*Landing*
EGTS	0	100	100	100	100	100
DEAP	50	100	100	10	0	100
SUGAR Volt	100	100	100	0	100	100

strong public sector support, she contends the industry will miss the opportunity to install these important technological breakthroughs and lower prospects to achieve neutrality through technological and efficiency improvements.[42]

Air traffic control efficiency

On an average day in Beijing in 2013, over 1,500 planes take off and land, many of which are late. China's biggest airports have the worst flight delays in the world, with only 18% departing on time. Chinese air traffic has grown 20-fold in 20 years, and Chinese airports are quickly becoming the busiest in the world. Air traffic controllers lack incentives to move airplanes efficiently, and even at best Chinese flights are kept much farther apart than in the U.S. or Europe. Flights in China are crowded into highly restricted routes, and alternate routes are prohibited by the Chinese military for security reasons. Planes stack up in bad weather, all inefficiently burning fuel. As air traffic problems worsen in China, systems to create more efficient air patterns are delayed by concerns raised by the military.[43]

While air traffic control systems are aging globally, the improvement of this infrastructure could be one of the most feasible, least expensive approaches to cutting total global emissions from aviation in the near term. Most experts agree that some basic fixes are required. The technology to improve existing air traffic control systems is available, and the problems associated with inefficient flight patterns could lessen overall aviation impacts by over 10%.[44]

Author of the report *The Future of Mobility*, Dr. Chris Carey states that present control systems are part of the cause of air traffic congestion worldwide. If they are improved, landing and takeoffs could be quicker, and planes could fly closer together and avoid headwinds.[45]

Airspace is managed by national organizations. Like China, each country sets its own priorities based on national interests. There has been surprisingly little international cooperation. Some of the problems include:

- Management in non-radar areas, where flights must be kept on highly consistent, limited flight pathways to ensure they can be tracked until they reenter radar tracked areas;
- Limited accuracy and time lines of weather information for long-range flight planning; and
- Limitations on airspace for political, security, or environmental concerns.

Large-scale investment in more integrated air traffic control systems will create more fuel-efficient routing. The goal is to create a more flexible, nimble use of airspace, known in the trade as FAU (flexible airspace use). FAU facilitates better coordination between the military and civil aviation authorities on the use of national airspace. This requires a definition of airspace policies, day-to-day allocation of airspace according to national needs, and real-time management of these day-to-day allocations.[46]

Regional and national systems are emerging in Europe, Asia, and the U.S. These systems have the capacity to triple the number of flights that air traffic controllers can handle. Of course, it is likely that this technology will simply just lead to more aircraft volume and undermine all efforts at emissions reduction. But, at present, this is not the assumption.

The Federal Aviation Administration (FAA) of the U.S. has a road map for transition to NextGen air traffic control technology. They estimate that by 2020, NextGen improvements will reduce air traffic delays by 41%. Better performance-based navigation in busy hubs, improved navigation equipment on airplanes, and immediate weather information are all part of the deployment of this system. The goal is to reduce total emissions and create a better flying experience according to the FAA and to cut costs for fuel for the airlines.[47]

Satellite positioning systems and high bandwidth communications systems are the technologies that are making this next generation of flight planning possible. Quieter, smoother descents, more options for routing long-haul flights, and use of prevailing winds will be the outcome. These technologies can be implemented relatively quickly if there is public–private cooperation. The world's commercial fleets will have to embrace the technology and purchase what is required, but the savings involved should make this transition the most attainable in the short term for lowering aircraft emissions.[48]

Alternative fuels

In 2008, the airline Virgin Atlantic made history. A jumbo jet in its commercial fleet flew powered, in part, by biofuel. Virgin Atlantic has sought to take leadership in the biofuel business and has challenged its competitors to stay in the game or fall behind. To maintain their leadership position, Sir Richard Branson held a press conference in 2011, announcing a breakthrough in aviation technology. Branson stated that Virgin's new fuel production process will recycle carbon monoxide to create fuel out of lost, unused energy, resulting in "doubling the use of oil."[49] He went on to state that the announcement of this new technology "is the most important announcement I have made in my lifetime."[50]

Virgin's biofuel technology partner, LanzaTech, is using gas-fermenting microbes to convert carbon-rich waste gases and residues from steel mills to aviation fuel.[51] The company suggests the process can apply to 65% of the world's steel mills, allowing the company in theory to meet nearly 20% of commercial aviation needs.[52] Virgin predicts its new jet fuel will reduce its aircraft carbon emissions by 50%. While Virgin had suggested the fuel would be operable in 2013, further testing and certification extended the process. LanzaTech has built two demonstration-level ethanol facilities in China, and more financing is needed to push the biofuel production process forward.

While Virgin is pursuing a unique conversion of carbon monoxide using ancient microbes that predate cyanobacteria and algae,[53] most of the aviation industry is pursuing the conversion of plant materials into biofuels to begin to lower their carbon impacts and lessen global dependence on fossil fuels. This endeavor is being advanced by governments together with industry across the globe. There is a clear mandate to innovate and produce alternative fuels, and the potential for biofuels is real. Research and development funds are flowing generously to academic institutions and to start up industries to launch this new industry.

Test flights are taking place regularly. Among many flights, Airbus A380 flew with a 40% blend of alternative fuel called gas to liquid (GTL) in 2008, an historic flight. Air New Zealand flew with biofuels derived from the Jatropha plant also in 2008.[54] As the excitement builds, the real barriers to achieving production at scale are also emerging. Creating all-new production facilities and amassing the appropriate food or forest product waste in sufficient quantities has still to be achieved. The biofuel industry has momentum. But like all new industries, it will need a great deal of support from both industry and government to scale up sufficiently to begin to meet even a small portion of global demand.

Biofuels can be produced from any renewable, biological carbon material. Crops that are rich in sugars (such as sugar cane) have been a primary source of biofuels for biodiesel production for decades. Corn has also been used extensively to create ethanol, which is used in nearly all gas stations in the U.S. These fuels would not be suitable for powering jets. The second generation of biofuels will be sourced from plants not used for food, all of which could be suitable for jet fuel. The principle behind plant-based jet fuels is simple: the carbon dioxide absorbed by the plant materials is released again to the atmosphere, making the fuels – in theory – carbon neutral. But the total life cycle of production is not equal for all biofuels, and researchers are carefully comparing the total carbon emissions for the production of each.

The use of land, forests, and wetlands to produce food stocks for the production of biofuels has come under increasing investigation. As more and more food crops are converted to fuel sources, prices of food commodities have started to rise. Millions of acres of land may have been converted to cropland because of the demand for more first-generation biofuels. According to one research report, 3.45 million acres of cropland, roughly the size of Northern Ireland, was required to produce 3% of the UK's fuels for transport in one year.[55]

The aviation industry has already acknowledged that food crops are not suitable for jet fuel production. The main biomass sources are forestry and agricultural residues, waste materials, and inedible energy crops. Second-generation crops under investigation include the following:[56]

- Jatropha is a plant the produces seeds that can produce fuel. It grows in difficult soil conditions.

- Camelina has a high lipid oil content making it suitable only for biofuel production. It is used in rotation with other food crops when land would otherwise remain fallow.
- Algae is highly touted as a major potential source for jet fuel. Its rapid growth rates and oil-rich production capacity makes it the possible winner for production capacity but only in the long term, according to experts.[57]
- Carinata is an oilseed crop in the mustard family suited to production in semi-arid areas.[58]

Pathways to produce aviation fuel have increased considerably since 2010. The original systems to produce biofuels were:

- Synthetic Fischer-Tropsch (FT), produced by biomass gasification;
- Hydrogenated esters and fatty acids (HEFA), produced from plant oils, animals fats, algae, and microbial oil; and
- Hydrogenated pyrolysis oils (HPO), produced from gasification of lignocellulosic biomass.[59]

Newer processes include:

- Alcohol to jet (ATJ), which produces alcohol from raw materials and then chemically synthesizes jet fuel;
- Direct sugar to hydrocarbons (DSHC), which is produced by fermenting sugar and hydrogenation;
- Lignin to jet fuel (LJF), a chemical-catalytic conversion of lignin converted to jet fuel.[60]

Jet fuel producers must develop fuels that can be mixed with existing diesel fuels on a "drop-in" basis in order to allow for a smooth transition to alternative fuel sources. Any other approach would require the industry to retool or redesign their engines, which would not be tenable.

New fuels have a rigorous testing regimen to ensure viability and safety. Tests must take place on the ground and in the air at takeoff and cruising speeds. Certification of fuels for commercial use is required. The Roundtable on Sustainable Biofuels (RSB) has emerged as a leading certifier setting global benchmarks, but there are over a dozen certifiers in Europe alone. The RSB is now authorizing third-party organizations, such as SCS Global, to perform third-party certification, auditing, and testing.

The European Commission (EC) set out a goal in 2011 for the production of 2 million tons of sustainable biofuels to be used in aviation annually by 2020. This is 3% of 2010 levels of jet fuel consumption in Europe – a modest projection in principle. But current reports indicate reaching this goal may be difficult. From a direct cost perspective, all biofuels will be significantly more expensive than fossil fuel–based jet fuels until 2020 and beyond. Even blends of only 5–10% biofuel will have uncompetitive prices.

The general principal is that the rising cost of carbon allowances should provide an incentive for the aviation industry to purchase biofuels, but this system has not proven itself in Europe where the carbon market is the most active in the world. The EU moved to tackle the oversupply of carbon in the marketplace at the end of 2013. A 2013 study led by the European Commission together with biofuels producers and the aviation sector, titled *Two Million Tons per Year: A Performing Biofuels Supply Chain for EU Aviation*, on Europe's capacity to meet its 2020 goals, which does not express the opinion of the EC, suggests that "no airline at present is in the position to voluntarily pay a biofuel premium in large volumes because the cost is too high, and all incentives in place are not sufficient."[61]

Europe is making progress with its hydrogenated esters and fatty acids (HEFA) biofuel industry, which is already producing biofuel for cars and other forms of transport. But the jet fuel industry is in its infancy. The industry will need financial incentives, sources of affordable sustainable feedstock, and scaled-up production plants. Feasibility for the production of jet fuels appears to be futuristic without heavy government support. For example, one researcher from Australia concluded that second-generation algae technologies to produce jet fuel would not become viable enough to break even until crude oil hits $1,000 per barrel. Feedstock storage is another issue. Even if the second-generation biofuel industry does not take up land that might otherwise be used for food production, the amount of waste materials from agriculture or forestry required to make viable quantities of jet fuel is beyond imagination, requiring the storage of enormous proportions.[62]

In the U.S., where the size and scale of agriculture dwarf the production in Europe, the U.S. Department of Agriculture has been investing in the deployment of first-generation biofuels for some time. The *Farm to Fly* program was launched in 2010 to serve as a catalyst for a successful U.S. biofuels industry. The economics of delivering jet fuel are far superior to the delivery of biofuels for automobiles. The largest 40 U.S. airports account for more than 90% of the jet fuel used by commercial aviation, a market worth $17–19 billion a year.[63] If producers can locate themselves near airports, the cost of fuel delivery would be low and improve the economics of the aviation biofuel business development program. It is likely that U.S. cities have more land available in the periphery of airports than Europe to build refineries and store adequate food or forest waste materials for production. Grants are backing biofuel research across the U.S. through its agriculture and forestry university programs, and regional solutions to producing biofuels based on locally available materials are being supported. Creative, economical solutions are on the table.[64]

The U.S. Federal Aviation Administration (FAA) established a Center for Excellence to study the potential of biofuels, and a study on the market costs of renewable jet fuels is remarkably optimistic compared to the studies in Europe. The system in place to support the purchase of new sources of alternative fuels was already established for the ethanol industry. While it is important to understand that the carbon emissions of ethanol are not superior to fossil fuels, the system of incentives to ethanol producers has worked wonderfully (some would argue too well). The system originated in the U.S. in 2005 as part of the Energy Policy Act, which mandated the production of ethanol from cornstarch. A progressive target of increased biofuel production and consumption was mandated by the federal government. A Renewable Identification Number (RIN) is assigned to each gallon of renewable fuel, and a subsidy is provided for every gallon of biofuel purchased to pay the difference between conventional and biofuels. As demand increases, the price on the alternative fuel lowers, until biofuels are at scale.[65]

The FAA has set a goal of 1 billion gallons of renewable jet fuel to be consumed in the U.S. each year from 2018 forward. Estimates for production are primarily based on the oily plant camelina, which is grown in rotation with wheat on 3–4 million acres of land that would otherwise be left fallow.[66] The use of tractors and fertilizer to farm camelina is built into the cost models and obviously contribute to a carbon footprint for these crops, but nonetheless the U.S. Navy estimates carbon emissions will be reduced by 80% compared to fossil fuels.[67] The U.S. military is setting out to use 50% renewable fuels by 2018, for a total of 0.65 billion gallons per year.

U.S. commercial airlines are mandated to use biofuels for 1.7% of their total projected jet fuel use in this same time frame, or 0.35 billion gallons per year.[68] Unlike in Europe, third-party studies predict that the industry will voluntarily purchase the renewable fuels because of the subsidies paid, which will be in the neighborhood of $2.86 USD per gallon of biofuel.[69] Under this system, the aviation sector would receive a $100 million USD fuel subsidy from the federal government to achieve goals for use of biofuels by 2018, reducing the use of conventional fuel by 1.7%.[70] But the system effectively creates a viable market and distribution system for biofuels in the U.S. that scales up over time, if federal support is not interrupted.

As new, cheaper systems for biofuels emerge, the subsidies will become lower and lower. While it is unknown at the time of writing how many new types of biofuels will be put into production at what cost, it is without argument that, as biofuels become increasingly economically competitive, they will contribute to a growing proportion of the total jet fuel requirements, if biofuel production can scale up at a pace that exceeds the growth rate of the U.S. aviation sector. Most experts agree that it may be as late as 2040 before biofuels have a real impact on total carbon emissions of the aviation sector, but the reality is that there is no real way of predicting the rate of advancement in this field. Airlines must also begin to disclose the percentage of their total fuel use that is composed of biofuels and display the certification system they are using.[71]

The frontier for biofuel production is China. In 2014, China was the second largest aviation fuel consumer at 20 million tons per year, with demand expanding 10% annually.[72] Early announcements on results have come from the Chinese company Sinopec, which projects it will produce 60,000 tons of biofuel a year from Jatropha.[73] The Virgin Atlantic project with LanzaTech is also producing on a test scale basis in China. In November 2013, the LanzaTech joint venture facility with the Shougang Jingtang Iron and Steel United Company and Tan Ming Group in China was certified by the Roundtable on Sustainable Biomaterials. It is estimated the process will reduce life cycle greenhouse gas emissions by 60%. Virgin Atlantic plans to start using the sustainable jet fuels on flights to and from China.[74]

As each system of biofuels scales up, there will be a healthy competition between airlines to include more biofuels. The major aircraft production companies, Airbus, Boeing, and Embraer, signed a memorandum of understanding in March of 2012 to work together on the development of drop-in, affordable aviation biofuels.[75] Virgin Atlantic is setting a standard for using an alternative fuel system that recycles existing gases from fossil fuels; the rest of the airlines are depending on the production of biofuel from waste agricultural products. Which will be the most cost-effective, receive the most beneficial subsidies from government, and how quickly the total scale of production advances will determine which system contributes the most to the goal to make air travel less carbon intensive.

Airport growth and environmental management

Nearly 3 billion airline trips transpired, either domestically or internationally, in 2012. This is the equivalent of one-third of the earth's total population taking one flight.[76] To be clear, 3 billion separate people did not fly in one year, as the statistic measures the number of passages, not unique passengers. But more and more members of our global society will travel in future, and each will travel more frequently. As a result, there is every expectation that air transportation will grow at about 6% annually, reaching 4 billion air passages by 2017 and at least 6 billion by 2020.[77] Figure 5.1 provides a visualization of how our globe is presently traveled using the top 7 airlines.

The monumental complexity of managing this number of flights and routes around the world is the responsibility of airports. As our earth becomes increasingly connected by air travel, airports have become the essential hubs of human activity,

FIGURE 5.1 Routes flown by the top 7 airlines by international passenger distance flown[78]

Ordered by scheduled international passenger kilometers flown in 2010 (Source: Wikipedia).

for both leisure and business. The geopolitical importance of seaports has always been well understood by historians. Kingdoms were built and defended around access to the sea. While airports may not have the same military significance partially because they can be built almost anywhere, they have extraordinary strategic significance in the world of global commerce and economic growth.

There is a well established link between air transport and growth in a nation's standard of living.[79] Countries that have recognized the strategic importance of airports are thriving economically, and those that are left out and are underserved by airlines are doing much more poorly. Aviation contributes approximately 2% to the gross domestic product of developed countries and has been leveraged by such nations as the United Arab Emirates, in particular the global hub of Dubai, to open up the nation to a broad array of investments. Nations that have not invested in airports have lost value. India, Nigeria, and the Philippines have aviation sectors that contribute only 1% of GDP, precisely because they lack good airport infrastructure.[80]

For example, the World Bank reports that Africa is the most underserved continent for air travel, with only 4% of the world's scheduled airline seats in 2010.[81] As of 2014, the lack of air connections at reasonable prices is a major impediment to the continent's economic growth. Experts on business in Africa who attended the Africa Accelerates annual conference at Harvard Business School in 2014 complained that travel in Africa can cost double or triple the price of air transport in Europe, America, or Asia. Without suitable airports, there are not enough routes to fly and infrequent connections. These problems have helped to hold the African economy back and were highlighted during the event as one of the most important problems that African countries must face to move their continent forward economically.

For Asia, Europe, and the Americas, the challenge of managing the global economy through these burgeoning strategic centers of business and leisure could not be greater. Airports must manage growth at rates of speed that would be difficult for any business. Their management is not private sector–led in most cases, precisely because airports have so much strategic national significance. Airport commissions, quasi-governmental organizations that report to mayors of metro areas, are the management bodies most often charged with the Sisyphean task of managing airport growth. Each time they complete an upgrade or expansion, another is needed. Their inability to meet these challenges is fairly evident to every passenger. The inconveniences are notorious and seem only to get continuously worse.

Few who travel by air could fail to notice the poorly planned, hodgepodge nature of many airports, especially in the U.S. Just as one new terminal or wing of an existing hub is completed, there is more construction, more disturbances to reaching gates, shuttles that travel strange routes between gates, and ever present concrete barriers and orange cones marking routes for buses carrying hapless passengers around ever expanding airport footprints.

But it is only fair to consider how difficult it is to actually manage and run an airport. Any business trying to handle 6% growth rates annually might have a problem keeping up. Few businesses have to manage this type of growth rate. And the planning and management of such massive construction projects in most societies takes years. The design requirements of airports are obviously specialized, and the size and scale of these huge construction projects must be planned within very restricted areas, often surprisingly close to major urban areas.

Aerotropolis is the new name for regional centers of growth driven by airports. Aerotropoli are urban centers, which can include office buildings, convention and exhibition centers, cultural attractions, leisure and recreation facilities, fitness facilities and spas, and medical wellness facilities in addition to the usual airport functions and duty-free shopping zones. They represent significant employment benefits to their locations. An increasing number of airports employ more than 50,000 workers. They offer hundreds of restaurants and retail outlets managed by concessionaires.[82]

Hong Kong is the leader in aerotropolis design. Hong Kong's SkyCity serves Hong Kong and southern coastal China. It is linked to efficient highways and express train connections to the city and Hong Kong's Disney theme park. A massive town near the airport houses 45,000 workers and their families with schools, churches, shopping and medical facilities. It is seamlessly connected to high-speed jet ferries to reach the booming Pearl River Delta. The ferries link the delta's major manufacturing centers to air transport, shuttling goods at high speed with efficiency for air freight.[83] (See Figure 5.2.)

FIGURE 5.2 Hong Kong Airport from planning documents[84]

The challenge of environmental management is as monumental as the size of these facilities. Each airport needs to plan carefully to avoid environmental damage to their sites, keep noise to a minimum, avoid destroying valuable ecosystems as they continue to grow across vulnerable landscapes, offer solutions to growing problems of toxic runoff and skyrocketing CO_2 and NO*x* air emissions, and lower their gigantic energy emissions footprints. Each facility must be planned as part of major urban areas, and yet there are still few examples of successful regional plans. The planning of airport facilities to meet the needs of a growing flying public takes time, and yet the rate of growth to meet those needs will continue to skyrocket.

The number of decision makers involved in every airport expansion program and the stakeholders they serve is almost beyond description. There are the government and municipalities, the corporate airline companies, the many businesses that depend on airports such as rental car businesses, and the concessions within airports including restaurants and retail. There is also an extraordinary array of workers involved in all kinds of professional roles, from engineers and pilots, to baggage handlers, to restaurant servers. The management of these complex business and infrastructure systems requires enormous expertise. Airport cities are business ventures that increasingly must be managed by private sector partners and specialized urban planners. Their challenge is to create efficient well managed hubs for transport that have also become entertainment centers, nexuses for global air freight shipping, and centers for a host of local and international businesses.

Expertise of managing these sprawling developments in future will have to include environmental management to a much greater degree than is presently advocated. Environmental research on the impacts of airports emerged around 2000 and has revealed a great many troublesome issues. More research is required to ensure the safety, health, and well-being of people living near these growing mega developments. The issues reviewed in this section are:

- Noise,
- Air quality,
- Waste management, and
- Storm water runoff and management.

Airport noise issues and solutions

Aircraft noise is the single most important environmental issue facing managers of growing airports and their aerotropoli. Noise can raise blood pressure, affect sleep, and disrupt digestion; it is linked to coronary heart disease and appears to be connected to negative impacts on cognitive development in children. It can cause aggressive behavior or a feeling of helplessness. The increasing frequency of aircraft movements has escalated impacts on communities beneath air flight paths. The aircraft industry continues to reduce aircraft noise with new technologies, but the benefits of such actions are largely offset by the growth of air traffic, as will be discussed.

Communities adjacent to airports have become ever more proactive in their efforts to protect themselves from unhealthy noise levels. Research on airport noise impacts on human health, though limited, is more definitive every year. Noise levels have been notoriously difficult to challenge because the measurement systems used average noise over a 24-hour period in the U.S., a system that is due for reform. While standards for noise are still evolving, efforts to mitigate the impacts of noise on residents by airports has become increasingly standardized in developed world areas where residents and local governments are organized and can demonstrate health impacts. Those communities that are not prepared to protect their soundscape can suffer, as the case study on the City of Inglewood demonstrates.

Los Angeles airport noise: case study

Harvard Extension student Kabibi Adunagow reviewed the case of noise standards and mitigation for the City of Inglewood, which is within 2 miles of the Los Angeles International Airport (LAX).[85] This minority community has grown from being an African American neighborhood in the 1960s–1980s to becoming a Hispanic majority neighborhood in the 2000s. These minority residents do not feel politically empowered to manage the difficult problems of the noise they face and are probably typical of many communities who live near airports.

Los Angeles International Airport (LAX) has been working on mitigating the impact of aircraft noise pollution on the community of Inglewood for over 3 decades. The LAX airport had 565 domestic flights daily in 2011 and 1,000 international flights weekly. The State of California set an acceptable level of noise at 65 decibels as part of their comprehensive Noise Control Act in 1973. Airports develop contour maps to identify and validate the boundaries of noise impacts to comply, and in areas that exceed these noise levels, LAX instituted a soundproofing program in Inglewood in 1992 for the installation of acoustically rated noise-reducing products in windows and doors, insulation in attics, and new or updated air conditioning systems. Since the program began, LAX has spent $167 million in Inglewood financed by FAA improvement grants and airport passenger facility charges and revenues.

Despite these investments, the number of complaints of airport noise has increased. LAX created a noise complaint line available 24 hours a day and found that the number of complaints are increasing by 20% annually. Author Adunagow[86] notes the increase in complaints aligns with the increase in volume of air traffic. In 2000, a Community Noise Roundtable was created by LAX to identify noise concerns in the surrounding communities. While over $100 million in noise pollution mitigation funding had been spent in Inglewood by 2011, only half of the residents have insulated their homes. Adunagow theorizes that the residents have little or no trust in the system, partly because of the long-term noise impacts they have had to endure.

As a result of their unwillingness to accept assistance, Inglewood residents are not protecting themselves or their children from the impacts of noise. Adunagow recommended that the community residents take part in Community Noise Roundtables, take advantage of mitigation soundproofing funds, and continue to register their concerns through the noise complaint lines. Without this participation, Inglewood residents are bound to suffer health impacts, especially among the children, that have been well researched.[87]

Research indicates that loud noise causes changes in blood pressure, affects sleep, and disrupts digestion. Specific research on airport noise pollution performed by the European Union suggests that living near an airport is a risk factor for coronary heart disease and stroke.[88] Airport noise especially affects children's health. A study on LAX airport in 1980 showed children in neighboring communities, such as Inglewood, had higher blood pressure than those living farther away. Studies in Germany found that airport noise at night harms children's nervous systems, disturbs dreams, and does not allow the body to rest properly.[89] Research in the UK[89] and New York[90] found that children living near airports lagged behind in reading by two months for every 5-decibel increase, even after socioeconomic differences were considered.

Noise measurement issues and solutions

The measurement of noise levels is not as straightforward as it may seem. In the U.S., average sound levels are measured in increments called Ldn (or DNL), which is the average equivalent sound level over a 24-hour period, with a penalty added for noise during nighttime hours between 10 p.m. and 7 a.m. During the nighttime, 10 decibels are added to reflect the impact of noise of local populations.[91] While this is a reasonable measure for constant noise, it is poor at measuring the impacts of loud infrequent noise events, such as jets flying over communities. This measure is nonetheless used by the FAA, even though the FAA's own research documents agree that the Ldn is a poor predictor of aircraft noise impacts.[92] For example, the Ldn system does not include sound frequencies below 1,000 hertz. A standard of 65 decibels, Ldn is the guideline for mitigation action. The EPA has long recommended that 55-decibel Ldn trigger actions to protect citizens from the impacts of noise, but the FAA until recently has cited that economic considerations must be the primary consideration for their guidelines, implying that research on human health is not yet substantial enough to merit changing how airports manage noise.

The FAA is reviewing the problem, at least since the beginning of the Obama administration. In 2013, Rebecca Cointin, Noise Division Manager of the FAA's Office of Environment and Energy, presented solid plans to respond to the questions of health effects caused by noise and consider the need for a new approach to measuring aircraft noise. She announced that the FAA should seek to reduce the number of people exposed to significant noise around U.S. airports, notwithstanding aviation growth, and provide additional measures to protect public health and welfare, with a goal of reducing the U.S. population exposed to significant aircraft noise around airports to less than 300,000 individuals. It appears all of the efforts of communities to draw attention to the problem are having an effect on the FAA.[93]

There has been a standoff between airports and neighboring communities in the U.S. because most communities lack input into the policies and decisions of their transportation agencies. One vivid story of the problem is well worth reading. Author Rae Andre wrote a compelling personal story about the problems her community faced fighting airport expansion near Lexington and Concord, Massachusetts, very historic American iconic communities, with their share of influential residents, who nonetheless could not arrest the expansion of their airport. She quickly learned that, even though noise levels exceeding FAA guidelines occurred 10 times per day, the total average noise level over 24 hours was well within legal limits. She soon waded into the arcane world of airport expansion regulations and learned how little she or any local person could influence the process. In response, she wrote a book *Take Back the Sky*, a well thought-out treatise on how communities should organize to ensure airport expansions take citizen concerns and needs into account. She suggests that independent state agencies that manage airport regulations and expansions, like Massport, redesign their governance to become more representative of local communities.[94] More inclusionary transportation governance will be essential, according to Andre, to ensure that the views of local citizens, who are impacted by airport noise, are part of airport expansion decisions.

One of the reasons administrators have been wary of including local viewpoints is that the response to airport noise can be anger. One of the symptoms of constant exposure to noise is aggressiveness, on the one hand, or among children a feeling of helplessness. Studies of noise complaints at airports by the FAA find that the large majority of complaints come from a few very angry individuals.[95] This may well be a result of their exposure to noise. On the flip side, a feeling of helplessness in towns such as Inglewood, California, may be the result of exposure to airport noise over generations. It is not without merit to consider that the minority residents living in Inglewood have been so affected by noise over their entire lives that they have no will to act or fight the problem, precisely because of the impact of noise on their health and mental well-being.

Airports can certainly respond and are increasingly doing so. Nearly all citizen reports on noise point out that managing single-event noise levels is crucial to effective noise control.[96] Improved measurement of noise could be a breakthrough approach to regaining an atmosphere of positive engagement.

Solutions to aircraft noise: noise management systems

Systems to alleviate the problem of growing airport noise include:

- Quieter aircraft,
- NextGen Air Traffic Control Systems, which can facilitate smoother takeoffs and landings,
- Land-use planning, which prevents building residential housing in noise-sensitive areas around airports,
- Fewer nighttime flights, and
- Soundproofing homes near airports.

Quieter aircraft

Aircraft today are 75% quieter than they were 50 years ago.[97] A great deal of effort has already transpired to reduce the noise coming from aircraft, and additional easy wins will be more and more difficult. New standards are set out by ICAO, and as new research and new aircraft are released, there are increasing improvements, but the percentage of improvements are decreasing.[98]

The Boeing Aircraft Company tests all aircraft to better understand the sources of noise. During the development of the Boeing 787, wind tunnel tests were employed to optimize wing and flap systems, allowing the manufacturer to identify the quietest possible airframe configuration.[99] Likewise, the Airbus 380 is also a much quieter aircraft that, according to the manufacturer, delivers unprecedented certified noise levels that satisfy ICAO standards and the noise requirements of international airports.[100]

Virgin Atlantic measures and reports on its total noise output. Its 45-aircraft fleet has been refreshed with newer, quieter aircraft, including 10 new Airbus 330s. As of 2016, the airline has 13 787s and 10 Airbus 330s in service.[101] They successfully reduced their airline's total decibels by 5% as of 2013 and plan to reduce total noise output by 6% more by 2020.[102] The total reduction in decibels will be 11% by 2020, based on their original 1982 benchmark. This airline's efforts are laudable and important and a very good example of what an airline can do. But despite the efforts of Virgin Atlantic, the rate of growth of the aviation sector will far outpace any one airline's efforts to put the quietest aircraft fleet in place. With the growth of aviation now projected to be 30% by 2017, more and more local communities will be affected by aircraft noise in future. ICAO estimates that by 2036 up to 34 million people will be exposed to aircraft noise over 55 Ldn.[103]

Improved navigation systems

The new flight navigation systems previously discussed, known as NextGen, have the potential to reduce aircraft noise. These systems are being deployed as rapidly as possible in the U.S. and worldwide to reduce airport congestion and flight delays. They use GPS satellite systems to allow air traffic controllers to have more ,precise information on aircraft movements, which will enable more efficient flight management. Aircraft will use more routes in positions that are closer together with more precise horizontal and vertical paths for departure and arrival. On the ground, taxi times will also be reduced. If aircraft are moving more smoothly and are well synchronized, pilots and air traffic control managers will be able to reduce the power required from jet engines. Simply put, there will be less need to step on the "gas" either for ascent or descent. Instead, there will be smooth patterns of trajectory. These smoother routes with less jet power will lower aircraft noise.[104]

The health benefits of NextGen navigation are being tested and have the potential to address noise impacts, but are hindered by a lack of funding. At the time of writing, the U.S. Environmental Protection Agency's (EPA) program to study the health impacts of airport noise, the Office of Noise Abatement and Control, has not been in effective operation since being defunded in 1981. After a long hiatus, the FAA began more research in 2010–2013 on a wide range of environmental health impacts of airports, including the issue of noise. Some of these documents have been used to understand the current state of research on noise and aircraft in this section.

While smoother flight paths may lower the decibels, there are concerns that additional routes for ascent and decent will threaten neighborhoods that have not suffered from air traffic noise impacts in the past. In Queens, New York, area residents complained the noise was unbearable when the new NextGen route emanating from New York's LaGuardia airport was launched.[105] Bayside, Queens, residents successfully lobbied New York Governor Andrew Cuomo to order a noise compatibility study and establish a community aviation roundtable with New York's Port Authority.[106]

Harvard Extension student Bryan Johnson of Phoenix, Arizona, found that in fact the implementation of NextGen aviation has caused a significant increase in noise complaints that were strongly related to spreading the ascent and descent of aircraft over a much wider number of residential areas. The flight paths that were once over an industrial area but were changed with little fanfare and without any introduction or public meetings.

> The Phoenix NextGen project optimized flight paths by changing their routes and locating them nearer the airport. Previous to its implementation, the city of Phoenix was commendably managing aircraft noise at Sky Harbor airport in that very few complaints were being received from the public. The community responded with immediate public comment to the flight path changes with a dramatic rise in aircraft related noise complaints. Complaints prior to the NextGen rollout averaged about 15/month with a dramatic increase in the volume of complaints to about 1,000/month (after implementation). Complaints have held at this level through the ensuing year to the present.
>
> The number of Sky Harbor flights hadn't changed from one day to the next with the NextGen implementation. The complaints filed were a direct result of flight path alterations that relocated aircraft over new locations within the community. With the implementation of NextGen flight path changes, turns were occurring in precise positions closer to the airport. This placed planes directly over densely populated residential areas. In some cases, these neighborhoods had not been previously exposed.
>
> The introduction of NextGen into Phoenix was poorly marketed and with no public opportunity to review the changes being made. The NextGen implementation was closely followed by a strong negative citizen outcry of being harmed by the additional aircraft noise. The city government then added its voice to the narrative by agreeing that significant harm to the public had occurred and implied that the cause of the harm was the FAA handling of the changes made to the flight paths. This city response

> cemented the strong negative public view of NextGen and framed the FAA as an unyielding organization.[107] In the year following NextGen implementation, the community complaint level has held constant at about 1,000/month (up from 15/month), indicating the depth of the negativity towards aircraft in the community. The irony of this is that an overall sound reduction likely occurred.[108]

This research confirms that the FAA needs to review its noise regulations and the implementation of its NextGen navigation program, or community concern will only grow in size and concern. Without public involvement in the growth of airports and their navigation systems, there will be angry citizen opposition. As of the writing of this book, citizens of Phoenix have sought to have their Congressional representatives develop alternatives to the present process that FAA uses for managing the introduction of its NextGen program. There is also a growing demand that the FAA noise regulations are based on different metrics, which according to a growing amount of research need to be revamped.

Land-use planning

Airports map the soundscape impacts of their runways and can anticipate where high-decibel impacts will transpire. These noise mapping procedures are now being used worldwide. The resulting maps illustrate noise contours over land, sea, and rural and urban communities. They are based on flight frequency, flight paths, and aircraft types. These noise maps help airport officials to analyze how to lower the noise impacts of flights over vulnerable communities that are closest to airports.

O'Hare Airport is an oft cited example of an airport that has had trouble managing its noise impacts on local residents. It began as a 4-runway airport in 1945. Community opposition began as far back as the 1960s when a new runway was built. Noise-related lawsuits began in the 1980s. As O'Hare expanded to become one of the U.S.'s busiest airports, residential and other incompatible land uses developed simultaneously around the airport. Since 1982, O'Hare and the O'Hare Noise Compatibility Commission began to acquire impacted homes and to pay for sound proofing of properties that were not acquired. It became the largest initiative of its kind, in response to noise complaints and lawsuits, in the U.S.[109]

Studies by the U.S. FAA are now performed to understand how airports can avoid lawsuits and expensive problems with neighbors by planning in advance for airport expansion and by carefully zoning property near anticipated runways for uses that are not residential or business but rather agricultural or industrial. These techniques are successful when there is enough available land not already occupied by communities vulnerable to excess noise. Many airports do not have that luxury and face neighbors who fight airport expansion and runways as soon as they are announced. The FAA has said that "allowing incompatible real-estate development around airports signals the first step toward closing the airport."[110]

Denver, Colorado Airport: case example

One example of an airport that has managed to use land-use planning as a means of avoiding noise impacts on residents is Denver International airport, located in Colorado, U.S. The airport it replaced, Stapleton, was overwhelmed by noise complaints and closed due to noise litigation. As part of the settlement, Denver was required to build a new airport by 2000 and close Stapleton permanently. Denver chose a remote, undeveloped location. It set stringent regulations working with surrounding communities to avoid any developments near present and future airport runways. All parties agreed to a master plan that provided a full build-out plan for the airport, with all communities involved in the process. Denver is a busy airport and transits tens of millions of passengers. It was developed to receive up to 1.5 million aircraft operations per year. Its noise contour maps were used to work with local communities to prohibit development anywhere where a contour indicated that a 65-decibel average sound level might be expected. Most of the neighboring cities have maintained the zoning restrictions, and noise complaints have been few. But as the city of Denver continues to expand, there are pressures from developers to build closer and closer to the airport. Residents are now beginning to oppose the build-out of the master plan and are using Congressional influence to allow for new developments near runways that have not yet been built.[111]

The conclusion from Denver's experience is that airports with the best intentions might still not successfully be able to use land-use planning as a means of controlling development near airports. Increased tax revenues alone are strong incentives for municipalities to change their view of agreements that originally prohibited residential development near airports. The fact is that airports represent immense commercial potential for developers of land. Controlling land development is very difficult in most parts of the world.

Night flights

Night curfews are advocated by citizen's organizations as a crucial approach to eliminating the most debilitating health effects of airport noise. As of 2012, 250 domestic and international airports worldwide imposed night operating restrictions, with 66% of these restrictions in Europe.[112] ICAO takes the position that removing night curfews would improve market access, alleviate slot problems, and contribute to economic development and trade. The problem is that airports cannot keep up with the demand for airline "slots," which allow airlines commercial access to airports.

In Europe, where noise problems are the most pronounced, airport congestion is also the highest. The key concern for European airports is the efficient allocation of scarce airport capacity. As bottlenecks increase, the annual and hourly capacity of the airport must be established in terms of available slots. Airlines must have slots to operate, and slot coordination is increasingly complex. While this book does not include a discussion of how slots are managed, it is clear from a review of the literature[113] on the topic that slots are highly valuable commodities in the world of air travel. There is increasing pressure to open up more slots, and the curfew rules once maintained by airports in Europe and other congested regions are increasingly difficult to justify from a commercial perspective.

Heathrow Airport has had limited night flights for 50 years, but in 2013 began the process of considering a revision of its policies. Cost–benefit policy documents indicate that at Heathrow, night flight restrictions impose costs on the airport, airlines, passengers, and the environment, and impact the operations of the airport in the daytime. Government documents state that the night flight ban at Heathrow has negative economic impacts on express freight and mail services and restricts early morning arrivals favored by high-value business passengers, especially those coming from Southeast Asia. Curfews also disrupt manufacturing supply chains that rely on access to last-minute shipments. The government recognizes the impacts of night flights and expects airlines to reduce and mitigate the noise through the use of best-in-class aircraft and best operating procedures. The difficulty of both managing impacts and permitting growing aviation traffic has weighed heavily on Heathrow and its neighboring communities. An independent airport commission was established that produced a report seeking to both allow growth and ban night flights. A third runway to allow for growth, $1 billion in community mitigation funds, and an independent board were all recommended, solutions that airports around the world will need to consider.[114]

The International Aviation Transport Association (IATA), a trade association representing the airline industry worldwide takes the position that night curfews have negative consequences for passenger airlines, travelers, and freight.[115] The airports and aviation organizations working on the problem appear to be unable to hold back the demand for more airport capacity. To manage the concerns of citizens, ICAO recommends a so-called Balanced Approach, including assessing current and future noise impacts and reviewing the costs and benefits of night flight restrictions.[116]

Soundproofing

Soundproofing may be the best solution in the long term. Thanks to the California Noise Control laws, neighbors to LAX in Inglewood were quite lucky to receive $160 million in soundproofing funds between 1992 and 2011. The total spent in the entire LAX noise affected area in this period was almost $1 billion to mitigate noise impacts on just over 30,000 dwellings.[117] This quite laudatory example should set a standard for how crowded aerotropoli regions of the world approach the noise problem in future. If they do not, they will face increasingly angry local communities who are seriously affected by noise pollution.

As airports grow worldwide, there will be an urgent need to address the issues of noise for residents near airports in very congested regions. Chicago O'Hare's Residential Sound Insulation Program offers sound insulation to eligible residences: both single owner–occupied homes and multiunit homes impacted by the

65-decibel-Ldn average trigger approved by the FAA.[118] In 2010 O'Hare approved 1,000 homes and buildings for sound insulation mitigation.[119] Similar efforts are under way in Europe and Australia. Throughout the world, it is important to consider that residents near airports frequently are lower-income, minority, or even homeless. Concerned architecture students from Rangsit University of Thailand won an architecture award in Bangkok by designing a house that used double-glazed glass walls and fiber insulation to deaden sound inside by 60% for people living near Bangkok's Suvarnbhumi airport. The students sought to assist residents living near the airport lives who lack funds to move away to improve their quality of lives.[120] This type of thinking will have to be scaled up in future.[121]

Many communities feel they are victims of airport growth, and there is a sense of helplessness and anger that is fueled by the intense odds they face and the lack of caring they perceive from airport authorities. The environmental management of airports is a growing field, which will have an increasingly important role in the future of travel. Experts in noise management will have to work closely with experts in community management and greatly improve skills in stakeholder engagement and involvement in decision-making processes. There also needs to be considerably more research on the specific health and cognitive problems caused by airport noise. This research will facilitate the work of experts in environmental management who will need to knowledgeably respond to neighborhood rage and problems of below-average reading skills in local schools, both of which are on the increase. Finally, there will need to be a multidisciplinary response to airport noise that increases efforts to help residents become more informed and involved in land-use planning decisions and obtain funding for soundproofing of homes and businesses.

Airport air quality issues and solutions

Airport air quality is regulated as part of national clean air laws across the world. Airports are usually monitored as part of regional air quality measurements and not measured as a distinct and separate source of emissions. For this reason, environmental management of air pollution from airports is a field that lacks data. Many airports do not account for the contribution of aircraft to their air pollution because they argue it is outside their scope of responsibility, even though airplanes are the most important source of air pollution generated by airports. Despite these specious arguments, experts in public health are striving to uncover the specific contribution of growing airports to local air quality because of the concern that airport air pollution is a growing threat to the health of neighboring communities, many of which around the world are disadvantaged communities with lower incomes.

A current study performed by Wolfram Schlenker of Columbia University and Reed Walker of the University of California, Berkeley, made a significant contribution to measuring the precise link between airport air quality data to public health in neighboring communities. The team estimated airport air pollution in airports in California but using real-time data. They sought to take this data and link it to the causes of air pollution, which they found is largely airport runway congestion. They then proceeded to review how this real-time data links to local hospital records, not only admissions but short visits due to respiratory complaints. This study was able to statistically link public health problems to the overcrowding of airports and the congestion of aircraft idling on runways.[122]

Schlenker and Walker's background research reveals that a large fraction of airport emissions come from airplanes and that most of these emissions stem from airplane idling.[123] Their data exposes the fact that runway congestion is a significant predictor of local air pollution levels. This important insight led to a series of conclusions. First, to address the impact of air quality near airports, the problem of backups and congestion, especially at peak flight rush hours, needs to be addressed by airport authorities. Second, flight patterns across the U.S. must be factored in, as weather as far away as the East Coast of the U.S. frequently causes significant delays in California, requiring national systems to arrest weather-related backlogs of flights idling on runways.

Flight delays in one part of the country or even the world have a domino effect on other airports. Aircraft arrive late and depart as quickly as possible, pushing off from gates to maintain on-time statistics. Congestion builds, as flights all seek to depart simultaneously. The more planes that push off at once, the more airplanes will simply have to wait on the runway to depart. These long lines are well-known to travelers and a real annoyance, but the increasing number of delays is startling and getting worse. Schlenker and Walker found that the average airplane taxi time increased by 23% between 1995 and 2007. They estimated that over 1 million airplane hours were spent idling on runways in the U.S. in 2008. Their conclusion was that the increase in delays and idling leads to significantly higher levels of ambient air pollution.[124] Schlenker and Walker suggest that inefficient queuing of airplanes needs to be regulated to prevent inordinate and growing taxi time and the idling of airplanes. They suggest that when congestion becomes too great, that aircraft are prevented from taxing, queuing, and idling. The solution to this problem is simply to stay at the gate until congestion diminishes.

The Schlenker Walker study measured both carbon monoxide and nitrous oxide levels surrounding three major airports in California and reviewed the levels of these gases as they corresponded with reported increases in airplane taxi times during crowded periods on the runways. The authors found a 23% increase in carbon monoxide within 6 miles of the Los Angeles International Airport (LAX) during periods of heavy runway congestion and backups. The real-time increases in carbon monoxide at LAX were statistically linked to a 33% increase in daily admissions for asthma problems, carefully studied to control all other factors, including access to health insurance.[125] The authors found that pollution levels near California's 12 airports is causing $1 million in hospitalization costs, or $70,000 per day.[126]

The authors suggest that pollution costs are real and quantifiable and can be incorporated into cost–benefit analysis used by airports to discuss how to manage growth. But these conclusions have not been fully accepted; recent studies on behalf of LAX found that the concentration of air pollutants associated with airports are not above EPA standards in neighboring communities.[127] This is an important debate because the health implications of airport air pollution, when measured, appear to be significant.

Accounting for the monetary value of the health implications of airport air pollution is one important means of ensuring that airports review not only the economic benefits of growing airports but the true costs to society. Comparing the costs of allowing aircraft to queue versus the cost to society of aircraft idling is an exercise that should be part of an airport's review of air pollution standards. But to date this has not transpired.

Air pollutants

Airport jet engines and other aviation-related activities produce:

- Carbon dioxide (CO_2),
- Nitrogen oxides (NO*x*),
- Carbon monoxide (CO),
- Sulfur oxides (SO*x*),
- Volatile organic compounds that are partially combusted hydrocarbons (VOC),
- Particulate matter (PM), and
- Hazardous pollutants (HAPs).

The sources of these emissions are as follows:

- Aircraft emissions
- Aircraft handling emissions
- Infrastructure sources
- Vehicle traffic sources

Aircraft cause emissions from their main engines and from their auxiliary power units, which power aircraft during ground operations. The ground support equipment, including power units, aircraft tugs, conveyor belts, forklifts, and cargo loaders, all have impacts, as do the aircraft fueling equipment and deicing equipment. There are emissions also from the airport buildings and from traffic associated with airports.

Solutions to aircraft emissions

Airports are on a pathway to measuring their emissions using systems that can capture the variability of air pollution sources. As opposed to the somewhat crude systems for measuring noise just described, air pollution and emissions are measured using modeling and with intent to capture variability according to changes in a variety of factors, such as weather or congestion of the airport. Computer models are now allowing for even more sophisticated emissions inventories. Graphic 2-D density grids and use of 3-D techniques are also being considered for sources, particularly aircraft.[128]

Zurich Airport in Switzerland is managing emissions from a holistic perspective, measuring both aircraft and ground operations emissions. Airports, including Heathrow, Seattle, O'Hare Chicago, and LAX, do not include responsibility for the operations of aircraft in their calculations of air quality. According to Zurich's studies, aircraft engines are the source of 89% of Zurich airport emissions.[129] They found that aircraft pollutants spread over a larger area than ground sources but that the contribution to pollutant concentrations of neighboring locations just over 1 kilometer away drops to 10%.[130]

In the future, discussions of the growth of airports, the management of air traffic, and systems for reducing idling of aircraft will be the most relevant approaches to managing air pollution. While reviewing airport ground operations is important, the reduction of these impacts will affect only 10% of total emissions from airports. Given aircraft emissions on site at airports are nearly 90% of the problem, airports need to move toward agreements with airlines that will reduce congestion and idling to address the real problems facing their neighboring communities.

Denver International Airport has been one of the innovators in this category and has deployed a system that keeps aircraft at the gates using auxiliary power units when they are experiencing a snowstorm. This system allows aircraft to remain plugged in at the gate until they are provided with "the quickest, most efficient route through their deicing process to the taxi runway."[131] Such systems need to be implemented to reduce runway congestion across the board at airports worldwide for all storm-related delays.

Airport waste management issues

Airports have an important role in municipal waste management and an outsized impact on how well states and regions meet waste management goals. But in many states in the U.S., airports do not monitor their waste and do not report on meeting waste management goals in a cohesive fashion.

Airports manage five types of waste:

1. Municipal solid waste, product packaging, bottles, cans, food scraps, and newspapers
2. Construction and demolition (C&D) waste, including concrete, wood, metals, debris, salvaged building components
3. Green waste, such as tree, shrub, and grass clippings
4. Food waste, food that is not consumed or food left over during food preparation
5. Deplaned waste, items removed from planes, including bottles and cans, newspapers, plastic cups, service ware, food waste, food soiled paper and paper towels

Each waste type requires different steps to ensure that waste reduction and minimization transpire. The Chicago Department of Aviation has set out a highly useful manual for the environmental management of airports that provides a set of guidelines and measurement tools for airport managers to benchmark their performance. Their *Sustainable Airport Manual* (*SAM*)[132] is the most comprehensive guide in the world and allows airports to score their performance and compare themselves with other airports via a standardized scoring system. Rosemarie Andolino, the Commissioner of the Chicago Airport Authority, has been a pioneer in airport green management, having initiated the Airports Going Green movement and overseeing the publication of the *Sustainable Airport Manual* and the implementation of Chicago airports' comprehensive environmental management program.

One of Chicago's strengths is the management of construction waste, a very important part of an airport's waste management program. The objectives of construction waste management (CWM) programs include the diversion of demolition debris from disposal in landfills or incineration facilities or redirection of recyclable resources back to the manufacturers for reuse. The management of airport surfaces using excavated soil to create runways and other landscaped areas can maintain and reuse excavated soil and land clearing debris to save funds and divert waste from landfills. A leader in the world of CWM, Chicago O'Hare recycles hundreds of thousands of tons of concrete, asphalt, bricks, scrap metal, lightbulbs, and landscaping waste. The city's ambitious program recycles 98% of its construction waste and has saved 20 million cubic yards of soil on site for reuse, saving well over $100 million in its earthworks program.[133]

Airports have a complex set of agreements with their resident retail and restaurant concessions, and airlines can be difficult to align into one system for waste management. But without cohesive, coordinated waste management programs, airports are not able to manage the waste stream of their facilities with adequate oversight. A study on Logan Airport in Boston by Harvard Extension student Ria Knapp (2010) reveals that the airport authority Massport does not manage the waste stream of lobby areas outside of security, the restaurants and retail outlets in the airport, or the majority of waste generated by airlines using the facility. As a result, Logan has no capacity to measure the solid waste generated by the facility as a whole. Logan reported the airport recycled 10% of its waste in 2010, but as Knapp rightly points out, this was based on only a small portion of the waste that Logan airport generates.[134]

Airline waste accounts for 50% of all airport terminal waste according to the Natural Resources Defense Council (NRDC),[135] while the FAA states that on average airplane waste is 20% of a municipal airport's solid waste.[136] Either way, about 40% of this waste could be recycled and often is not.[137] Airlines have high waste volume at hubs and low volume at spokes, making it difficult to manage a policy across the board that is cost-effective or that can be economically justified. For this reason, most airlines are still not recycling unless the airport authority or municipality requires it, and even if it is required under law, airlines may not be meeting municipal standards in many locations. Airlines prefer a centralized system according to the NRDC report because it provides a cost-effective system that they do not have to manage. Centralized waste management policies save funds for the municipality, the airport, and all of the airport tenants, and they greatly improve both waste minimization and recycling rates.

Solutions to aircraft waste

San Francisco Airport (SFO) took over waste management responsibilities for its terminal tenants in 2004. SFO increased the rate of recycling from 51% in 2002 to 75% by the end of 2011. The city has set a goal of 85% recycling rate by 2017 and Zero Waste (90% or more) by 2020. Waste reduction goals are being reached with the following approaches:

- A clause in all food concessionaire lease agreements at Terminal 2 requiring biodegradable food-ware
- Annual waste characterization study to understand the composition of the solid waste streams and to evaluate progress in recycling
- Reduced water bottle waste by providing drains in pre-security checkpoints
- Installed electric hand dryers in restrooms to avoid paper towel waste
- Partnering with contractors to achieve over 90% construction waste recycling
- Monitoring of custodial staff and tenants to ensure proper segregation of waste at collection points[138]

While many airports are not yet managing their facilities via a centralized system, the options for airports that do are much greater. They can work with all of the facilities on their properties to set goals, a critical part of the process of achieving effective environmental management. Airports that set goals with a wide range of property representatives are more likely to meet the type of recycling and waste minimization goals achieved by SFO. In Portland International Airport, Portland, Oregon, U.S., the airport authority undertook a waste characterization study to understand how to target their efforts. They found that deplaned waste was 40% of their waste stream. The airport committed to providing specialized recycling support to individual airline tenants and to coordinate communication between airlines and their ground service caterers to increase recycling rates.[139]

Airlines in the U.S. are lobbying airports to take over the task of managing waste. Through their industry association, they are publically asking for more support for recycling in airport facilities, and on a global level they are seeking fewer regulations. In Europe, airlines face waste regulations that make it impossible to compost, and in Australia waste is quarantined. The lack of infrastructure and legal restrictions for recycling of deplaned waste are primary obstacles to more recycling. Suppliers that can reuse their materials or offer recycled materials, such as coffee cups, are clearly needed. And airplanes need well designed systems for attendants to manage recycled goods. Certainly any passenger can see that the attendants are limited in their capacity and facilities to sort waste, but the response has been extraordinarily slow given how long recycling has operated effectively in most communities on land.

On a global scale, the management of aircraft waste is only at the earliest stage of assessment. An IATA program was begun to address concerns that airlines are not recycling sufficiently, and in 2012 a global working group was formed. The project undertook waste audits using a standard methodology and estimated that passengers worldwide are generating 4.43 million tons of solid waste in 2013, based on 3.1 billion passengers. Food and beverage was 18.5% of total waste, and unopened bottled water almost 5%.

In some countries, solid waste treatment is still underfinanced and not available to a broad segment of the local population. In these cases, recycling is still not common, and airports are not facilitating the recycling of any waste, including deplaned waste. The lack of consistency of waste treatment systems around the world does inhibit airlines from meeting basic standards for recycling. But as the industry continues to grow, new airports must and can afford to include the management of solid waste not only for their ground-based facilities but for airlines.

One concept for waste management that is suited to large projects is the application of vacuum waste collection and recycling systems, where the waste is transported along underground pipelines into containers and connected to underground railway systems. This type of system can be remotely monitored, reduces the use of trucks, and offers easy collection systems for the processing of mixed waste, organic waste, and paper. Still a very new concept, being implemented in locales such as Finland, these pneumatic piping systems for waste are perfect for new build facilities such as airports and could be part of airport design in future.[140] Vacuum waste collection and recycling is part of an exciting new generation of intelligent design solutions for greener airports.

Storm water runoff issues and solutions

Airports have acres of impermeable concrete surfaces with outsized storm water runoff impacts far beyond similar facilities, such as large shopping malls or schools. Runoff concerns include not just cars, but aircraft and the chemicals used to manage aircraft. These chemicals, particularly those used during winter months in cold climates, have toxic impacts on local waterways when not adequately managed. Airports need environmentally sensitive, low-impact "hardscape" management for their buildings and runways to lower their total impacts on nearby ecosystems.

Airplane de-icing is one of the most environmentally harmful procedures used by airports. De-icing fluids are composed primarily of ethylene or propylene glycol. These compounds cause the death of fresh water ecosystems by depleting dissolved oxygen in water and essentially suffocating life that needs oxygen to survive while fostering bacteria that can grow in abundance in waters that have almost no oxygen.[141] Runoff from the application of these chemicals was documented by the U.S. Environmental Protection Agency (EPA) to have caused fish kills, growth of biological slimes, die-off of aquatic life, impacts on wildlife, birds, and cattle, as well as impacts on workers exposed to the chemicals in airports.[142] In 1998, a coalition of local and national environmental groups filed a complaint against

Baltimore-Washington International Airport for discharges into Sawmill Creek. The Natural Resources Defense Council, together with local citizens' organizations, demanded compliance with federal laws that protect waterways, claiming that the airport was not moving to reduce its toxic runoff. According to the NRDC at that time, airports across America were allowing massive amounts of chemicals to run unimpeded into the nation's waterways.[143] The EPA began to review the problem after the NRDC lawsuit and estimated that 212 U.S. airports discharged approximately 28 million gallons of glycol solutions annually into local waters without treatment between 1990 and 2000. EPA recommended that these effluents be reduced by 87% from 1990 data and 62% as of the year of their study, which was 2000. Airports had already begun to act once it was clear that deicing would no longer be unregulated. Lawsuits were settled with NRDC, and mitigation efforts began across America.

The final EPA Airport De-icing Effluent Guidelines were not published until 2012. U.S. airports with over 10,000 departures are now required to collect 60% of their deicing fluid, while smaller airports are required to collect 20%.[144]

Solutions to storm water runoff

Solutions to the problems of de-icing runoff are as follows:

- Alternative de-icing and anti-icing materials
- Improved equipment and alternative de-icing technology
- Capture and treatment
- Recovery and recycle

Alternative solutions include potassium compounds that have lower biochemical oxygen demand (BOD), which is the property of most de-icing agents in use that cause depletion of dissolved oxygen in fresh water ecosystems.

Improved equipment is varied and can include more accurate spraying devices that reduce the amount of runoff, vapor steam devices that melt the ice before de-icing agents are applied, and vacuum sweeper trucks that remove the solution from areas after spraying.

Capture and treatment is the primary technique recommended for large airports. Airports have choices for the capture of these chemicals. De-icing pads are highly recommended to capture the runoff. These pads are surrounded by specialized drains designed to capture the toxic fluids and treat them separately from other storm water runoff sources. Containment of the chemicals makes the system easier to manage, and the chemicals can be stored in tanks to maintain the runoff until it is treated. This prevents de-icing fluids from entering into the storm water treatment systems of neighboring municipalities, which now charge extra fees to airports for processing of de-icing fluids. Some airports will manage their own chemical processing, while others simply pay a fee to their municipalities to do so.

Recovery and recycling of the fluids are viable options but expensive, according to EPA analysis. Nonetheless, in Munich, Germany, the international airport saves $2.6 million each year by recycling de-icing fluid, which allows them to avoid the cost of treatment. Munich has reduced its runoff by 70%. Munich uses a centralized de-icing pad near the end of the runway with a collection system that avoids mixing the de-icing fluid with other storm water sources.

With the combination of these 4 approaches, total discharges from U.S. airports could be reduced to 70% efficiency, or 4 million gallons a year, a substantial improvement that could surely secure a higher quality of life for local residents and the ecosystems upon which they depend.

Future of environmental management of airports

Travelers passing through Changi International Airport in Singapore are given royal treatment. The airport, which managed 53 million passengers in 2013, provides relaxed, speedy service with wonderful airport attractions, such as free movies and a butterfly garden, between flights. Singapore is moving quickly toward a third runway and two more terminals by 2017. A "bubble-shaped glass complex will sprout between the existing terminals, providing shops, gardens and a waterfall."[145]

The *New York Times* reports that $115 billion is committed to airport construction across Asia-Pacific, 45% more than North America or Europe. Airlines in the region carried 510 million people on 3,270 aircraft in 2007. Those numbers increased by 200% by 2013. Growth rates are estimated to continue at 6–7% annually in the region until 2017 and then moderate to 5%. Low-priced airlines are springing up, like Cebu Pacific in the Philippines and Lion Air in Indonesia, carrying 25% of all passengers in the region.[146] Rapid growth is the paradigm of the future. The question is can environmental management keep pace?

In the U.S., Atlanta Hartsfield-Jackson International Airport had as many visitors as Disney World, Graceland, and the Grand Canyon combined as of 2008. The largest concentration of hotel rooms on the U.S. West Coast surrounds Los Angeles International Airport. The emerging aerotropolis concept requires region-wide planning. There is little that an airport will not require in the future, and it will certainly have to have some of the most sophisticated environmental management teams in the world, given the many challenges involved. This profession should not only be stressed, there should be a worldwide demand for training and a global consortium of experts to guide this process.

Conclusion

The field of aviation is fortunate to have many visionary leaders and researchers who are committed to getting accurate data. The international organization, ICAO does outstanding work measuring impacts, working with scientists around the world, and giving unusually candid estimates of how well environmental management procedures are affecting overall outcomes. Measurements of total impacts are science based, and efforts to lower total cumulative impacts are transpiring in every corner of the industry, largely unrestricted by the questions of boundaries, though not entirely. Noise indicators are still measured according to highly antiquated systems that do not respond to high volume, irregular high-volume incidents, and low frequencies. And air pollution indicators have been averaged according to systems that are not responding to clear problems with congestion and runway idling. These problems need immediate attention.

Of course, there are some economically and politically challenging steps to be made to create a more efficient and accurate system – a system that does not fulfill the predictions of the skeptics but rather builds on the dreams of the optimists. The monumental elephant in the room is aviation carbon emissions and the necessity of establishing a market-based mechanism that can manage global emissions trading for the aviation industry. The aviation "emissions gap" threatens to be 7.8 billion tons of carbon between 2020 and 2040, an enormous volume to offset internationally through either emissions allowances or offset credits. (See Figure 5.3.) The global advocacy community organized to encourage ICAO to create these mechanisms suggests that there is sufficient supply and that new projects are likely to be available by 2021 thanks to the Paris COP 21 Agreement,[147] an important point discussed further in the Conclusion.

The hard facts are that the next generation of aircraft from Boeing and Airbus will lower the per-passenger impacts of flight by approximately 6% by 2025, and the overall emissions from flights are doomed to increase as fleet sizes continue to grow.[148] Alternative fuels are an extraordinary option and are presently dominating the investment scene with some $100 billion[149] committed to their creation. But

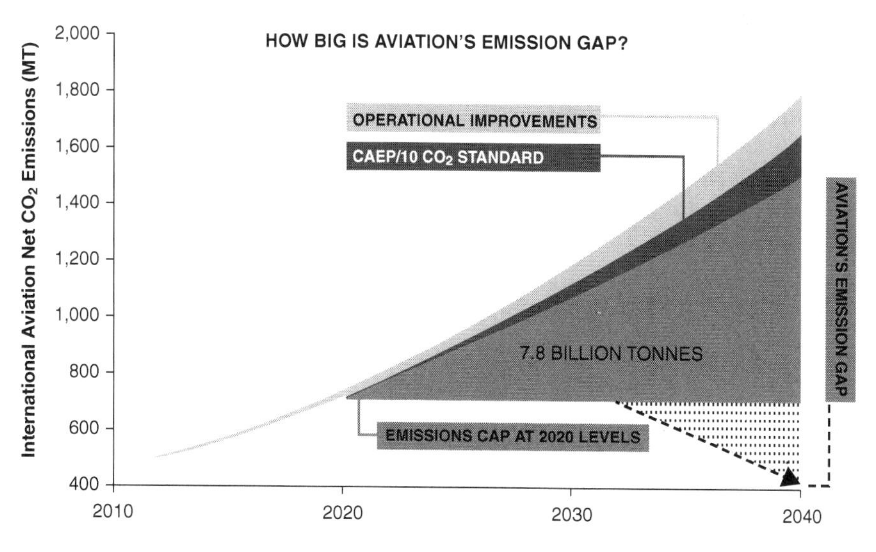

FIGURE 5.3 Projected carbon emission gap for aviation after cap set in 2020[150]

by all accounts these fuels face market limitations that could limit their use to just 1–2% of the total fuel mix for a generation. Business opportunities are there, such as the innovative LanzaTech system, which employs some of the earth's most ancient microbes to convert waste gases from steel mills to biofuel. With such innovation on offer, there will be continuing public private interest and burgeoning investment in biofuel solutions, particularly in the U.S. and China – but the question will be whether these can ever reach sufficient scale.

One alternative is greater investment in the research and development on hybrid engine models. The next generation of aircraft might actually cruise with zero emissions, according to Harvard Extension student Tania Fauchon's research, which would be far better than just trading the growing carbon emissions burden from aviation by buying out the global market for carbon credit permits.

Given the limited solutions to fully reverse the impacts aviation has on our atmosphere, a genuine discussion of curtailing the growth of the aviation industry through lowered consumer demand might need to be taken more seriously. Most of the major stakeholders in the tourism field do not care to broach this topic at all. But academics are ready to declare that in fact behavior change of consumers, government tax incentives to lower travel for business, short-haul flight limits for pleasure, and restraints on airport development might be necessary to lower travel by air.[151] If travelers were encouraged to increase their length of stays when on vacation and lower the number of frequent short trips for business using teleconferencing instead, the industry's carbon footprint could be reduced by an astounding additional 43% – more than all the technological solutions combined![152] It is high time more discussion begins on how to change the voracious use of air travel and find ways to make that palatable to business, governments, and consumers. It is likely that technological solutions and market-based mechanisms will not suffice due to the many limitations involved with both options.

Airports have become the earth's global transportation hubs directly built into the DNA of Asia's next generation of economic growth. The monumental complexity of managing these facilities needs to be reduced. Management needs to become driven by business-style decision makers, who must make certain there is systematic inclusion of local residents in decisions regarding expansion and top-notch environmental specialists and planners on their core management teams. Regular consultations with councils who represent citizens, businesses, and local leaders will help to alleviate the growing rage against airport growth.

Noise management must be reformed in the U.S., and new systems for noise measurement need implementation worldwide. Growing airports need strict zoning, which prevents residential or highly populated business zones within the airport's loudest noisescapes. And soundproofing for those who cannot escape the loudest airport noise should be paid for via standard airport operations fees.

Airports face increasing pressures to limit their impacts on local residents in Europe and America. It is unknown how this will proceed in Asia, Africa, and Latin America, but it is a fair assumption that all citizens should be protected because many people living near airports are minorities or disadvantaged economically.

Airport air quality impacts need further study, with air quality research that is real-time and not averaged over long periods of time. These studies should help to account for the costs of hospitalizations of those who are impacted by air pollution, and the true cost of airport expansions to society should be built into the equation for airport growth. Systems to prevent aircraft from idling for unnecessary periods of time, especially during storm delays, must be implemented.

Airport waste management is in its infancy. Centralized management of airport waste would raise the bar. By centralizing waste management, San Francisco Airport increased recycling rates for the total facility from 50% to 75% in 10 years and expects to achieve Zero Waste (90% or less) by 2020. Airports that argue they cannot take responsibility for their concessionaires need to revise their policies. While there are many impediments to aircraft waste recycling, which causes the lion's share of the problem at airports, the companies involved need to implement action plans for waste recycling and reduction and work with ICAO to demand that airport infrastructures in Asia and around the world install advanced waste management systems with proper mechanisms for deplaned waste to be recycled.

The runoff from the de-icing fluids used to treat aircraft in cold climates has endangered fresh water ecosystems, suffocating oxygen-dependent creatures with millions of tons of effluents, which remained unregulated in the U.S. until 2012. Efforts to recover and treat this waste are now on track, with a goal of collecting 60% of the fluids. In Europe, recycling has been successful and has been shown to be cost-effective. This option should be considered as the most preferable approach by airports that are already collecting the fluid separately for treatment.

There are challenges of enormous proportions in the field of aviation. But the solutions are dynamic, the capital is ready, and the scaling-up of the aviation system should simply include the cost of managing its impacts. As with much of the travel industry, the real cost of doing business must not be ignored at a time when our climate and the earth's ecosystems and landscapes are in much need of proper respect and management. And as the footprint of the industry continues to grow at nearly double digits, vulnerable citizens must be protected from the impacts of noise, waste, waste water, and air pollution.

Notes

1 The UN Framework on Climate Change set a 2-degree Celsius cap benchmark in Copenhagen in 2009, which is considered by most scientists as the basis for analyzing the extent to which the global community is on track to meet long-term temperature goals. Subsequent meetings have indicated that this goal will not be met. Research shows there will be devastating affects if the climate warms by 4 degrees Celsius, such as the inundation of coastal cities, increasing risks for food production, unprecedented heat waves in many regions, exacerbated water scarcity in many regions, increased frequency of high-intensity tropical cyclones, and the irreversible loss of biodiversity, including coral reefs, according to the World Bank (November 2012, *4 Degrees, Turn Down the Heat: Why a 4 Degree Warmer World Must Be Avoided*, Potsdam Institute for Climate Impact Analysis Research and Climate Analytics)

2 Atmosfair, 2013, *Atmosfair Airline Index*, Atmosfair, Berlin 2013

3 Cohen, Scott Allen and James E.S. Higham, n.d., Eyes wide shut? UK consumer perceptions of aviation climate impacts and travel decisions to New Zealand, *Current Issues in Tourism*,Vol. 14, Issue 4, pages 323–335, page 12
4 Higham, J. E.S. and Scott Allen Cohen, 2011, Canary in the coalmine: Norwegian attitudes towards climate change and extreme long-haul air travel to Aotearoa/New Zealand, *Tourism Management*,Vol. 32, Issue 1, pages 98–105
5 Ibid., page 8
6 Ibid, page 10
7 *Travel and Leisure*, http://travelandleisure.com/articles/eco-travel-guide Accessed October 1, 2016
8 Gössling, S. and Upham, P., eds., 2009, *Climate Change and Aviation, Issues, Challenges and Solutions*, Earthscan, London
9 Cohen, Scott Allen and James E.S. Higham, Eyes Wide Shut? page 17
10 Gössling, Stefan, 2011, *Carbon Management in Tourism, Mitigating the Impacts on Climate Change*, Routledge, International Series in Tourism, Oxon, page 102
11 Ibid.
12 Oxford Economics, n.d., Aviation the Real World Wide Web, Overview, Oxford Economics, Oxford, UK http://oxfordeconomics.com/my-oxford/projects/128832, Accessed October 1, 2016
13 Gössling, *Carbon Management in Tourism*, page 85
14 Scott, Daniel, Paul Peeters and Stefan Gössling., April 2010, Can tourism deliver its "aspirational" greenhouse gas emission reduction targets?, *Journal of Sustainable Tourism*, Vol. 18, No. 3, 393–408
15 Carey, Christian, Georgina Santos, and Ziaoyu Yan., 2010, *Future of Mobility*, University of Oxford, Smith School of Enterprise and the Environment, Oxford, page 56 http://library.uniteddiversity.coop/Transport/Future_of_Mobility.pdf, Accessed October 1, 2016
16 Lee, D.S. Ling L. Lim, and Bethan Owen, August 2013, Mitigating Future Aviation Co_2 Emissions – "Timing Is Everything," Manchester Metropolitan University, Centre for Aviation Transport and the Environment, http://cate.mmu.ac.uk/news/mitigating-future-aviation-co_2-emissions-timing-is-everything/, September 20, 2016
17 Lee, D.S., L.L. Lim, and B. Owen, March 2013, Bridging the Aviation CO_2 Emissions Gap: Why Emissions Trading Is Needed, Manchester Metropolitan University, http://cate.mmu.ac.uk/wp-content/uploads/Bridging_the_aviation_emissions_gap_010313.pdf; Centre for Aviation Transport and the Environment, Manchester Metropolitan University, http://cate.mmu.ac.uk/news/bridging-the-aviation-co_2-emissions-gap-report/, Accessed September 20, 2016. Note that emissions are measured using a 2006 baseline.
18 IPCC WGH AR5 Summary for Policymakers, 2014, *Climate Change 2014: Impacts, Adaptation and Vulnerability*, Summary for Policy Makers
19 Kyoto Protocol, http://kyotoprotocol.com/, Accessed September 20, 2016
20 International Civil Aviation Organization (ICAO), 2013, 2013 Environmental Report, Destination Green, Environment Branch, ICAO, Montreal, Canada
21 Centre for Aviation, Transport and the Environment, Manchester Metropolitan University, http://cate.mmu.ac.uk/, Accessed September 20, 2016
22 A Sustainable Future for Aviation: The Future of International Aviation Emissions Regulation Could Be Determined in the Next 12 Months, n.d., PWC, http://pwc.co.uk/sustainability-climate-change/publications/a-sustainable-future-for-aviation.jhtml, Accessed September 20, 2016
23 Lee, D.S. et al., *Bridging the Aviation CO_2 Emissions Gap*, page 2
24 Campos, Pamela, 2016, Why aviation's carbon must be capped, and how to do it, Environmental Defense Fund, August 31, http://blogs.edf.org/climatetalks/category/aviation/, Accessed September 20, 2016
25 Emissions Caps and Allowances, European Commission, http://ec.europa.eu/clima/policies/ets/index_en.htm, Accessed September 20, 2016

26 A Sustainable Future for Aviation: The Future of International Aviation Emissions Regulation Could Be Determined in the Next 12 Months, n.d., PWC, Summary page, http://pwc.co.uk/sustainability-climate-change/publications/a-sustainable-future-for-aviation.jhtml, Accessed September 20, 2016
27 EU ETS 'stop the clock' proposal passes crucial test in the European Parliament but faces legal challenges from European airlines, 2013, *Greenair*, February 27, http://greenaironline.com/news.php?viewStory=1665, Accessed September 20, 2016
28 Lee D.S. et al., *Mitigating Future Aviation CO2 Emissions*
29 Rock, Nicholas, Peter Amana, Shariq Filani and Alexandra Gordon, January 7, 2014, European Union: Aviation Emissions: The ICAO Outcome and Its Impact on the EU Aviation Emissions Trading Scheme http://mondaq.com/x/284726/Aviation/Aviation+Emissions+The+ICAO+outcome+and+its+impact+on+the+EU+aviation+emissions+trading+scheme, Accessed October 1, 2016
30 Reyes, Oscar, September 27, 2016, Climate Con: Why a global deal on aviation emissions is a really bad idea. https://newint.org/features/web-exclusive/2016/09/27/climate-con-a-new-global-deal-on-aviation-emissions/ Accessed October 8, 2016
31 ICAO Assembly climate change outcome hailed by industry but seen as a missed opportunity by environmental NGOs, 2013, *Greenair*, October 6, http://greenaironline.com/news.php?viewStory=1763, Accessed April 8, 2014
32 ICAO assembly achieves historic consensus on sustainable future of aviation industry. http://icao.int/Newsroom/Pages/ICAO-Assembly-achieves-historic-consensus-on-sustainable-future-for-global-civil-aviation.aspx
33 UN aviation agency adopts historic global aviation climate agreement https://edf.org/media/un-aviation-agency-adopts-historic-global-aviation-climate-agreement Accessed October 8, 2016
34 Aviation Transport Action Group (ATAG), November 2010, Beginner's Guide to Aviation Efficiency, ATAG, Geneva, Switzerland, page 9
35 Khan, Aamir, n.d., *Boeing Strategic Analysis*, https://academia.edu/4626722/Boeing_Strategic_Anlaysis, Accessed September 20, 2016
36 Future Technology, 2013 Environment Report, Boeing, http://boeing.com/aboutus/environment/environment_report_13/3_1_designing_future.html, Accessed September 20, 2016
37 Cleaner Products, 2013 Environment Report, Boeing, http://boeing.com/aboutus/environment/environment_report_13/3_2_cleaner_products.html, Accessed September 20, 2016
38 Aerospace: EADS and EOS – Study demonstrates savings potential for DMLS in the aerospace industry, n.d., EOS, http://eos.info/press/customer_case_studies/eads, Accessed September 20, 2016
39 Oliveira Fernandes Lopes, Joao Vasco, November 2010, Life Cycle Assessment of the Airbus A330-200 Aircraft, Dissertacao para obtencao do Grau de Mestre em Engenhaira Aeroespacial, Instituto Superior Tecnico, Universidade Tecnica de Lisboa, Lisbon, Portugal, page 89
40 Verghese, Vijay, Airbus vs Boeing: the big question, n.d., *Smart Travel*, http://smarttravelasia.com/AirbusVsBoeing.htm, Accessed September 20, 2016
41 Fauchon, Tania, December 2015, *Hybrid Aircraft Powerplant, How Can They Help the Aviation and Tourism Industries Reduce Their GHG Emissions?* Unpublished Manuscript, Environmental Management of International Tourism Development class paper, Department of Sustainability and Environmental Management, Harvard Extension School, Cambridge, MA
42 Ibid.
43 Crowded skies, frustrated passengers, 2013, *The Economist*, August 8, http://economist.com/news/china/21583273-military-control-airspace-and-risk-averse-culture-threaten-cripple-chinas-rapid-growth, Accessed September 20, 2016
44 Carey et al., *Future of Mobility*, page 66
45 Air traffic control improvements are key to cutting aviation emissions in the short term, finds new Oxford study, 2010, *Greenair*, February 10, http://greenaironline.com/news.php?viewStory=752, Accessed September 20, 2016

46 Flexible use of airspace, http://eurocontrol.int/articles/flexible-use-airspace, Accessed October 1, 2016
47 Federal Aviation Administration, June 2013, *NextGen*, Office of NextGen, Washington, DC
48 Carey et al., 2010, *Future of Mobility*, page 64
49 Sir Richard Branson announces biofuel breakthrough – press conference, n.d., Virgin Atlantic, http://virginatlantic.digitalnewsagency.com/stories/6005/videos, Accessed September 20, 2016
50 Ibid.
51 Technical Overview, n.d., LanzaTech, http://lanzatech.com/innovation/technical-overview/, Accessed September 20, 2016
52 World first low costs fuel to be developed for Virgin Atlantic, n.d., Virgin Atlantic, http://virginatlantic.digitalnewsagency.com/stories/6005, Accessed September 20, 2016
53 Ibid.
54 http://biofuelstp.eu/air.html#demo_flights, Accessed April 24, 2014
55 Phelan, Ben, 2013, Biofuel crops: food security must come first, *The Guardian*, August 29, http://theguardian.com/environment/2013/aug/29/biofuel-crops-food-security-prices-europe, Accessed September 20, 2016
56 Air Transport Action Group, May 2009, *Beginner's Guide to Aviation Biofuels*, enviro.aero, page 4, Accessed September 20, 2016
57 Ibid.
58 Agrisoma Announces Closing of Series A Round, 2014, Agrisoma, September 26, http://agrisoma.com/agrisoma/index.php?pageID=109, Accessed September 20, 2016
59 Maniatis, Kyriakos, Michael Weitz, and Alexander Zschocke, eds, 2013, *2 Million Tons Per Year: A Performing Biofuels Supply Chain for EU Aviation*, August 2013 update, Brussels, Belgium, kyriakos.maniatis@ec.europa.eu, page 12
60 Ibid., pages 18–19
61 Ibid., page 21
62 What happened to biofuels? 2013, *The Economist*, September 5, http://economist.com/news/technology-quarterly/21584452-energy-technology-making-large-amounts-fuel-organic-matter-has-proved-be, Accessed September 20, 2016
63 U.S. Department of Agriculture (USDA) in conjunction with Airlines for America and the Boeing Company, January 2012, *Agriculture and Aviation: Partners in Prosperity*, USDA, Washington, DC, page 10
64 Ibid.
65 Nastu, Paul, 2010, BA, Bombardier announce biofuel plans at Farnborough, *Environmental Leader*, July 22, http://environmentalleader.com/2010/07/22/ba-bombardier-announce-biofuel-plans-at-farnborough/, Accessed September 20, 2016
66 Winchester, Niven, Dominic McConnachie, Chrisoph Wollersheim, and Ian Waitz, March 2013, *Market Cost of Renewable Jet Fuel Adoption in the United States*, Massachusetts Institute of Technology, Cambridge, MA, page 6
67 Nastu, BA, Bombardier announce biofuel plans at Farnborough
68 Ibid., page 8
69 Winchester et al., *Market Cost of Renewable Jet Fuel Adoption*
70 Total life cycle emissions from biofuels are not 100% carbon neutral.
71 Hanmel, Debbie, January 2015, *Aviation Biofuel Sustainability Scorecards, NRDC Issue Brief*, Natural Resources Defence Council (NRDC), Washington, DC
72 Aviation biofuel goes into commercial use in China, 2014, *Alexander's Gas & Oil Connections*, February 14, http://gasandoil.com/news/2014/02/aviation-biofuel-goes-into-commercial-use-in-china, Accessed September 20, 2016
73 Biofuels International, http://biofuels-news.com/display_news/5281/Sinopec_company_prepares_to_produce_green_jet_fuel/, Accessed September 20, 2016
74 Virgin Atlantic hails RSB certification of LanzaTech's Chinese venture to convert waste gases into sustainable jet fuels, 2013, *Greenair*, November 25, http://greenaironline.com/news.php?viewStory=1791, Accessed September 20, 2016

75 Accelerating availability is key pillar of industry's sustainable growth strategy, 2012, Airbus, March 22, http://airbus.com/presscentre/pressreleases/press-release-detail/detail/airbus-boeing-embraer-collaborate-on-aviation-biofuel-commercialisation/, Accessed September 20, 2016
76 IPK International, 2013, *ITB World Travel Trends Report 2013/14*, Prepared on behalf of ITB Berlin by IPK International, Munich, Germany, page 5
77 Annual updates on air travel found in press releases from IATA, http://iata.org, Accessed October 1, 2016
78 Map: James Cheshire, Mapping the World's Biggest Airlines, http://spatialanalysis.co.uk/2012/06/mapping-worlds-biggest-airlines/, Accessed September 20, 2016
79 Perovic, J., 2013, *The Economic Benefits of Aviation and Performance in the Travel and Tourism Competitiveness Index*, The Travel & Tourism Competitiveness Report, 2013, *Reducing Barriers to Economic Growth and Job Creation*, Jennifer Blanke and Thea Chiesa, eds, *World Economic Forum*, Davos, Switzerland, pages 57–59
80 Ibid.
81 Christie, I. et al., 2013, *International Bank for Reconstruction and Development*, The World Bank, Washington, DC, page 47
82 Kasarda, John D., ed., 2010, *Global Airport Cities*, Insight Media, http://aerotropolis.com/files/GlobalAirportCities.pdf, Accessed September 20, 2016
83 Ibid., Accessed May 2, 2014
84 @SOM, no date, Skidmore, Owings and Merrill LLF, the master planner
85 Adunagow, Kabibi, December 2011, *Los Angeles International Airport Noise Pollution: A Case Study of the Impact on the City of Inglewood*, Unpublished Manuscript, Environmental Management of International Tourism Development class paper, Department of Sustainability and Environmental Management, Harvard Extension School, Cambridge, MA
86 Ibid.
87 Ibid.
88 European Commission, January 2015, *Science for Environmental Policy Thematic Issue: Noise Impacts on Health*, Issue 47, European Union, Brussels, Belgium
89 Noise as a hazard: Medical professionals talk about the effects of night-flights on the psyche and body of a human being, April 21, 1999, AReCO, http://areco.org/499noise.htm, Accessed September 20, 2016
90 West, Larry, 2016, What are the health effects of airport noise and pollution? *About News*, August 8, http://environment.about.com/od/pollution/a/airport_noise.htm, Accessed September 20, 2016
91 Andre, R., 2004, *Take Back the Sky*, Sierra Club Books, San Francisco, CA, page 36
92 Ldn and Lden Calculator, Noise Meters Inc., http://noisemeters.com/apps/ldn-calculator.asp, Accessed September 20, 2016
93 Mestre, Vincent, Paul Schomer, Sanford Fidell, and Bernard Berry, June 2011, *Technical Support for Day/Night Average Sound Level (DNL) Replacement Metric Research*, Final Report, USDOT/TITA/Volpe Center, Report Number: DOT/FAA/AEE/2011–02, page 14
94 FAA Noise Research, n.d., Federal Aviation Administration, http://aci-na.org/sites/default/files/cointin_-_final_-_noise_research_v3.pdf, Accessed May 6, 2014
95 Andre, R., *Take Back the Sky*, page 158
96 Mestre et al., *Technical Support for Day/Night Average Sound Level*
97 Ibid., page 105
98 IATA Position on Noise-Related Operating Restrictions, n.d., IATA, http://iata.org/policy/environment/Documents/paper-on-operating-restrictions-august-2013.pdf, Accessed September 20, 2016
99 ICAO Secretariat, 2010, Aviation Outlook, *ICAO Environmental Report*, page 22, http://icao.int/environmental-protection/Documents/EnvironmentReport-2010/ICAO_EnvReport10-Outlook_en.pdf, Accessed September 20, 2016

100 Boeing 787 Dreamliner, https://en.wikipedia.org/wiki/Boeing_787_Dreamliner, the best overview of the development and testing of the aircraft, Accessed October 1, 2016

101 Eco-Efficiency, Airbus, http://airbus.com/innovation/eco-efficiency/design/, Accessed September 20, 2016

102 Virgin Atlantic, https://en.wikipedia.org/wiki/Virgin_Atlantic, Accessed October 1, 2016. The best summary of information on the airline

103 Aircraft Noise Management Strategy, 2013, Virgin Atlantic, http://virgin-atlantic.com/content/dam/VAA/Documents/sustainabilitypdf/Final_VAA_Aircraft_Noise_Management_Strategy.pdf, page 14, Accessed September 20, 2016

104 ICAO Secretariat, 2010, Aviation Outlook, *ICAO Environmental Report*, Overview, page 2

105 Performance Based Navigation (PBN), http://faa.gov/nextgen/update/progress_and_plans/pbn/ General Updates on Next Generation navigation now called Performance Based Navigation by FAA, Accessed October 1, 2016

106 Bayside residents say noise from low-flying jets is ruining quality of life, 201, *CBS New York*, August 26, http://newyork.cbslocal.com/2013/08/26/bayside-residents-say-noise-from-low-flying-jets-is-ruining-quality-of-life/, Accessed September 20, 2016

107 Queens case continues to evolve per the following references: Avella Applauds Governor Cuomo for Directing Port Authority to Conduct Noise Studies an Establish Community Roundtable to Address Increase in Airplane Noise over Queens, March 24, 2014, https://nysenate.gov/newsroom/press-releases/tony-avella/avella-applauds-governor-cuomo-directing-port-authority-conduct, Accessed October 1, 2016; Shastri, Veda, n.d., Noise Pollution Affects Jackson Heights, http://projects.nyujournalism.org/goinggreennewyork/the-roar-above/noisepollution/, Accessed October 1, 2016

108 Wasser, M., 2015, Sound and fury: Frustrated Phoenix residents are roaring ever since the FAA changed Sky Harbor flight paths, *Phoenix New Times*, March 4, http://phoenixnewtimes.com/news/sound-and-fury-frustrated-phoenix-residents-are-roaring-ever-since-the-faa-changed-sky-harbor-flight-paths-6654056, Accessed September 20, 2016

109 Johnson, Bryan, December 2015, *Aviation Noise: More Than an Annoyance*, Unpublished Manuscript, Environmental Management of International Tourism Development class paper, Department of Sustainability and Environmental Management, Harvard Extension School, Cambridge, MA

110 Ming Li, K., Gary Eiff, John Laffitte, and Dwayne McDaniel, 2007, *Land Use Management and Airport Controls, Trends and Indicators of Incompatible Land Use*, Partnership for Air Transportation Noise and Emissions Reduction, Massachusetts Institute of Technology, Cambridge, MA, page 2

111 Ibid., page 4

112 Ibid., pages 40–42

113 ICAO Secretariat, 2010, Aviation Outlook, *ICAO Environmental Report*

114 De Wit, J, and G. Burhhouwt, 2008, Slot allocation and use at hub airports, perspectives for secondary trading, European Journal of Transport and Infrastructure Research, Volume 8, http://dare.uva.nl/record/1/296183, Accessed October 1, 2016

115 Airports Commission, July 2015, Airports Commission: Final Report, London, UK, page 10, https://gov.uk/government/uploads/system/uploads/attachment_data/file/440316/airports-commission-final-report.pdf, Accessed October 1, 2016

116 IATA Position on Noise-Related Operating Restrictions, n.d., IATA

117 Worldwide Air Transport Conference (ATCONF), Sixth Meeting, 2012, International Civil Aviation Organization, http://icao.int/Meetings/atconf6/Documents/WorkingPapers/ATConf6-wp008_en.pdf, September 20, 2016

118 IATA Position on Noise-Related Operating Restrictions, n.d., IATA

119 O'Hare Noise Compatibility Commission, http://oharenoise.org/residential_program.htm, Accessed September 20, 2016

120 O'Hare Sound Insulation Programs, http://flychicago.com/OHare/EN/About Us/NoiseManagement/SoundPrograms.aspx. Landing Page for information on Chicago O'Hare's on-going noise mitigation programs, Accessed October 2, 2016
121 Sound-proof house for people living near Suvarnabhumi Airport, Suvarnabhumi Airport Thailand, http://airportsuvarnabhumi.com/sound-proof-house-for-people-living-near-suvarnabhumi-airport/, Accessed September 20, 2016
122 Ibid.
123 Schlenker, Wolfram and W. Reed Walker, October 2012, *Airports, Air Pollution and Contemporaneous Health*, NBER Working Paper No. 17684, National Bureau of Economic Research, Cambridge, MA, pages 1–3
124 Ibid., page 3
125 Ibid., page 1
126 Op. cit., page 31
127 Ibid., page 32
128 Tetra Tech, June 2013, *LAX Air Quality and Source Apportionment Study*, Vol. 1, Los Angeles World Airports Environmental Services Division, Los Angeles, CA
129 International Civil Aviation Authority (ICAO), 2011, *Airport Air Quality Manual*, Montreal, Quebec, 4.5
130 Airport Local Air Quality: Zurich Airport Regional Air Quality Study, 2013, Zurich Airport, http://zurich-airport.com/~/media/FlughafenZH/Dokumente/Das_Unternehmen/Laerm_Politik_und_Umwelt/Luft/2013_LocalAirQuality_E_final.pdf, page 4, Accessed September 20, 2016
131 Airport Local Air Quality: Zurich Airport Regional Air Quality Study, 2013, Zurich Airport
132 Scott Morrissey, 2016, personal communication regarding Sustainability Initiative for Denver International Airport, Project name: Aerobahn Deicing Surface Management System
133 Chicago Department of Aviation, 2013, *Sustainable Airport Manual*, City of Chicago, http://airportsgoinggreen.org/sustainable-airport-manual.aspx, Accessed October 2, 2016. Manual is continuously updated and found on this site
134 Federal Aviation Administration, April 2013, *Recycling, Reuse and Waste Reduction at Airports a Synthesis Document*, Washington, DC, pages 24–25
135 Knapp, Ria, December 2010, *Terminal T? Exploring Ways to Improve Trash & Recycling Management at Logan International Airport*, Unpublished Manuscript, Environmental Management of International Tourism Development class paper, Department of Sustainability and Environmental Management, Harvard Extension School, Cambridge, MA
136 Ibid., page 17
137 Federal Aviation Administration, *Recycling, Reuse and Waste Reduction*, Page 3
138 Ibid., page 3
139 Ibid., page 43
140 Ibid., page 37
141 Honkio, Katariina, 2009, The future of waste collection? Underground automated waste conveying systems, *Waste Management World*, January 7, http://waste-management-world.com/articles/print/volume-10/issue-4/features/the-future-of-waste-collection-underground-automated-waste-conveying-systems.html, Accessed May 16, 2014
142 Gray, Larry, December 2013, *Review of Aircraft De-icing and Anti-Icing Fluid Storm Water Runoff Control Technologies*, Rensselaer Polytechnic Institute, Hartford Campus, Connecticut, pages 4–5
143 U.S. EPA, August 2000, *Preliminary Data Summary, Airport Deicing Operations (Revised)*, Office of Water, EPA, Washington, DC, pages 10–16
144 Environmental Groups Sue BWI Airport for Toxic Discharges, March 16, 1998, U.S. CAW Activities, http://us-caw.org/lawsuit3.htm, Accessed September 20, 2016
145 Airport Deicing Effluent Guidelines, Environmental Protection Agency, http://water.epa.gov/scitech/wastetech/guide/airport/index.cfm, Accessed September 20, 2016

146 Wassener, Bettina, 2013, Singapore leads pack as cities prepare for an influx of fliers, *The New York Times*, December 30, http://nytimes.com/2013/12/31/business/international/singapore-leads-surge-in-airport-construction-across-asia-pacific.html?_r=1&, Accessed September 20, 2016
147 Ibid., Accessed May 21, 2014
148 Flightpath 1.5 degrees, *Frequently Asked Questions*, http://flightpath1point5.org/, Accessed May 2, 2016
149 Carey et al., *Future of Mobility*, page 65
150 Fauchon, *Hybrid Aircraft Powerplant*
151 How Big Is Aviation's Emissions Gap?, Flightpath, http://flightpath1point5.org/, Accessed October 2, 2016
152 Dubois, Ghislain et al., 2013, Tourism Sensitivity to Climate Change Mitigation Policies, Lessons from Recent Surveys, in *Tourism, Climate Change and Sustainability*, eds. Maharaj Vijay Reddy and Keith Wilkes, Earthscan, Routledge, London and New York, page 169
153 Gössling, *Carbon Management in Tourism*, page 102

6

TOUR OPERATORS

Exploring and importing customers worldwide

Introduction

> Explorations guide, Richard Avilino realized as far back as primary school he wanted to make a career out of his passion for the bush. Richard grew up along the Thamalakane River in Maun, Botswana. Some of his most memorable moments as a boy were watching wildlife coming down to the river to drink. Richard applied for and completed his guiding qualification with the Botswana Wildlife Training Institute. In 2005, Richard heard about Wilderness Safaris. "I was drawn to Wilderness after hearing about the high standard of guide training." After contacting Wilderness and being hired as a guide he was not one bit disappointed.[1]

Wilderness Safaris was founded in 1983 and built around the passion of its partners for protecting and conserving the ecosystems where they worked. In the 1980s, Botswana was a little known safari destination and not a hub for safari companies. As a mobile safari tour operator in Maun, Botswana, they immediately were faced with competing with well financed international safari hunting companies, which controlled large tracts of the renowned Okavango Delta.[2] The early days of their business were not easy, and their profits were small. The founders lived on small salaries. In the 1990s, the firm gained access to valuable concessions (15-year leases) to wildlife tourism areas in Botswana. They had realized the future for safari tourism lay in obtaining fixed concessions on wildlife-rich lands for photography, not hunting. In a gutsy business move, they borrowed funds to offer the Botswana government what they viewed as the "true value" of the wildlife-rich land that had previously been leased by hunting groups for small fees.[3] Because of their business foresight, Wilderness Safaris went on to win the concessions it bid for in Botswana and took a leadership position in the safari business in the region. The vision of this company is based on a strong understanding of how to maximize the enjoyment of their clients in pristine landscapes, conserve the value of the ecosystems where they work, develop the potential of a local, trained workforce, hone an efficient approach to operations, and ensure that local communities benefit from tourism and conservation.

Wilderness Safaris is now a public company, listed on the Botswana Stock Exchange as well as the Africa Board of the Johannesburg Stock Exchange, with $89 million USD in gross revenues reported in their 2015 annual report. Their net profit is 7% of gross revenues in the same year,[4] significantly up from 2% in 2013.[5] After recovering from the 2009–2011 global economic recession, which hit the luxury travel trade hard, the company nimbly restructured the business in the face of the downturn, consolidated management where possible, reduced operating costs in part by significant investment in solar energy systems for their safari camps as part of their commitment to long-term sustainable practices, and concentrated on markets where currencies were strongest for their product, in this case the U.S. as opposed to Europe.[6]

Wilderness Safaris operates on a quadruple bottom line (QBL) system, which exceeds the usual concept of triple bottom line to accommodate reporting on environment, economics, and social outcomes.[7] A fourth bottom line has emerged that measures business success beyond economic results, environmental management efficiencies, and labor and employment outcomes. The fourth bottom line is in essence a reinvestment in the future through ethical payments for positive growth in ecosystem and sociocultural health. Wilderness Safaris defines their responsibilities in four categories, known as the 4Cs: "commerce, conservation, community, and culture." In each case, their consolidated annual report provides clear measurements of how they are achieving goals in each category.[8]

The company operates in 8 countries: Botswana, Namibia, Rwanda, Zambia, South Africa, Zimbabwe, Seychelles, and Kenya. It manages its markets via the traditional tour operator model by selling its products through a dedicated network of tour operators and travel agents around the world. It also owns its own ground and air transfer business, manages all of its own reservations through regional offices, and owns safari camps. This makes it a vertically integrated company that employs 2,600 people, owns 32 aircraft, and owns or manages 58 destinations with 1,000 beds. A Chief Sustainability Officer works for the company at the same level as the Chief Operating Officer, making it clear from the organizational chart alone that conservation priorities and sustainability rank as high as operational realities for the company.[9]

Wilderness Safaris owners are committed to the conservation of ecosystems, local employees, and community partners. For example, in 2013 the company spent $782,000 USD on conservation initiatives, which represents a half of one percent of their gross turnover and an astonishing 24% of their net profits.[10] Over one-half of its employees received ongoing training in its diverse regions of operation, and nearly $1 million USD went to community partners in the form of contracts related to land leases and services rendered.[11] In 2015, Wilderness Safaris paid $4 million in the form of concessions, which consist of rentals, royalties, or profit shares with wages paid into local communities, constituting a major contribution to local economies.[12]

Its low working-capital requirement and lighter costs for maintenance of hotel infrastructure allow it to invest more in the quadruple bottom line. This prize-worthy approach to building equity in their people and places is partially driven by their founder's values. But now that the company is public, the firm will increasingly have to prove its worth to shareholders. The company is committed to building "conservation economies" in Africa.

Tour operators like Wilderness Safaris offer a model that is in direct contrast to the trends within the industry, which are increasingly driven by online marketing dollars. Their willingness to invest profits in both the conservation of ecosystems and the well-being of local people sets them apart and makes them an important exception to the rule for the industry, which is highly worthy of analysis. They are not only successful in helping to preserve the destinations where they work, they are pushing the boundaries of what a profitable company should invest in by creating a fourth bottom line and doing this at a scale that far exceeds most ecotourism operations, which largely have failed to emerge beyond small owner–driven enterprises. They are genuine specialists with knowledge and capacity that cannot be replaced or underpriced. And their commitment to their internal operations and the willingness to invest in the skills of their own team make them formidable competitors in this service industry.

A brutal restructuring of the tour operator marketplace is under way, and only the savviest will survive. Tour operators are travel businesses that procure services from a wide variety of suppliers, which requires business contracting with restaurants, hotels, camps, guides, and transportation services including buses and small charter aircraft. The pricing they negotiate with both suppliers and local talent form the foundation of the cost of doing business for these firms. They can seek to invest in quality by paying for above-average costs for their suppliers and talent, or they can seek to cut costs and increase volume as much as possible. Most tour operators are moving to ever more aggressive online marketing, forcing them to cut overhead, push supplier pricing down, and keep a cap on the cost of talent.

Wilderness Safaris operates from the ground up. They invest in their suppliers and their agents who sell them and have created a long-term strategic partnership of committed customers and local suppliers that all believe in investment in the 4th bottom line, which will help them to protect the future of the destinations they serve, even as global tour operators are cutting overhead and seeking to pay less for more. Tour operators were once proud specialists and experts on destinations who could tout their expertise, and some still use this formula, but the new version of tour operating offers a selection of trips based on comparisons of thousands of options, rates, reviews, and rankings online.[13] This has changed the revenue structure of tour operating permanently and threatens to drive investment in local suppliers and talent down to perilous degrees, and certainly precludes investment in the long-term health and well-being of the destination.

Companies, such as Wilderness Safaris, break this mold. Their investment in pristine wilderness and local people may well be the secret to help them to keep ahead of the online marketers and their algorithms. They retain a close connection to the extraordinary ecosystems and peoples they represent, something international operators can only pretend to achieve. Their value-added offering is likely to become increasingly valuable as global destinations become overrun with volume driven in part by aggressive online marketing and poor management by destinations. Sustainable tour operations that support conservation with significant dollars may well be the leading allies in the quest to preserve the most important assets on earth, as the pressures of tourism become greater and greater on vulnerable ecosystems, cultures, and monuments.

Their Value Chain Model includes rural development, community income, conservation projects, and economic contributions. (See Figure 6.1.)

Environmental impacts of tour operators – key issues

Tour operators have five major environmental management and conservation areas of engagement that were initially outlined by the Tour Operator Initiative of the United Nations Environment Programme, which was launched in 2000 to present good practice to the overall tour operator community and explore new tools to address environmental, cultural, and socioeconomic issues.[15] Some of these tools have become standard issue sustainability practice for all global corporate activity while other areas, such as Product Development and Management, Supply Chain Management, and Destination Stewardship are of particular relevance strictly to the tour operator community.

Digital travel purchasing models will have a transformative effect on the travel industry and tour operators in particular in the next 20 years. These trends are accelerating, and there are concerns that the aggressive pricing comparisons and discounts the online travel agencies (OTAs) offer have removed any incentive for tour operators to include sustainability in their offer.[16] The different business models for tour operators will be discussed in depth here in order to provide insight into how the restructuring of the market may affect future environmental impacts of the industry.

Tour operator business models defined

Tour operators are the glue that holds together the tourism industry. Together with the transportation industry, they manage the delivery of tourists worldwide. The consumer demand for travel is global, but the delivery of tourism services is local. In general, a small number of tour operator companies manage the packaging of the services for tourists, while a large number of suppliers deliver components of the package on a local basis. Tour operators remove the complexity of purchasing so many products from small suppliers when traveling and the headaches of

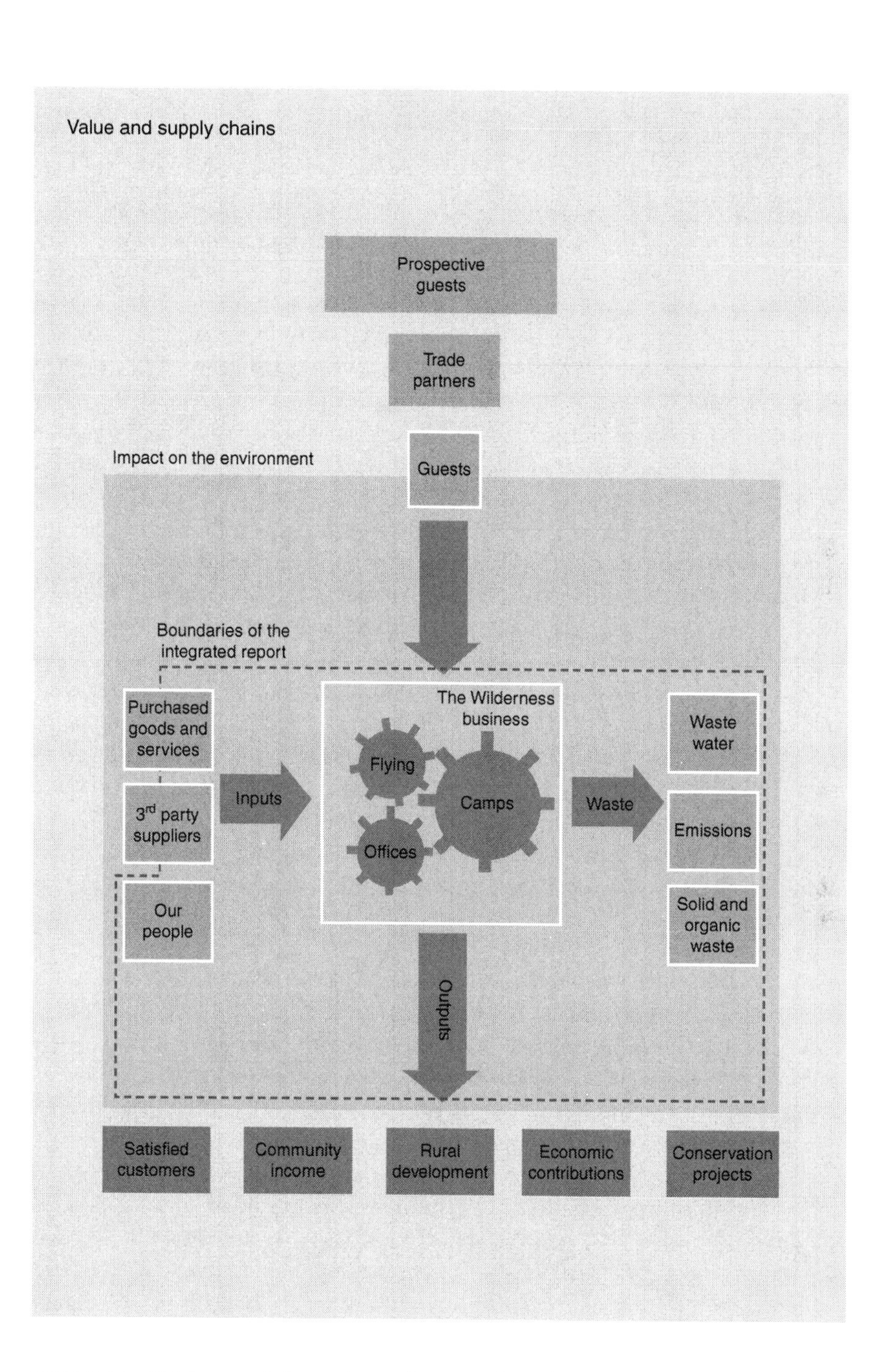

FIGURE 6.1 Wilderness Safaris value and supply chains[14]

getting the best prices. There are two parts to the tour packaging business: travel agencies and tour operators. Tour operators combine two or more travel services (e.g., accommodations, attractions, restaurants) and sell them either through travel agents or direct to consumers.

Travel agencies sell travel services directly to consumers, but they do not package these services. Either they purchase packaged tours from tour operators, or they purchase single services, such as hotels or tickets, for consumers without packaging the services. Online travel services are now common and known as online travel agencies (OTAs), which include Expedia and Travelocity. OTAs are quickly taking the place of storefront travel agencies. Travelers now book hotels and airline services either through OTAs or directly online from airlines and hotel companies. This is causing a decline in travel agencies worldwide.

There are two types of tour operators. The *outbound* tour operators focus on regional and international travel for their customers who live in the countries where they are selling travel. The *inbound* tour operators manage the services for visitors in the countries where travelers are arriving. The outbound tour operator often contracts the inbound tour operator to manage all services within one region. The inbound operators offer good local business relationships, capacity in local languages, payment for services in local currencies, and management of all local negotiations and contracting. The outbound tour operator specializes in marketing, branding, excellent website services, customer service, and quick response times to tourist needs in their home markets. Inbound operators often remain anonymous to the consumer and simply manage the business for the brand name outbound company.

This symbiotic relationship between inbound and outbound tour operators has worked well for decades, but new market dynamics are sweeping aside the old business relationships that were once the foundation of tour operating. Online purchasing platforms are so inexpensive to construct in nearly every country in the world, there are fewer barriers to entry for inbound operators who seek to sell their services directly to consumers in international markets.

To illustrate this shift, consider Wilderness Safaris, which is an inbound tour operator. While they maintain important business relationships with outbound tour operators, they are also fully capable of selling their own services directly to the online consumer. As such, they have representation in their most important source markets to get a larger percentage of the total package price. The larger size of these companies gives them a more competitive edge to explore every option for capturing a larger percentage of the retail price. This may not always be the case for smaller tour operators around the world, who continue to depend on the large outbound operators in source markets. But there is no doubt the basic outbound–inbound model is fracturing.

Tour operating is generally a low-margin business. Considering the number of parties being contracted for each tour, from local guides to transport services, to hotels, every member of the supply chain is often squeezed to create price-friendly packages. Tour operators block the availability of services with their vendors for reduced pricing with guaranteed payments for volume bookings. As online booking becomes easier and easier, the cost differential between a package tour and a service booked directly is disappearing. Tour operators often seek to prove their value by showing that their packages are cheaper than traveling independently. To achieve the best package price, outbound tour operators often strong-arm their local suppliers to reduce overhead cost in order to deliver one competitive package price.

When looking at the issues related to how well tour operators can manage sustainability, the Catch 22 is that they frequently work with two goals simultaneously: cutting the margins of all their suppliers while simultaneously seeking to boost sustainability standards worldwide. They are also burdened with a great many diverse suppliers, which are highly difficult to monitor effectively, unless an expensive system is put in place. According to research, large tour operators in Europe seek the least demanding steps to achieve environmental performance of their suppliers. The European operators surveyed faced a daunting task to review their supply chain globally, with one respondent suggesting it would take over 5 years for their business to evaluate its 10,000 accommodation suppliers.[17] Only the largest firms in Europe, with revenues exceeding $1.5 billion annually, sought to meet sustainability goals and report on their goals and objectives in a consistent fashion.[18]

Without the support of their international tour operator clients, local operators and vendors struggle to meet any sustainability standards because they do not have the profits in their pockets to invest in basic efficiencies, not to mention reporting systems. Local suppliers frequently complain they are left with no choice but to operate almost at cost or turn down large tour operator business and the security this represents. If the local hotelier or transport company decides to accept the low-overhead pricing offered by a tour operator, they must cut staffing, goods, and services. Trained staff is essential to achieve better management of environmental impacts on the local level. Partnerships between the tour operator and their suppliers could allow for many more environmental management procedures to take place worldwide. The success of these partnerships depends on profit margins for local suppliers, who are often left out of the discussion and forced to cut their costs to the bone and remain silent partners without options to improve their sustainability performance.

A wide range of cases in the tour operating world demonstrate this point, while there is a minority of cases where genuine collaboration between international operators and local suppliers results in sustainability results. But these successes are difficult to measure. Only the largest firms deliver consistent sustainability performance reporting, while most small and medium-sized international firms tout their philanthropic projects without providing a broader context of their firm's performance. Many firms simply cannot manage a sustainability agenda because the staffing and talent are not there at any level of the company. Sustainability students of the author who come from the tour operating industry have related the brick walls they encounter when proposing sustainability initiatives because their companies simply do not incorporate sustainability in their planning and supplier contracting. There are solutions to this, which are now discussed.

European tour operating models

The tour operating market is so consolidated in Europe that 70% of clients now book under only 5 brands.[19] The United Kingdom has become dominated by just 2 travel groups, TUI and Thomas Cook. The UK tour operating market represents a "huge share" in the total turnover of the tour operator and travel agency market in Europe, equal to $69 billion before the economic crisis, or one-third of Europe's total tour operator economy.[20] Thomas Cook has over 20,000 employees, 45 aircraft, its own television channel, and many well-known sub-brands under its umbrella. This model of tour operating sets a very different standard and enables a very different type of decision making than what is typical in the Americas, as will be reviewed in this chapter.[21]

The low profitability of the tour operating industry and travel agencies is one reason for the increasing consolidation of the business. With net margins at an average of 5%, the tour operating industry in Europe has margins that are far below the average double-digit profitability of the hotel sector due to the increasing cost of doing business (including airfare, local transport, and accommodations). European tour operators have streamlined and increased volume to maintain competitiveness[22] Within the EU27, medium-sized and large tour operators in Belgium, Germany, Spain, and the Netherlands dominate the tour operating world, earning 65% of the gross revenues for the region.[23] These operators have used their market position to achieve preferred pricing arrangements with their suppliers, giving them an increasing market advantage,[24] and larger companies like TUI are now using an All-Inclusive strategy, which keeps all the business in-house, contains costs, and allows them to offer a unique selling proposition to their clients.[25]

U.S. tour operating models

The U.S. tour operator business model is very distinct from Europe, so much so that it can be difficult to compare these two marketplaces. Even though volume in the U.S. travel market is three times larger than the outbound market in Europe, U.S. tour operators are predominantly small enterprises that earn less than $10 million per year.[26] Historically, U.S. tour operators survived on booking volume tours to easy-to-package destinations such as Florida, Las Vegas, or the Caribbean. But with modern online booking, U.S. consumers can easily do this themselves. In general, the mainstream U.S. tour operating business is not growing or integrating into a large-scale, all-service business as it has in Europe.

Even though large tour operators have not thrived in the U.S. market, a highly diversified and vibrant niche tour operator industry has. Adventure travel was spearheaded in the U.S. in the 1960s and 1970s and became the pioneering model for niche market tourism. Early adventure companies such as Mountain Travel and

Sobek set the standard for companies that attracted the large outdoor adventure market in the U.S. These companies created a whole new tour operating business model. The founders of Mountain Travel and Sobek, which later merged, are emblematic of the early outdoor adventuring market. They came from the Western U.S. where outdoor sports in extraordinary parks such as Yosemite and the Grand Canyon offered terrific experiences that could be packaged and sold. These entrepreneurs have historically sought to avoid traditional business models and buttoned-down business-suit management techniques. They have lived the life of outdoor adventurers themselves, taking on first ascents of mountains and then descents of wild rivers. To this day, they view themselves not as business people but as pioneers of outdoor lifestyles that they themselves enjoy.[27]

Their idea of selling outdoor adventure sold quickly and morphed into a wide variety of niche market concepts. Small, specialized tour operators began to take off in every form with creative leadership from both the adventure and the environment communities in the 1980s and 1990s. By the end of that decade, the U.S. had become the largest outbound market in the world for both adventure and ecotravel.[28] These entrepreneurs built businesses based on their interests in wildlife and cultural preservation, mountaineering and rafting, biking, and walking. Many of the founder owners of ecotourism companies were very concerned about the environmental conservation of wildlands and the well-being of the local people affected by their tours. It was this movement that fostered the founding of The International Ecotourism Society in 1990, whose charter company founders included Wildland Adventures and International Expeditions, companies I interacted with extensively and learned from. I visited ecotourism companies worldwide between 1991 and 2002 and learned how they developed their businesses and how they managed them on the ground to both respect and contribute to the local places and people their clients were visiting.

Interest in rugged outdoor backpacking and river rafting also grew enormously in the 1980s and 1990s, powering an active market for backcountry tours, adventure travel magazines, and outdoor gear, which helped to drive an increasingly robust marketing environment for adventure travel.[29] The adventure market is now tracked by the Adventure Travel and Tourism Association, which reports that 63% of adventure tour operators reported an average of 17% growth in revenues in 2011, with hundreds of North American tour operators offering specialized products for niche markets.[30]

While ecotourism and adventure travel was taking off in the U.S., another vibrant niche market also emerged. Nonprofit organizations, which are more numerous in the U.S. than anywhere else in the world, began to use travel as a means of building loyal membership bases and raising funds. Their market is now called educational travel. Large NGOs, including the Smithsonian, World Wildlife Fund, and The Nature Conservancy, established partnerships with North American tour operators to offer specialty trips to their members. This phenomenon grew rapidly in the 1990s, with membership organizations of many kinds marketing to a wide variety of affinity groups. Zoos, museums, alumnae associations, and even public radio and television organizations all developed tours for their supporters in partnership with tour operators. This allowed niche market U.S. tour operators to develop packages for markets with like interests, a model that operates at smaller volumes but tends to yield good returns and allows the tour operator partners to design trips that are focused not on cutting costs but on providing exceptional experiences for individuals interested in art, wildlife, culture, and local ways of life. Small group tours thrived, helped to support universities and nonprofit organizations, and attracted well-to-do Baby Boomers, who were thrilled to support both their favorite nonprofit and find like-minded friends on tours. The Non-Profits in Travel Conference, formed in 1987, spurred the growth of this market by creating a marketplace where small vendors offering tours could meet with nonprofits and create exciting new offerings for educational travel.[31]

All of these American niche markets have now morphed into the concept of so-called experiential tourism, which is sweeping the marketplace and attracting a wide range of demographics and age groups.[32] Experiential tours yield "authentic experiences" according to the current buzz in the marketplace, allowing travelers to connect with the people and cultures of places, with positive impacts on the environment. For tour operators to achieve this kind of authenticity, they must rely on excellent delivery of service from both their inbound operator and local tour guides, who must be very plugged into the local culture and environmental milieu.

The remaining U.S. tour operator community has about 20% of the U.S. market and appears to be shrinking. Mainstream U.S tour operators are on a collision course with cruise lines, which can offer trips to Europe or the Caribbean for half the price. With market erosion a reality, traditional U.S. tour operators are seeing a future with diminishing success.[33]

Hybrid business models

The experiential travel movement combines adventure, eco, and educational travel into one niche market. One group of international tour operators has taken this combination of travel experiences to a new level, fueled in large part by a youthful market that seeks to travel with their peers and experience countries around the world at prices that they can afford. These companies removed the inbound tour operators from their supply chain, eliminated the minibus almost entirely, and created trips without frills that delivered more authentic experiences. To manage their growing businesses, they built networks of low-rent international offices with locally hired youthful managers. By cutting out the cost of the "middleman," these companies jumped forward in the marketplace, charging highly competitive rates. They outcompeted the small niche tour operating model pioneered in the 1990s by marketing directly to consumers, using digital marketing and social media platforms, and putting local guides in the position of "experiential gurus" who help travelers to live like the locals.

This new hybrid business model is forging ahead, with G Adventures out of Canada and Intrepid of Australia now part of PEAK Adventures a joint venture with

TUI of Germany, as the leading examples of this tribal movement of youth travel. They play on a global stage, gross hundreds of millions a year, are based primarily in English-speaking countries, and, contrary to what might be expected, are not dependent on the U.S. market. Both trade heavily in the Commonwealth markets, where youth have fewer educational debts and are able to take long, adventurous trips around the world before going ahead with more traditional career pursuits.

Business models in transition: destination online

The travel industry was one of the first industries to show solid growth in online purchasing in the late 1990s. There is natural marriage between travel and tourism and online platforms. Dr. Michael Frenzel, the former supervisory chairman of TUI and chair of the World Travel and Tourism Council (WTTC), states that the digital economy is driving the growth in travel worldwide, creating a new ecosystem. This ecosystem is helping consumers to customize their travel experience, similar to the way they buy shoes or other types of consumer goods.[34]

Hotels, destinations, tours, and cruises all benefit from a visually enriched sales platform, and customers have quickly learned to demand sites that give as many angles and pictures as possible. In raw numbers, travel sales online and via mobile devices grew by 10% in 2012, and 40% of total travel sales were made online in the same year in the U.S.[35] Notably, only 24% of global travel consumers are now booking with physical travel agencies, a clear signal of the decline of this industry.[36] In Europe, 36% of travel bookings take place on the web, but online behavior varies vary greatly between countries in Europe, with the UK the most active and Spaniards and Italians the least likely to purchase online.[37]

Online booking will continue to expand and replace all other forms of doing business. The entire travel industry is closely watching the changes in how travelers book, generating market research reports that look at every angle. Mobile phone booking is growing by 50% a year as of 2014 in the U.S. and Europe.[38] Google entered this market in early 2016 with its destination program that allows mobile phone users to instantly compare flight and hotel prices, for flexible dates, itineraries, and attractions, all on one screen by simply adding the word "destination" to a mobile search.[39]

The OTAs have strengthened their market share and are speeding up the trend toward booking online. In 2014, it was estimated by an industry report that OTA bookings would grow 5% annually worldwide, with figures double or triple that pace in the emerging economies where mobile phone bookings rule. Expedia began stating on its website in 2014 that it is the largest travel company in the world.[40] Its multiple brands now enable travelers to do research, planning, and booking. In 2016, Expedia carried 307,000 bookable properties, 475 airlines, and 15,000 activities worth nearly $68 billion in gross booking.[41] Market consolidation is reducing competition to two dominant players in the U.S. – Expedia, which owns Travelocity, Orbitz, and hotels.com, and Priceline, which owns booking.com. In 2015, these two companies control 95% of the U.S. online booking market.[42]

Suppliers are battling the trend with a great deal of revenue at stake. Hotel companies, cruise lines, and airlines are very competitively offering online purchasing and spicing up the offers with discounts and loyalty programs. Hotel websites managed 26% of all online travel bookings in 2012, while OTAs managed just over 10%. The battle over market share on all the different digital devices is also in full swing. Research shows that desktops are still where the majority of consumers purchase travel, but researching on a tablet or mobile is very common.[43]

Social media is also driving sales and a larger conversation about the quality of travel experiences. Forty percent of travelers in 2012 stated that social network comments influenced their travel planning. This connection to peers is growing widely worldwide and is particularly powerful in Asia.[44]

The digital and mobile phone travel booking revolution will fuel the growth of independent travel in the next century and will grow at the expense of the tour operating business community. The overall growth of the market may compensate for any percentage losses the tour operator community experiences, as long as they offer a digitally savvy purchasing platform. Delivering a solid set of predictions on how the travel industry and tour operators will be affected by changes in global travel patterns, digital purchasing, and demographics may not be entirely possible in this period of transition. However, there are some key takeaways:

1 The U.S. Millennial generation, born between 1980 and 2000, have a stronger desire to travel than other generations, by a 23% margin. They appear to be more socially conscious and want more experiential travel options.[45]
2 Online booking systems will help the overall travel economy to grow even more rapidly than in 2013. With growth from China and other emerging countries, such as Indonesia, Vietnam, Colombia, and Egypt, the total travel market will continue to expand at rates well beyond the global GDP, creating exceptional opportunities for tour operators familiar with the needs of these travelers.[46]
3 The new international travel community is digitally savvy and will demand more services online, through their mobile devices especially, and expect the best deals via online comparison shopping. They will avidly mine social media sources to review travel options. They will expect more pricing transparency and more pricing options to be based on an "uncoupling" of services, with cost breakdowns for each option.
4 The experience-sharing social media economy will continue to grow. Global customers, especially U.S. Millennials and Asian travelers, will share their travel experiences and ratings extensively, and provide instant recommendations to their peers. These instant ratings will have an influence on travel offerings, forcing rapid responses to customer concerns.

Traditional tour operators outperform online firms in most measures of sustainability initiatives. In early reviews of the sector by academics, none of the online tour operators had information on the sustainable performance of their suppliers, nor did they adhere to any sustainability initiatives.[47]

Given the fact that tour operators have generally found that managing their supply chains for sustainability is challenging from a labor and expense perspective, there are increasing concerns that sustainability will be priced out of the market or driven under by OTA comparison shopping. This remains to be seen. In an industry survey of European industrial sectors in 2005, tour operators in general were the least likely to produce sustainability reports compared to such sectors as pharmaceuticals, forestry, and construction, which were the most likely.[48] From this point of view, tour operators have not been a stellar industrial sector even before the rapid rise of OTAs. It is therefore inaccurate to suggest that online marketing is at the root of the problem, but it is a factor that needs much more research and scrutiny.

Online travel rating and environmental screening systems

The experience-sharing social media economy has reinvented travel decision making by revamping how tour operators reach customers. The social media phenomenon, TripAdvisor, also calls itself the largest travel website in the world, which offered links to booking tools that check hundreds of websites reaching 350 million average unique visitors monthly in 2016.[49] They provide a powerful new tool to manage customer outreach and a place to manage environmental information of tour operators and their suppliers. The game-changing effect of TripAdvisor has put them in a leadership role of offering ratings not only of quality and service but also of environmental performance.

In 2013, TripAdvisor launched its GreenLeader program, which categorizes the environmental performance of its hotels and B&Bs in the U.S. and Europe according to a set of criteria easily downloadable from their website in survey form. All properties are required to have a number of basic systems in place:

- Energy tracking system
- 75% energy-efficient lightbulbs
- Current and active towel and linen program
- Recycling at least 2 types of waste
- Staff training on green practices
- Guest education on green practices.[50]

Scores for GreenLeader properties are broken down into bronze, silver, gold, and platinum, with required responses in categories of energy, water, purchasing, waste, site management, and innovation and education. A third-party auditing program has been established, and properties must be ready to submit all documentation to auditors. At launch in early 2013, the program had 1,000 properties, and by July 2014 that number reached 6,000 properties.[51] Travelers can sort their hotels according to green practices, and as of July 2013, 50,000 travelers had used this search engine.[52]

Two of the largest OTAs, Expedia and Travelocity (which have now merged), feature green hotel search engines, both with thousands of properties listed, and therefore have made advances on listing green attributes of their suppliers since the academic surveys, previously referenced, were performed. The screening mechanisms are managed slightly differently by each – Expedia via the nonprofit Sustainable Travel International and Travelocity by means of approved third-party certification systems under the Global Sustainable Tourism Council (GSTC) system. The market interest in these systems is growing according to controlled research,[53] but much more research is needed to determine the exact value of these green rating systems in terms of influencing customer purchasing behaviors.

With the growth of the digital purchasing economy, opportunities for consumers and tour operators to screen their travel choices according to green or sustainability factors could increase if OTAs find there is consumer interest. Online screening of the sustainable management of hotel properties will need years of work to perfect. The best systems depend on third-party visits, which are costly. TripAdvisor's GreenLeader program is completely free and offers a wide variety of support programs and affiliations. Properties of all sizes receive excellent benefits for reporting and access to millions of potential site visits. The value proposition is attractive and will certainly help to enroll many properties that may never have considered environmental management before.

But the next steps for online environmental management reporting may be more difficult. Third-party auditing is really the only route to ensuring that properties will meet the sustainability criteria they are reporting on. TripAdvisor's GreenLeader is offering spot-checks for now, a cost-containment strategy. Full-fledged certification systems offer a range of options from self-reporting to third-party certification, and it is likely TripAdvisor will have to create this type of tiered system to become a more credible player in this arena. With the system as it is, experts fear there are inadequate assurances that the thousands of GreenLeader properties are providing sufficiently verifiable information.[54] TripAdvisor's generous offer for free searchable green listings with consumer ratings will bring many businesses under the sustainability tent for the first time, but if the information is not verifiable, the validity of the system for all involved is at risk.

Environmental management of tour operations

As explored in this chapter, the business model for tour operators varies greatly in different regions. Europe follows a more consolidated model of ownership, while North American tour operators seek to own as few assets as possible. Inbound operators around the world use different strategies of ownership as well. Some operate with few assets, while others own their own vehicles and lodges. Decisions about how to manage the costs of investment in sustainability for operations vary according to how much the company is willing to invest in assets that it may or may not own. Tour operators that own more of their own assets and are more vertically integrated are more likely to be proactive about environmental management in order to achieve savings for their companies.

A Tour Operator Supplement for the Global Reporting Initiative was created based on the original work done by the Tour Operator Initiative, which provides performance indicators for each of these key areas, a highly useful way for students of the industry to review how well a tour operator is succeeding in measuring these key areas.[55] The five categories of review are:

1. **Product Development and Management:** Product managers can work with their suppliers, including restaurants, activities, hotels, and transport companies to determine which have environmental management systems in place and evaluate their environmental performance to date. They can help improve supplier footprints taking advantage of monitoring systems for tour operators such as Travelife – (more later in the chapter).
2. **Internal Management:** The labor practices of the tour operator include health and safety policies for staff at headquarters and destinations, training and education available for staff and contractors, and policies to minimize the environmental impacts of the office.
3. **Supply Chain Management:** Product and contracting managers can lay out the terms of contracting to include improvement in sustainability measures.
4. **Destination Stewardship:** Tour operator leadership can work with their suppliers to contribute to the environmental management and conservation of ecosystems and local cultures in destinations by working with local experts and NGOs.[56]
5. **Customer Relations:** Sales and marketing tools are available to influence customer awareness of environmental responsibilities and inform on all aspects of the company's efforts to environmentally manage their company, which discuss their work to support suppliers and specify their efforts to conserve local destinations.

Many tour operators have been very slow to adopt these basic environmental management procedures, but the difference between larger firms and small firms may be the most important predictor of capacity to manage a tour operation with environmental and social goals incorporated into operations.[57] Smaller firms do not have a great deal of money to save by instituting environmental management standards and often use philanthropy as a means to give back to local people and destinations.

Environmental management goals are more deeply embedded in the DNA of the large, consolidated European brands than the niche market American ones. TUI AG has won numerous awards for environmental management, with a plan that is so comprehensive it will be presented in depth in a case study later in the chapter. The methodical environmental management systems implemented by EU tour operators are surely influenced by how many physical assets the companies own and how much they have to gain by achieving environmental efficiencies.

With the young hybrid firms, the enthusiasm is there, and so is the revenue. But the more informal business cultures beloved by adventure travel business owners and youth culture ethic hinder the adoption of rigorous procedures. With young, sustainability-minded Millennials joining hybrid travel companies worldwide, tour operators have an outstanding opportunity to use their energy and interest to institute sustainability measures and policies. But there is little distinction between philanthropy and more rigorous corporate social responsibility (CSR) in their publicity, and third-party data is not present on the websites.[58] Few travel and tourism students have the opportunity to study the differences between philanthropy, environmental management, and sustainable development technique, and are therefore largely unaware of the difference and are happy to promote what the companies say is best without looking at the underlying differences. Tour operators working in the adventure travel trade in general conflate good deeds with caring for the environment. Leadership, staff training, and consistent, transparent reporting on the impacts of their efforts to assist in destinations would make a considerable difference and in future could be the key to transforming the sector into a powerful sustainable development player.

Product development issues and solutions

In theory, tour operators can deeply influence the services of local providers by selecting local products and services that are sustainably managed. Tour operators also have a responsibility to assess the sustainability of the destination they are working in.[59] Efforts to define and manage the process of selecting sustainable tourism products for tour operators can be daunting because they work with a large number of suppliers on an international scale. While global initiatives such as the Tour Operator Initiative and the GRI tour operator reporting guidelines seek quantification of triple bottom line procedures for product selection with a review of the key environmental, economic, and social issues faced in each destination,[60] there are many barriers to achieving these goals. In an analysis of the process of adopting sustainability goals for tour operators in the Netherlands, the researchers found that even with the incentive of a law that requires sustainability reporting in Holland, tour operators with high interest still chose small new products that could be developed with local NGOs, as opposed to reviewing their entire portfolio. The Dutch law's vague criteria and lax requirements for reporting made it possible for these operators to stress good deeds over consistent product selection using CSR criteria.[61]

These early efforts at achieving sustainable product development were ambitious, considering the extremely high standards early designers were hoping to achieve, the industry's lack of full involvement in evaluating these standards, and the perceived conflict between health and safety standards and sustainability criteria.[62] The GRI standards allow tour operators to use anecdotal examples and report on successes but never thoroughly review their overall product line, except in a few exceptional cases. The volume of products tour operators are responsible for and the wide variety of social and environmental issues involved generally hinder more thorough approaches to product development that would meet higher standards for sustainability. But the internal culture within travel companies seems highly resistant to quantification and monitoring, even while marketing such projects is given the highest priority. This is a question of leadership, risk evaluation, internal management, and training – all issues that travel companies may understand in future to be essential to the preservation of their own businesses.

Exceptions to this rule can be found in countries that provide consistent governmental support and university cooperative research partnering, such as Costa Rica and Australia. The EcoTourism Australia Ecocertification Program (a GSTC-recognized standard) has successfully applied three levels of certification to tourism operations within Australia as of 2013 for 10 years, with over 600 tour operators certified at one of the three levels, which are progressively strict.[63] Australia's tour operator sector is dominated by micro, small, and medium-sized enterprises, and the majority of operators certified fall into these categories. Current statistics indicate that certified businesses garner $1 billion in revenues, or 2% of Australia's total tourism GDP.[64] This is the longest running program in the world. It is a public–private program that benefits from an accreditation for quality and that helps consumers to understand that environmental sustainability is closely linked to receiving a quality product and experience.[65] Ecotourism Australia, which manages this program, effectively seeks benefits for their certified members, such as financial incentives, tax breaks, and priority access to public lands from government, arguing in their policy documentation that increased participation in their program boosts national tourism product credibility.[66]

In the Netherlands, a research organization, the Center for Sustainable Tourism and Transport based at NHTV Breda University of Applied Sciences, produces scientific publications to support the sustainable development of tourism. This Centre offers support to tour operators, including its award-winning CARMACAL carbon footprint assessment tool and other science-based tools to assist tour operators with managing their impacts.[67] This type of research center, especially if government funded, can offer substantial support to tour operators in future seeking to find the right tools to manage their global impacts.

Internal management issues and solutions

Offices of tour operators do not differ in any way from other industry offices. Like all industry, tour operators must seek to reduce the use of energy, the unnecessary use of paper and packaging, the reduction of water use, avoidance of contamination of water supplies, and the promotion of the sustainable use of transport.[68] Tour operators must operate in similar ways to all socially responsible businesses and be certain that their staff receives appropriate mental health and well-being policies. Travel benefits may seem to be so exotic that they can replace other more consistent policies of training and education. Running a company with hundreds of employees requires any firm, wherever it is located, to protect their labor force from overwork and such problems as sexual harassment – even when the team is in the field.

Another distinctive consideration for tour operators is how much they should limit staff travel. On the one hand, a tour operator needs employees who are selling products they have experienced and destinations they understand. On the other hand, a heavy use of travel to reward and educate sales and marketing staff creates a large carbon footprint for the company. Roughly every overseas trip a staff member takes doubles their personal annual carbon footprint.

For a number of years in the 2000s, tour operators investigated and implemented carbon offset programs to manage their corporate carbon footprint. Excellent work was done to calculate office and travel emissions, helping operators to understand how to lower their impacts.[69] Many purchased carbon offsets from credible vendors for their offices. Often they implemented voluntary offset programs for their customers simultaneously. While carbon offsets remain available, many questions have been raised related to their effectiveness in credibly offsetting the impacts of travel. Carbon offset programs were called an effort to "rearrange the deck chairs on the Titanic" in the results of a major industry event in Berlin.[70] While this may be a shortsighted approach (see Chapter 5 regarding questions of aviation's emissions gap), the argument has never been fully settled among industry experts.

Harvard Extension School graduate student Meghan Henry performed a small survey of educational institutions in the U.S. that offer tours to determine how many are aware of environmental management tools, are implementing them, or have interest in doing so. She found that one-third of the 9 organizations she surveyed were monitoring their carbon footprints. While 78% would like to implement environmental standards for their tours, the same percentage found that environmental monitoring would be too difficult. She concluded, "Most have not implemented any sustainability standards due to costs, time, issues with controlling supply chain management, and lack of consumer interest."[71]

Henry, who worked for Harvard's educational tour program, is hoping to influence this. She found that few of the organizations were aware of existing monitoring systems and that there was great potential to share good practices among themselves, particularly the efforts of National Geographic tours, which "include $5–$12 in carbon offset fees in the trip price and calculate the carbon impact of each of their tours."[72]

Since the economic downturn in 2009–2010, the tour operator community by and large has not used the carbon offset model because their industry leadership was not fully in support of it. It was one less expense, and customers were very unlikely to pay for the offsets, so much so that major airlines such as British Airways were forced to scrap their voluntary carbon offset programs.[73] Discussions of how to reduce the carbon impacts of travel largely fall on the question of reducing the impacts of air travel. (See Chapter 7.) Fortunately, the new CARMACAL carbon calculator for tour operators can enable an easier and more seamless planning tool for tours large and small, which includes the entire supply chain, 25 modes of travel, and some 550,000 accommodations worldwide.[74]

For tour operators, the most important step they can take to reduce the carbon impacts of their office operations would be to reduce management and staff travel. Their best alternative is to use Voice Over the Internet Technology (VOIP) for meetings and training of marketing and sales staff. Knowledgeable sales representatives from different regions of the world can also easily be on tap for customer relations via VOIP, eliminating the need to fly sales and marketing staff around the world to experience and understand their travel products.

Supply chain management issues and solutions

Tour operators drive a great deal of business in destinations around the world, purchasing from transport providers, restaurants, hotels, attractions, museums, protected areas, and monuments. They hire local guides and work with local communities to offer services to clients. They have market power to influence local supply chains. Tour operators are well-placed to help their suppliers to meet environmental management and sustainability standards because they are the contractor of their suppliers' services. Any serious attempt by a tour operator to improve sustainability requires the implementation of a sustainable supply chain management framework and can have important implication as well for quality control.[75]

A tour operator's main contracted products and suppliers are as follows:

- **Accommodations:** Hotels, bed and breakfasts, self-catering (serviced) apartments, campsites, cruise ships
- **Transport to and from destinations:** Public transport, airports, scheduled air carriers, air charters, scheduled sea trips, chartered sea trips, bus trips, cruises
- **Catering and food and beverage:** Restaurants, bars, grocery stores, farmers, fishermen, local markets, bakers, butchers, food wholesalers
- **Ground transport:** Car rentals, boat rentals, fuel providers, gas stations, bus and minibus rentals
- **Ground services:** Agents, inbound operators in destinations
- **Cultural and social events:** Excursion and tour providers, sports and recreation facilities, shops, and factories
- **Environmental, cultural, and heritage resources of destinations:** Public authorities, protected site managers, private concessionaires, and owners[76]

The management of a tour operator's supply chain can be complex, and it is very helpful to use an existing system that assists tour operators and their global team to assess their suppliers. The Travelife Sustainability System (a GSTC-recognized standard) was designed for this purpose by the tour operator industry. Its web-based solutions are practical and affordable for small and large companies. A Travelife tour operator begins with an engagement stage that introduces the appropriate management procedures via an appointed sustainability coordinator. In Stage 2, a Travelife Partner reports on requirements using a self-reporting form. In Stage 3, the company's compliance is evaluated by an auditor.[77] The tour operator sets out a compliance plan not only for its in-house operations but also for its suppliers. The suppliers reviewed are transport, accommodations, excursions, local partners and representatives, guides and group leaders, and destinations. For example, excursion operators are asked to select suppliers that use locally produced goods, recommend local guides, use local restaurants, and visit local craft centers as part of their tour operating procedures. Accommodation suppliers are asked to use automatic devices to switch off air conditioning and control heating, reduce towel changes, and switch off lights when guest rooms are vacated as part of their responsibility to lower energy and water use in hotels.

A baseline assessment of a tour operator's suppliers is a fundamental step to set priorities. Many departments must be involved to establish a supply chain policy, including:

- Contracting director and managers,
- Country and destination managers,
- Human resources,
- Legal advisors,
- Marketing director, and
- Internal communications/training unit.

Suppliers will need time, training, support, and possibly funds to implement a tour operator's goals. While the requirements of a program such as Travelife program are not at all onerous, technical support is often required. Travelife offers training around the world to help suppliers meet tour operator goals. They have increasingly reached out to suppliers in emerging economies who may not have adequate exposure to environmental management principles and practices. But there are barriers to success, and one third-party study of Travelife found that Health and Safety inspections can trump sustainability concerns, such as "the overuse of pesticides, over chilling of foods, the use of disposable rather than reusable plastic, overwrapping food mandatory," all in contradiction to Travelife's recommendations. Cleanliness standards set by health and safety officials also often undermined efforts to reduce chemical uses, and ambient lighting standards were at times set at such a high brightness level they caused the denial of installations of more efficient appliances. These issues resulted in disagreements between Travelife auditors and local health and safety officials.[78] Around the world, there is often a lack of congruence between government standards and sustainability goals. Efforts to bring health and safety standard regulations in line with sustainability goals will be essential. Only joint initiatives can solve this problem. If governments work with Travelife and tour operators to help pay for research into new standards, health and safety and sustainability standards could be integrated.

Certainly local operators frequently do not have the same municipal and governmental services common in developed countries, and bureaucratic approaches can block progress. At the same time, local suppliers may not have the services required to meet international sustainability standards. For example, outbound tour operators must recognize that there may be no systematic recycling programs available at all in the destination, and the demand to meet benchmarks to lower solid waste impacts may not be feasible. Meeting sustainability standards cannot simply be made the responsibility of the local suppliers, particularly in small communities where there is little governance and no support. Tour operators must effectively partner with their local suppliers to begin the long process of effectively advocating for better local services, while at the same time improving efficiencies, managing resources appropriately, and developing the capacity of local managers to benchmark environmental and social management goals as a joint enterprise.

Destination stewardship

Destinations are the physical and socioeconomic assets that make tourism possible. The concept of destination stewardship implies that tour operators, as well as all other industry users of the places tourists visit, have a responsibility to help conserve their fundamental ecological, social, and cultural qualities.

Guidelines for the responsibility of the travel industry to conserve destinations were released in 2013 by the Global Sustainable Tourism Council. The GSTC standards require that destinations to promote sustainability standards for enterprises that are consistent with their hotel and tour operator standards.[79] In advance of that process, the United Nations Environment Programme (UNEP) outlined specific responsibilities for tour operators as follows:

- Influencing environmental planning and management through involvement in public policy decisions on infrastructure, protected areas, watersheds, and community livelihoods
- Promoting local products by supporting the production of food, crafts, and other locally made products
- Supporting local charities and NGOs[80]

The responsibility of tour operators to protect destinations can be very complex, and the level of responsibility of the industry is still largely unclear. The problems are not always as simple as managing the environmental footprint of a hotel or transportation supplier. Much deeper exercises in measuring the cumulative impacts of tours visiting vulnerable ecosystems will be needed in future.

Guatemala: case example

Harvard Extension School graduate student, Rod Santos, of Guatemala, found dramatic evidence of the cumulative impacts of tourism on Lake Atitlan, a region favored by a large number of tour operators in Guatemala. He investigated the causes of the 2009 blue-green algae outbreak. Blue-green algae is a sticky toxic algae that forms in vast beds that can spread for miles. He found that travelers were bringing in excess plastic inorganic waste, which local people did not know how to dispose of. According to the paper, there is only one municipal dump in the 224-square-mile watershed, which covers 15 municipalities and a total of 200,000 people. During the past 23 years, the population in the lake's watershed area has doubled with a great deal of additional waste being generated by tourists and residents.[81] In addition, local people were doing much more laundry in open basins, cleaning sheets and towels for overnight visitors in traditional ways. High levels of phosphates and nitrogen were overflowing into the lake, substances known to cause blue-green algae outbreaks. But even after the outbreak, there was no action. He writes,

> The Guatemalan Ministry of Environment and Natural Resources published an action plan in November 2009, enumerating steps necessary for the environmental remediation of Lake Atitlan. The plan calls for 310 million quetzals, equivalent to nearly 40 million dollars for the cost of putting it into action, mainly to erect wastewater and solid waste treatment. However, at the beginning of the rainy season in 2010 the country was devastated by a series of heavy rains, and reconstruction efforts throughout the country will likely use all available funds, leaving Lake Atitlan waiting for the necessary investment to build treatment plants.[82]

His conclusions were that local people lack the education and wherewithal to fix the problem. He recommends more support from the tourism community.

> Tourism may bring funds but those funds must generate financial support at the municipal level to afford education and infrastructure improvements. More importantly, tourism must be an example and driver for sustainability and environmental stewardship in order to ensure future use of the lake and its resources to continue generating income from tourism.[83]

On occasion, tour operators do become involved in questions of how to deliver the necessary public services, such as waste treatment, to local people. UNEP identified a case in Turkey where several tour operators worked to prevent the dumping of solid waste in sand dunes near an important archeological site. The tour operators made financial contributions to the Turkish government to develop a waste separation scheme and a new landfill 30 kilometers inland. Training sessions on the management of solid waste and waste separation were held for 200 managers and staff from local hotels. The collective action of tour operators made this possible.[84] But in general, the international outbound tour operator industry is often hesitant to involve themselves in questions of environmental management of destinations and will require more incentives and support from global and national institutions to become a part of the destination preservation equation. (See Chapter 3.)

Customer relations

Tour operators work with their customers on a daily basis to prepare them for their travel. A communications plan for travelers can help tour operators develop a strategy for improving their customer's interest in sustainability of travel.

Tour operators have quite a few venues to spread the word:

- Company websites and print catalogues
- Predeparture packages of information
- Discussion with salesperson
- Discussion with operations team who are managing details of the travel experience
- Arrival and welcome briefings
- Tour guide information
- Destination and attraction information
- Flight videos and magazine materials
- Post-trip follow up information

Surprisingly, many tour operators do not choose to use sustainability as a primary message in their communications with their travelers. The theory that sustainability builds brand equity is still not holding water. Ogilvy Earth publications on green markets put it this way:

> While we have been relatively good at getting people to believe in the importance of more sustainable behaviors, practices, and purchases, we have been unable to convert this belief fully into action.[85]

The primary problem this report identifies is that the so-called Green Gap makes it difficult for many corporations, selling to mainstream consumers, to make a successful business model out of green products and services.[86] While small, committed tour operators, working with the educational community, may highlight their sustainability credentials, many more mainstream companies do not precisely for this reason.

Consumer market behavior

Tour operators manage their businesses according to what the marketplace demands. For the time being, consumers rank sustainability only as the 7th most important factor out of 8 when booking a tour. For 22% of clients, sustainability is much more important and ranks in the top three factors for deciding on a holiday. In emerging countries, such as India, Brazil, and Russia, sustainability ranks higher, with 26% of customers looking for a balanced approach to ecological, social, and economic issues.[87] But there is a notable gap between interest in sustainability factors and actual booking behavior, making these statistics notoriously unreliable. This has been called the Green Gap by a variety of authors including my own publication on the topic in 2004.[88] Most recently researchers found that while 22% of sustainability-minded tourists consider sustainability important when booking a holiday, only one-third of that 22%, or 6%, of the total market actually books according to the sustainability criteria they claim to be looking for.[89]

Until consumers demand a higher level of responsibility, it is unlikely tour operators will do more to demand and develop product that is environmentally and socially sound. In the Americas, where small, niche tour operators are the dominant business model, the industry is even less likely to adopt criteria for operations than their larger Euro counterparts. While they manage what often appears to be a socially and environmentally responsible model of operations and work on a scale where real relationships with local suppliers can help them to monitor their impacts, few companies seek to report according to globally accepted standards. Because the companies are small, profit margins are not large, and they prefer to use simpler models for investment in local causes, avoiding complex sustainability systems.

The consumer market, even if interested in sustainability, has not responded to any one sustainability standard. What is encouraging is that tourists are well-informed about the important aspects of sustainable tourism and understand the importance of ecological, social, and economic topics. They see the role of environmental protection and the involvement of local communities, and they support the provision of locally sourced products. These attitudes vary, with different nationalities responding to different norms. For example, the Swiss are very concerned about ecology, while travelers from Brazil and India are concerned about poverty. Despite their concern, none of the nationalities are willing to pay a substantial premium for the inclusion of any sustainability component, with a willingness to pay at just 1.5% of the total tour price for conserving the planet and sustaining local peoples.[90] And thus, small and medium-sized tour operators are free to publicize their level of responsibility and promote their responsibility without any genuine comparative data. Why? Because the market is not expecting to see travel companies producing reports that allow for a genuine comparison of environmental performances. The supposition is that good anecdotal stories sell, and this may well be true given the market's hunger for short bits of compelling content, not long boring reports. But as the adventure, eco, and hybrid tour operator community matures, and their founding owners seek to pass on a legacy that is quantifiable, it will be important for the next generation of owners to achieve consistent environmental reporting. TUI sets the standard for this, and their reports offer a model of tour operator sustainability reporting. While smaller operators may suggest that such reports would be too costly to achieve, in fact establishing cloud-based reports working with local staff, is becoming more affordable every year and could well be established by tour operators seeking to establish and verify long-term goals for lowering impacts.

TUI: case study

TUI Group is Europe's largest tour operator group. They host 30 million customers from 27 countries, are listed on the London Stock exchange, and are composed of three sectors: the tour operator, TUI hotels and resorts, and cruises. What is highly unique about this conglomerate, with so many brands and divisions, is how well it describes its goals and objectives, both as a business and as a company that aims to be a leader in sustainable development in the industry. Their economic goal to reach gross revenues of 1 billion Euros by 2014–2015 was met in 2015.[91] The company is leveraging its hotel and cruise sector's contribution up to half of all corporate revenues, while previous contribution levels were once well less than a third. Clearly, tour operating is being downgraded within a company that is the largest tour operators in the world. While not eliminating the tour operating model, TUI is quickly moving to be a provider of what they call content (in their case hotels and cruises) that will give them a more secure position in this rapidly changing marketplace, which requires that they increasingly reduce costs and develop efficiencies in their home offices due to the growing influence of digital purchasing.[92]

TUI's strategy for sustainable development is comprehensive and can be found in their biannual report, summarized here.[93] Their commitment to sustainable development is front and center on their corporate home page. Their materials outline in detail the corporate approach to environmental management, corporate governance, internal team communications, social commitment, conservation policies, and commitments to their employees. TUI does not use anecdotal approaches to tout its successes. The company performs a "precise analysis" of sustainability factors that affect their business areas, and they take a systematic, holistic approach. They use a triple bottom line strategy and measure their performance in all categories. Their goal is to avoid and reduce negative impacts on the natural and social environments where they operate.

TUI outlines its responsibilities clearly and strategically. Its corporate governance is connected to sustainability goals at all levels, with human resources, marketing, environmental management, investor relations, and group strategy and development all sitting on their Corporate Responsibility Council.

The entire company has measured and set goals for sustainability. The company uses International Standard Organization (ISO) 14001 for all of its environmental reporting, due to its worldwide acceptance and applicability. Under their environment goals, they present the problem of climate change up front. The company has a comprehensive plan to reduce their climate impacts by first and foremost avoiding and reducing their greenhouse gas emissions along their entire value chain. They employ efficiency-enhancing technologies and conserve resources in five different divisions: airlines, water transport, ground transport, administrative buildings, and hotels. They measure their emissions group-wide and report with the Climate Disclosure Project. Their carbon-heavy divisions, airlines and cruise, have taken a very proactive approach. The company is purchasing the newest, most carbon-friendly crafts. They openly discuss the urgent challenge posed by the soot, sulfur, and nitrogen oxide emissions of their cruise ships with plans for a new generation announced. They also discuss the problem with bunker fuels and reference conversion to diesel fuels worldwide. (See Chapter 7.) TUI also uses carbon offsets to invest in the climate protection foundation *Myclimate*, which invests in renewable energies, energy efficiency, and reforestation worldwide.[94]

The company commits to a thrifty use of water, with water saving taps and showerheads installed in most hotels they own. They have incorporated environmental and social minimum standards into contracts with accommodation and excursion suppliers. For their "differentiated hotels," which are fully owned by TUI, they are expecting every hotel to have an environmental management standard, with 43% achieving this by the end of 2014. They have set average benchmarks of 24 kilowatt-hours and 400 liters of water per person per night as a commitment for these hotels and reached 513 liters of water and 239 kilowatt-hours by the end of 2014. They expect all of their hotels to invest in local socioeconomic improvements in destinations and work with donor partners, such as GIZ, to leverage investment in tourism destinations.[95]

The company has set a goal to purchase more fruit and vegetables locally, in order to cut transportation and packaging costs. It presently has 170 of its hotels purchasing 45% of their food from the local region and 80% from the same country. They now have a system that allows them to track and influence the environmental impact of their key suppliers and begin to monitor social indicators such as local employment and procurement. Its labor policies encourage diversity and gender balance, with training available for young staff to move ahead. TUI also stresses its responsibility to local societies in countries where it works worldwide, taking stands on human rights, training for young people, and protecting children from sex trafficking.

TUI also is developing a biodiversity strategy in cooperation with the Global Nature Fund. It is taking some strong stands on conservation, such as supporting the global opposition to the Serengeti Highway, a Tanzanian governmental project supported by the Chinese to create a highway through Serengeti National Park in order to export minerals and oil from Central Africa to Asia via Dar es Salaam, the capital of Tanzania. At the initiative of TUI, other major global tourism organizations followed suit.

TUI takes its communications responsibilities seriously and seeks to inform customers about its sustainable development goals and projects. It is offering more excursion products that focus on encounters with local people and is building its relationships with local communities to offer sensitive products that provide information about local cultures and local environments. It offers sustainability training to new hires in 74% of its businesses, especially with customer representatives who need to explain corporate goals consistently to visitors from throughout the world.

The company has impressive resources and can allocate more to sustainable development initiatives by virtue of its size. But the corporate ethics and goals that are expressed in these projects are noteworthy because they do not seek to avoid tough subjects such as climate change or sex trafficking. They focus on issues that are not sexy, such as sulfur oxide emissions of their ships. There is a clear commitment to transparency. But most importantly, the company measures and reports on its efforts and covers all of the important commitments tour operators can make to an improved planet.

Conclusion

Tour operators are functioning in a world where their market is eroding and being influenced by digital sales and marketing environments that they do not control. This lowers incentives for the tour operator community to invest in sustainability management programs. Large European conglomerates like TUI have created a vertically integrated product marketing, sales, operations, and supply chain system that enables them to effectively manage their global environmental footprint and sustainability communications, possibly more effectively than any other travel company in the world.

In the North American market, most niche businesses and nonprofits in travel have not embraced concerns about carbon reporting or other fundamental environmental impacts of their companies. The internal cultures of these companies are not fostering the concept of rigorous sustainability procedures, and their employees are not in the position to question this, nor are they trained to do so. Most tour operators have found taking on projects with high publicity appeal, together with NGOs, takes care of their responsibility to make their supply chains more responsible. And consumers do not perceive the difference. But this is a misperception that is being actively fostered by the leadership of some of the most important tour operators in the business, who clearly believe that customers and the global community do not care about sustainability metrics.

Online procurement review systems, such as Travelife and CARMACAL, are ideal to help tour operators large and small with the process of benchmarking their supply chains. Such systems are built into the booking procedures of tour operators and facilitate the use of sustainability criteria in the review and contracting of suppliers. These online procurement review systems are the most convenient and targeted approach to improving tour operator sustainability. But because one size indeed does not always fit all, customized systems could be created without enormous cost, especially with the support of universities.

For the traveler seeking to buy authentic responsible travel, the best advice is to shop local by purchasing from inbound tour operators, such as Wilderness Safaris. Companies that transparently and accurately report on their sustainability metrics. Inbound operators are now able to reach out to the global community via online marketing platforms, cut out international buyers, and sell directly to the consumer. They can resist online marketing pressures and deliver a higher percentage of their returns to local people. Local tour operators see immediate returns when they invest in their own countries and cultures because of their location in destination countries. They have every reason to preserve their own ecosystems, cultural heritage, and more motivation to invest in local communities and help the labor pool to become more educated. They are able to deliver a much higher-quality experience for their customers and their staff, and together with both their employees and the many residents who supply services to them, they learn how to achieve an increasingly beneficial tour program that meets the needs of local populations and protects local ecosystems. Local operators are not automatically better or greener. They must report on their triple bottom line results, using third-party auditors, or work with local certifiers who can achieve that on their behalf. But they are under less pressure to lower prices if they deliver on the ground, have a dedicated team of global agents who sell them, and market direct to consumers.

Undervaluing the people and ecosystems that supply an authentic tourism experience is a shortsighted mistake, and local tour operators see the consequences of such actions immediately. They design and deliver based on real feedback from the field. Sustainability and the protection of the environment may seem more remote and less important for many international tour operators and OTAs looking to cut costs, but they will see consequences in the future if they do not embrace a more concerted approach to partner with their suppliers and local governments to protect and benefit the iconic destinations and local people they depend on.

Sustainability initiatives resoundingly prove to be real "sellers" for the stellar universe of specialist inbound operators. Award-winning companies like Wilderness Safaris are proving that solid investment in local people, conservation, and genuine partnerships with local communities can generate thriving business models that create long-term equity in the business of conserving and sustainably developing the planet. And they have the proven ability to not only promote content about their company but measure their success. Given that digital marketing will only help these companies to thrive and that an increasing number of the world's travelers are seeking to travel to emerging destinations, the future for sustainable inbound tour operating not only looks positive, it appears to be the most important business model for sustainable tour operating to generate the most extensive benefits worldwide.

Notes

1 Get to Know Richard Avilino, http://wilderness-safaris.com/blog/posts/getting-to-know-richard-avilino, Accessed October 2, 2016
2 Our History, http://wilderness-safaris.com/about/history, Accessed October 2, 2016
3 Russel Friedman, Founding Partner of Wilderness Safaris, http://safaritalk.net/topic/6729-russel-friedman-founding-partner-of-wilderness-safaris/, Accessed August 1, 2013, vetted by Russel Friedman and edited for this account via personal communication, May 3, 2016
4 Wilderness Holdings, February 28, 2015, *Integrated Annual Report*, Gaborone, Botswana, page 110
5 Wilderness Safaris, February 2013, *Consolidated Annual Report*, Gaborone, Botswana, page 1
6 Russel Friedman reviewed account, personal communication May 3, 2016

7 Elkington, John, 1998, *Cannibals without Forks*, New Society Publishers, Gabriola Island, BC, Canada
8 Wilderness Safaris, *Consolidated Annual Report*, page 1
9 Ibid., page 11
10 Ibid., page 26
11 Ibid., page 14
12 Wilderness Holdings, *Integrated Annual Report*
13 Benckendorff, Pierre J., Pauline J. Sheldon and Daniel R. Fesenmaier, 2014, *Tourism Information Technology*, 2nd Edition, CABI, Oxfordshire, UK. Provides full array of digital travel technology advancements.
14 Wilderness Holdings, *Integrated Annual Report*, page 11
15 Tour Operators' Initiative on Sustainable Tourism Development to Present Good Practices at World Travel Market in London, http://unep.org/Documents.Multilingual/Default.asp?DocumentID=180&ArticleID=2678, Accessed October 2, 2016
16 Tourism today- big power- minimal responsibility http://travelmole.com/news_feature.php?c=setreg®ion=1&m_id=s~T_b_rs~m&w_id=31532&news_id=2020225, Accessed October 2, 2016
17 Budeanu, Adriana, 2009, Environmental supply chain management in tourism: the case of large tour operators, *Journal of Cleaner Production*, Vol. 17, 1385–1392
18 Van Wijk, Jeroen and Winifred Persoon, 2005, A long-haul destination: sustainability reporting among tour operators, *European Management Journal*, Vol. 24, No. 6, 381–395, 390
19 Ibid., page 76
20 Ibid., page 72
21 Ibid., page 76
22 Ibid., page 82
23 TOURISMlink, 2012, *The European Tourism Market, Its Structure and the Role of ICTs, Brussels*, page 28
24 Ibid., page 87
25 Ashton, Jane, 2014, *Lecture for Harvard Class on International Development of Sustainable Economies*, Harvard University Extension, Cambridge, MA
26 Tour Operators Have Steadily Lost Market Share to Online Travel Agencies, http://hospitalitynet.org/news/4042583.html, Accessed October 2, 2016
27 Our Heritage, http://mtsobek.com/heritage, Accessed October 2, 2016
28 EplerWood International, January 2004, *A Review of International Markets, Business, Finance and Technical Assistance Models for Ecolodges in Development Countries*, International Finance Corporation (IFC)/GEF Small and Medium Enterprise Program, page 4
29 Ibid., page 27
30 Adventure Travel Trade Association (ATTA) et al., October 2012, *Adventure Tourism Development Index 2011*, ATTA, Seattle, WA, page 3
31 Educational Travel Consortium, http://travelearning.com/content/index/about_etc_conferences?subCat=, Accessed August 6, 2013
32 Consumer Travel Editors Get Personal, http://travelweekly.com/Travel-News/Travel-Agent-Issues/2013-Travel-Editors-Roundtable/, Accessed August 6, 2013
33 Tour Operators, Unite!, http://travelweekly.com/Arnie-Weissmann/Tour-operators,-unite!/, Accessed October 2, 2016
34 Summarized from speech at the World Tourism Forum, April 17, 2013
35 NewMedia TrendWatch, http://newmediatrendwatch.com/markets-by-country/17-usa/126-online-travel-market?start=1, Accessed October 2, 2016, a good landing page for up-to-date new media trends
36 IPK International, 2011, *ITB World Travel Trends Report*, ITB Berlin, Germany, page 5
37 TOURISMlink, *The European Tourism Market*, page 49
38 Carroll, Bill and Lorraine Sileo, 2014, *Online Travel Agencies, More Than a Distribution Channel*, PhoCusWright White Paper, New York

39 Google Destinations lets you plan a trip on your phone, no app required, http://mashable.com/2016/03/09/destinations-on-google/#xwGAxhJdS5qr, Accessed October 2, 2016
40 Expedia inc. Global Network of Brands http://expediainc.com/expedia-brands/ The company has changed its claim, it now states it is one of the largest online travel companies in the world. This landing page lays out its services and sub-brands, Accessed October 2, 2016
41 Expedia, Inc. Overview, http://expediainc.com/expedia-brands, Accessed October 2016
42 Analysis of Major Online Travel Agencies, https://cloudbeds.com/articles/analysis-of-major-online-travel-agencies-otas/, Accessed October 2, 2016
43 Ibid.
44 IPK International, *ITB World Travel Trends Report*, page 27
45 Machado, Amanda, June 18, 2014, Changing travel, *The Atlantic Magazine*, http://theatlantic.com/international/archive/2014/06/how-millennials-are-changing-international-travel/373007/, Accessed October 3, 2016
46 The CIVETS market (Colombia, Indonesia, Vietnam, Egypt, Turkey and South Africa), rated as among the next emerging markets to rise quickly which results in more active travel throughout these regions. These statistics are constantly changing. https://en.wikipedia.org/wiki/CIVETS, Accessed October 3, 2016
47 Wijk and Persoon, A long-haul destination, pages 381–395
48 Ibid., page 392
49 Fact Sheet, TripAdvisor May 2016, https://tripadvisor.com/PressCenter-c4-Fact_Sheet.html, Accessed October 2, 2016
50 Why Become a TripAdvisor GreenLeader?, https://green.tripadvisor.com/survey/about, Accessed October 2, 2016
51 TripAdvisor Puts Green Hotels in the Spotlight, http://edie.net/news/6/TripAdvisor-puts-green-hotels-in-the-spotlight, Accessed October 2, 2016
52 Posts from the "Green" Category, http://tripadvisor4biz.wordpress.com/category/green/
53 Kuminoff, Nicolai, V., Congwen Zhang and Jeta Rudi, 2010, Are travelers willing to pay a premium to stay at a "green hotel"? *Agricultural and Resource Economics Review*, Vol. 39, No. 3, Northeastern Agricultural and Resource Economics Association
54 TripAdvisor – Yet Another Shade of Green http://travelmole.com/news_feature.php?c=setreg®ion=1&m_id=s~T_b_rs~m&w_id=10148&news_id=2012492, Accessed October 2, 2016
55 Global Reporting Initiative and the Tour Operators Initiative, November 2002, *Tour Operators' Sector Supplement, for Use with the GRI 2002 Sustainability Reporting Guidelines*, Global Reporting Initiative Secretariat, Amsterdam, The Netherlands
56 UNEP, 2005, *Integrating Sustainability into Business, a Management Guide for Responsible Tour Operators*, United Nations Environment Program Tour Operator's Initiative, Paris, France
57 Wijk and Persoon, A long-haul destination, page 390
58 For example, the Planeterra Foundation website, which is the nonprofit foundation for G Adventures, one of the largest hybrid tour operators in the world, makes no reference to triple bottom line metrics or third-party audits for any of the sustainability and value chain projects listed.
59 UNEP, *Integrating Sustainability into Business*, page 22
60 Global Reporting Initiative and the Tour Operators Initiative, *Tour Operators' Sector Supplement*, page 10
61 Van der Duim, Rene and Ramona van Marqijk, 2006, Implementation of an environmental management system for Dutch tour operators: an actor-network perspective, *Journal of Sustainable Tourism*, Vol. 14, No. 5, 449–472
62 Baddeley, J. and Xavier Font, 2011, Barriers to tour operator sustainable supply chain management, *Tourism Recreation Research*, Vol. 36, No. 3, 205–214

63 Ecotourism Australia Certification Programs, http://ecotourism.org.au/products.asp?mode=search, Accessed August 7, 2013
64 Ecotourism Reaps Green Tourist Dollars, http://travelweekly.com.au/article/Ecotourism-reaps-green-tourist-dollars/, Accessed October 2, 2016
65 Ibid.
66 Tourism for the Future, Strategic Discussion Paper, Ecotourism Australia, http://ecotourismaustralia.files.wordpress.com/2012/08/tourism-for-the-future-2012-edition-final.pdf, Accessed March 26, 2014
67 Center for Sustainable Tourism and Transport, http://cstt.nl/goals, Accessed October 2, 2016
68 Tourism for the Future, page 17
69 The author worked with the firm Native Energy in this period and helped tour operators to do these calculations.
70 IPK International, *ITB World Travel Trends Report*, page 21
71 Henry, Meghan, December 2013, *The Educational Tourism Industry and Environmental Sustainability: Starting the Discussion*, Environmental Management of International Tourism Development, Department of Sustainability and Environmental Management class paper, Harvard Extension School, Cambridge, MA, page 15
72 Ibid., page 10
73 British Airways to replace passenger carbon offset scheme with new fund to aid UK carbon reduction projects, http://greenaironline.com/news.php?viewStory=1320, Accessed October 2, 2016
74 Carmacal- Carbon Calculator Travel Industry https://climateneutralgroup.com/en/carmacal-carbon-calculator/, Accessed May 3, 2016
75 Schwartz, Karen, Richard Tapper and Xavier Font, 2008, A sustainable supply chain management framework for tour operators, *Journal of Sustainable Tourism*, Vol. 16, No. 3, 310
76 Center for Environmental Leadership in Business, 2004, *Supply Chain Engagement for Tour Operators, Three Steps toward Sustainability*, Tour Operators Initiative for Sustainable Tourism Development, Madrid, Spain
77 Travelife Certification, http://travelife.info/index_new.php?menu=certification&lang=en, Accessed October 2, 2016
78 Baddeley and Font, Barriers to tour operator sustainable supply chain management, 205–214
79 GSTC Destination Assessment, http://gstcouncil.org/sustainable-tourism-gstc-criteria/gstc-early-adopter-destinations/766-global-sustainable-tourism-criteria-for-destinations-gstc-d-version-10-november-2013.html, Accessed October 2, 2016
80 UNEP, *Integrating Sustainability into Business*, pages 48–54
81 Santos, Rodrigo, December 2010, *Tourism and Environmental Management in Lake Atitlan*, Harvard University Extension, ENVR E118, Cambridge, MA
82 Ibid., pages 14–15
83 Ibid., page 15
84 UNEP, *Integrating Sustainability into Business*, page 48
85 Bennett, Graceann and Freya Williams, 2011, *Mainstream Green: Moving Sustainability from Niche to Normal*, Ogilvy & Mather, New York, page 13
86 Ibid., page 15
87 IPK International, *ITB World Travel Trends Report*, ITB Berlin, Germany, page 21
88 EplerWood International originally published about this in March 2004, http://eplerwood.com/publications.php, since published with coauthors, e.g., Epler Wood, M. and Xavier Font, 2007, Sustainable Tourism Certification Marketing and Its Contribution to SME Market Access, in *Quality Assurance and Certification in Ecotourism*, CABI, Oxfordshire, OX and Cambridge, MA, Major marketing firms such as Ogilvy Earth now also uses the term, https://assets.ogilvy.com/truffles_email/ogilvyearth/Mainstream_Green.pdf, 2011, Accessed October 2, 2016
89 Ibid., page 21

90 IPK International, *ITB World Travel Trends Report*, ITB Berlin, Germany, page 21
91 TUI Group, *Annual Report 2015/5 Upgrade*, Hanover, Germany
92 Ibid.
93 TUI Group, 2014, *Sustainable Holidays Report*, Hanover, Germany
94 About Myclimate, http://myclimate.org/about-us/portrait/, Accessed July 28, 2014
95 TUI Group, *Sustainable Holidays Report*

7

THE CRUISE INDUSTRY

Empire of the seas

Introduction

In February 2013, the Alaska Senate passed a measure, HB 80, to roll back cruise ship waste water discharge standards. The new bill weakened the previous, voter-backed legislation of 2006, which had banned the discharge of untreated or treated sewage and other waste waters in State marine waters without permit.[1] With the House approval already in place, Governor Sean Parnell signed the bill, which took effect on March 1, 2013. This concluded yet another round in the debate over how cruise ships manage their effluent in pristine waters, where salmon still roam wild and are dependent on clean water.

Lawmakers in Alaska were flooded with comments, many in opposition to HB 80, the bill that some say the cruise line industry won through threats and intimidation. In March of 2010, cruise line industry representatives on a panel at the Cruise Shipping Miami event in Florida, with Governor Parnell in attendance, threatened to pull cruise ships from Alaska if the state did not ease up on pollution regulations and taxes.

The comments made by Steve Kruse, CEO of Holland America, sent a message directly to Parnell and Alaska's legislators.

> "In so much as the diminishing financial returns and a punitive regulatory environment makes a given destination or region less viable, then ships will move," Kruse said. "Our assets are moveable, they're designed to be moveable, and we are very good at moving our ships if the conditions necessitate doing so." Kruse added, "And remember, that 17% decline in Alaska doesn't mean the ships aren't operating full. They are. They're operating full somewhere else, because the ships can move."[2]

The outcomes of this drama are evocative of a battle transpiring in ports around the world. The questions raised go far beyond pollution of port waters. The cruise industry's economic influence gives it power that not many other members of the leisure industry hold. Cruise ships are moveable assets of enormous value, which can be transferred to other destinations. Tens of thousands of jobs can be on the line. Cruise lines do not hesitate to make the economic value of their trade explicit and routinely threaten to walk away from ports if they do not receive the deals they are looking for.

In Alaska, the cruise industry was able to move the State toward a set of laws that facilitate industry growth and allow more ships in Alaskan ports, at less cost. Whether or not this is at the expense of Alaskan waters and taxpayers remains controversial. In March of 2010, right after the Miami conference, Governor Parnell launched a bill to lower the "head tax" for cruise passengers visiting Alaska from $46 to $34.50, a fee that had been established in 2006 as a means of helping Alaskan ports manage the cost of infrastructure that cruise ships require. This head tax was confronted head-on via a lawsuit filed by the Alaska Cruise Association in 2009.[3] In April of 2010, a settlement was released as soon as the State moved to reduce the tax.[4] Alaska has now removed all impediments to cruise growth, by way of scaling back its environmental regulations and the cost for passenger visits in line with what the industry required. It is not often that a major industry so baldly pressures the governor of an American State in a public forum, especially one as economically prosperous as Alaska. But it is precisely the brassy business model of the industry that keeps it very profitable and perpetually controversial.

These tactics are counterproductive in the 21st century. As the industry expands, the assumption that cruise lines can command compliance from states and nations needs to be revisited. Regulatory systems are tightening in response to real concerns from well versed nations, and the industry will benefit from working within a set of norms that protects social and environmental resources. While the cruise industry continues to respond to environmental challenges and has set out to resolve many of the issues by academics, legal scholars, and national regulatory agencies,[5] this chapter suggests that a full restructuring of the legal and regulatory systems that apply to cruise on an international level would be the best solution to bring a level playing field to the industry and provide ports around the world with a consistent system to manage.

Environmental impacts of cruise lines: key issues

The cruise industry faces enormous challenges on land and sea to comply with a growing set of regulations. As the industry expands at a pace far beyond the rest of the leisure industry, with ambitions to operate in small and large ports worldwide, there must be a new harmonized approach to regulations, monitoring, and enforcement that is based on passenger ships operating within coastal waters.

The industry is fully prepared to argue that it is operating well within legal limits worldwide, using the norms laid out for ships in the maritime industry.[6] A wide range of issues require a broad set of disciplines to evaluate this claim. Science-based environmental concerns have been raised in the Caribbean, Baltic, and Adriatic Seas, which are already on the cusp of ecological collapse if they do not initiate stringent oversight. There are the sociopolitical challenges and the sociocultural questions of managing cruise arrivals for local communities that are increasingly outnumbered by their cruise visitors. At the bottom of this set of issues are the legal questions that dictate whether there is to be genuine progress on environmental management of cruise lines in the future. The industry faces a wide variety of critics, partly because of their pugilistic business style and partly because they respond on an incremental basis to legal challenges. It is critical for all students of the industry to grasp that the cruise industry will not stand down and develop a more *realpolitik* set of tactics to reach a more holistic set of solutions.

To respond to the wide variety of NGO efforts to alert regulators and passengers of the need for more oversight,[7] the cruise line industry has created a one-stop-shop website (cruiseforward.org) that addresses many of the issues raised by the NGOs. This web presentation seeks to reassure the public and regulators that "every aspect of the cruise experience is regulated."[8] At present, there is a patchwork of regulations, largely put in place by the most powerful nations with strong regulatory systems to protect their waters, with a weak international legal system that has no enforcement outside of national waters. Most maritime legal requirements and environmental management protocols are not designed to meet the challenges cruise lines represent, and the many local governments faced with a billion-dollar industry at their door are not ready to negotiate effectively with the cruise industry's powerful and legally armed local and international associations, nor is it reasonable to expect them to be. For this reason, international fixes are urgently required.

Cruise ships have a footprint that is very different from the rest of the shipping industry. The thousands of passengers they carry have effluents that are of a magnitude that cannot be compared with container ships and tankers. According to scientific testimony given in the U.S. Congress, one large cruise ship will produce approximately 300,000 gallons of sewage on a 1-week cruise.[9] Even with advanced waste water treatment, 4,000 gallons of sludge can be produced per day after treatment.[10] Cruise ships are also estimated to produce 8 tons of solid waste per week from a moderate-sized ship, or 24% of all solid waste produced by vessels worldwide.[11] And 25,000 gallons of oily bilge water is produced weekly by an average cruise ship, which collects on the bottom of a vessel.[12] All of these effluents and solid waste by-products of their operations need to be carefully monitored by ports and by governments overseeing their national waters.

With 23 million passengers sailing worldwide on cruise ships in 2015 and a growth rate of 21% since 2010,[13] the paradigm of cruising is no longer a regional issue; it is a global phenomenon. (See Figure 7.1.) More well versed, well trained managers are needed who can understand the issues at stake. Dialog to improve regulations is crucial on an international level to protect ports and vulnerable marine ecosystems wherever the industry is active.

FIGURE 7.1 Ports of call for the cruise line industry 2015 forward[14]

Cruise lines bring hundreds of thousands of visitors into cities around the world and will bring a rapid growth of passenger numbers in nearly every case. Stockholm, Sweden, saw a 12% average growth per year in passenger numbers beginning in 2007 with 500,000 passengers arriving by 2015.[15] On a cruise route in the Baltic Sea, which in total received 5 million cruise passengers in the same year, Sweden is one of the first countries to react with both solid research, regional cooperation, and scientific baselines.[16] The Baltic (Figure 7.2) needs protection from sewage and gray water discharges because of its unique ecology that makes it prone to eutrophication – an ecological vulnerability of enclosed seas that, lacking adequate circulation, are harmed by an excess of nutrients that cause toxic algae blooms.[17]

This ecological weak point is shared by many of the waters where cruise lines prosper, including the Adriatic Sea. (See Figure 7.3, page 230.) Stresses on the pristine waters of the Inside Passage of British Columbia are also escalating due to the heavy cruise passenger traffic, according to Canadian scientists.[18]

Cruise lines thrive in narrow scenic seas, where they can make frequent stops in accessible ports. This requirement puts them in waters that are much closer to land than the shipping industry, which normally plies open oceans, except when in port. Under maritime law, cruise lines can legally discharge sewage and gray water of millions of passengers 12 miles from land.[19] The Baltic Sea region, represented by the Helsinki Commission (HELCOM), drew up a regional strategy to respond to the growing challenge of sewage from cruise lines by establishing a special area for sewage treatment for passenger ships, which now regulates sewage discharges and ensures that adequate port treatment facilities are available.[20,21] The black and gray waters emitted from these heavily populated ships contain high levels of nutrients,

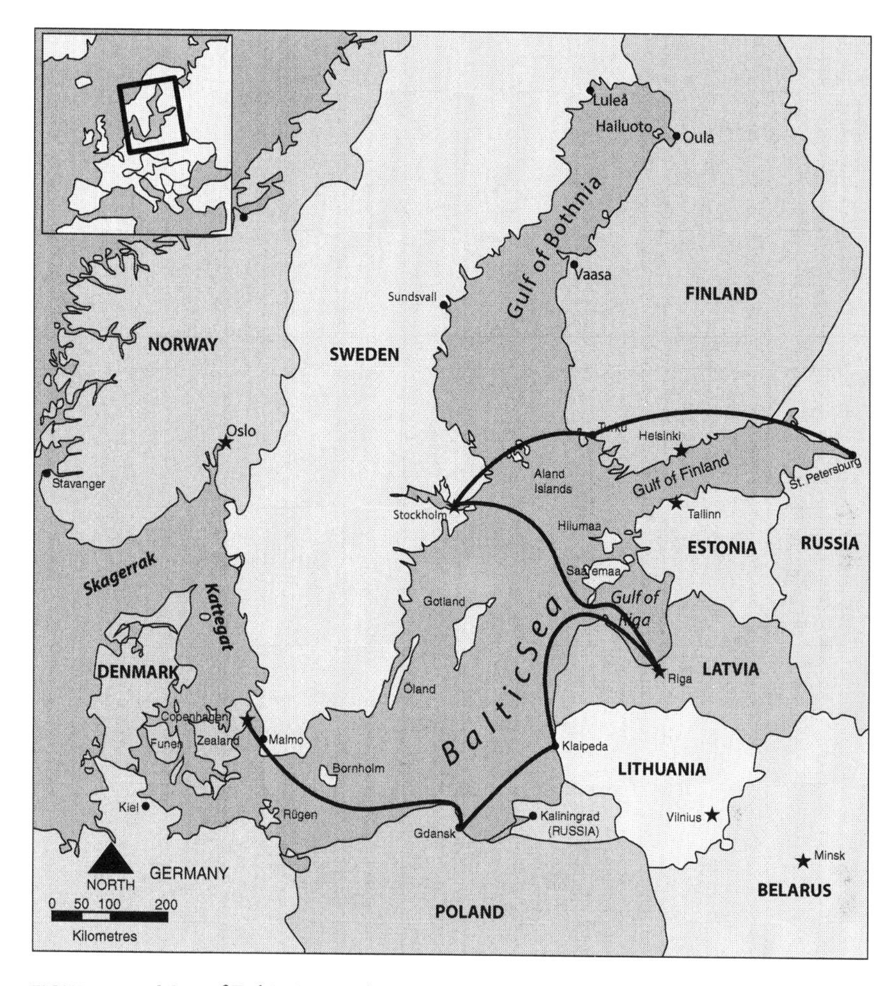

FIGURE 7.2 Map of Baltic Sea with standard routes for cruise lines

pathogens, residual levels of organic material, and cleaning chemicals, a problem recognized and regulated by the US EPA in 2013.[22]

Nations such as Sweden and the U.S. are in the vanguard of a process that will need to transpire around the world. Special maritime environmental regulations that respond specifically to cruise line impacts on coastal ecosystems will be required and should be considered for ports and vulnerable seas across the board. Many countries, without robust regional commissions similar to HELCOM and with vulnerable seas to protect, are struggling to respond. In Croatia, cruise passenger numbers have grown over 400% between 2002 and 2012, threatening the Adriatic Sea.[23] As a protected arm of the Mediterranean, the Adriatic was once considered a paradise for divers but is now being invaded by algae due to a lack of natural flushing sea currents and runoff from land. With 5 million cruise passengers arriving in the Adriatic in 2012, it is easy to see how the nations bordering this enclosed sea, pictured in Figure 7.3, need to monitor and review this growing industry which has arrived on its shores with great rapidity. But the environmental regulatory system of the region is not as strong as that of the Baltics, Canada, or the U.S.[24]

Uniform standards for discharges that take into account the seas and coastlines where cruise lines operate are needed. As it is now, the industry faces a patchwork system of scrutiny from state governments such as Alaska, from national governments such as the U.S. and Sweden, and from regional bodies such as HELCOM to protect them from the intensity of cruise operations and the resulting discharges

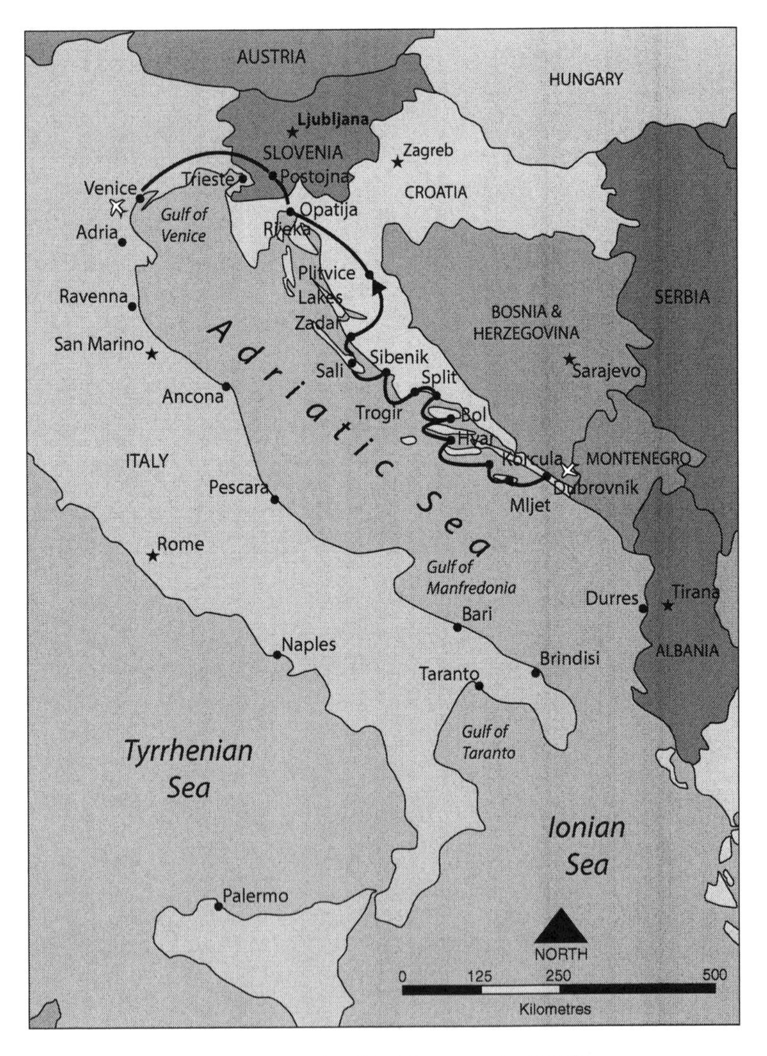

FIGURE 7.3 Adriatic seaports on cruise ship itineraries in 2014[25]

that need monitoring. In all cases, there are genuine, science-based concerns. The Inland Passage in Canada and the U.S., the Adriatic Sea, the Baltic, and also the Caribbean, where cruise lines are the most active, are all vulnerable.

Each type of cruise pollution should be considered for systematic updating to take into account the special characteristics of environmental management of large passenger ships; these include:

- **Sewage or black water:** Water that originates from toilets and medical facilities.
- **Gray water:** Waste water from sinks, showers, galleys, laundry, and cleaning activities on board.
- **Oily bilge water:** Water containing solid wastes, lubricants, pollutants, oil, and chemicals.
- **Ballast water**: Water, which can contain oily substances, invasive species, and hazardous waste, used to stabilize the vessel and compensate for changes in the ship's weight as fuel and supplies are used.
- **Solid waste:** Glass, paper, cardboard, aluminum, steel cans, and plastics, which can be non-hazardous or hazardous.
- **Hazardous waste:** Photo processing, dry cleaning, and equipment cleaning waste; medical waste; batteries; fluorescent lights; spent paints and thinners.
- **Air pollution:** Contaminants generated by the cruise ships' diesel engines that burn high-sulfur-content fuel, known as bunker fuel, which produces sulfur dioxide, nitrogen oxide particulates, carbon monoxide, carbon dioxide, and hydrocarbons.

Cruise line industry business model and management

The cruise line business model relies on affordable, safe, and entertainment-filled vacations for families at prices that few land destinations can beat. The industry, based in the U.S., out of Florida, has depended on the Caribbean Sea for decades to create exotic travel destinations that can be easily reached on a 10-day itinerary, with short and fuel-efficient distances between stops. In 2010, 43% of all travelers on cruise trips went to the Caribbean. Alaska is the second most important destination to the industry.

The industry has an average annual growth rate of 9.5%,[26] well beyond the average growth of tourism internationally, which is 4%. A $30 billion market in 2011, the industry is expanding internationally rapidly. While growth beyond the U.S., Canada, and the Caribbean is a major part of the cruise world's business strategy, it still depends on North American departures for 60% of its market.

The industry offers recreation and entertainment services in a package at prices superior to any other part of the leisure industry. New ships and larger ships are essential to this strategy. The new megaships now include Broadway musical venues, dozens of upscale restaurants, and large water parks, all available daily with no additional charges paid on board. High-tech cruising includes already ubiquitous

Wi-Fi access and crystal-clear cell phone reception. Onboard, plasma and LCD touch screens provide up-to-date information throughout the ship, and radio frequency identification technology (RFID)–enabled wristbands help keep track of family members in these 20-story, 1,100-foot long ships. iPad-based menus allow for no-fuss ordering in the numerous venues, and interactive wine menus also help pair meals with a complementary wine selection. Royal Caribbean's *Splendour of the Seas*, a megaship, is the first in the world to offer an iPad in every cabin that announces daily events and activities. These iPad systems also offer a vast array of educational options with complimentary courses available for enrichment.[27]

So much entertainment in one package is achieved by nearly doubling ship sizes. New economies of scale bring more revenue per ship, while keeping cabin prices affordable. The new megaships, launched each year since 2009, offer a 6,300-passenger capacity, with the Royal Caribbean International's *Oasis of the Seas* and its twin, *Allure of the Seas*, topping the list as the largest cruise ships in the world. In 2010, the industry added enough new cabins to jump passenger capacity by 10%.[28] Between 2000 and 2013, 167 new ships were launched, double the number that set sail in the 1990s, and 4 times more than those launched in the 1980s.[29]

One corporation, Carnival, dominates over 50% of the market, and Royal Caribbean, its closest competitor, has over 25% market share. Each of these two brands operates many sub-brands. Carnival operates 9 sub-brands, including Princess, Cunard, P&O Cruises Australia, AIDA Cruises, and Holland America. The Carnival target market varies according to the sub-brand from youthful travelers to older retirees. The Princess line targets budget-minded, younger travelers in their child-raising years with a fleet of 17 ships and 270 destinations throughout North America, the Caribbean, South America, Australia, and the South Pacific. An all-inclusive, 10-day Caribbean cruise on the Princess Line will cost a reasonable $1,100 per person, plus air fare to Florida, which can be commonly purchased at approximately $250 from most U.S. cities. For a grand total of $1,350 per person including airfare, this price point is doable for many families of four. Competing with these prices on land in the Caribbean is difficult. For example, a Caribbean trip to St. Thomas with overnight stays of 10 days will cost a minimum of $2,500 with airfare included, nearly double what it costs to take a cruise with the Princess line. If this sounds similar to a press release for the cruise lines, in a way it is. Understanding how cruise lines are tapping into the hunger for exotic travel at affordable prices, with 24/7 digital access, entertainment, business, and real enrichment capacity, is crucial to recognizing that these ships and their corporate owners are working a business model that is vibrant, cutting-edge, and not likely to disappear.

Huge capital investment is required to enter the industry, and this prevents the threat of new market entries underpricing the existing players or diverging from industry business practices. The high concentration ratio of just a few players in the field reduces competition and increases the likelihood of collaboration, which is attained through their professional association.[30] The Cruise Line Industry Association (CLIA) is a highly effective and organized player with 15 offices around the world,[31] which organizes the industry, provides research, and represents cruise's interests in capitals around the world. It emphasizes the mantra, "Our Industry, One Voice." While growth rates of U.S. passengers have been steady with 15% growth since 2010, global markets are the industry target in the long term.[32]

Expansion in the Mediterranean Sea is viewed as a no-brainer with its equally affordable itineraries featuring famous cities all within reach on one cruise. In 2013, Princess Cruises premiered a new megaship, *Regal Princess*, with an inaugural cruise to Venice. A twin ship, *Royal Princess*, was launched in 2014.[33] The capacity of both ships is over 3,000 passengers. While the industry gears up to double its capacity to serve Europe with additional megaships coming online, there are increasing uneasiness and tension in European ports that speak to local socioeconomic and environmental concerns. Venice is in the crosshairs with residents organizing protests and waving "No Big Ships!" banners. Cruise ships bring nearly 2 million passengers to Venice on 650 ships annually. Protestors comment that the ships are twice the size of the Palazzo Ducale (Doge's Palace) and twice as long as the Piazza San Marco, obscuring and overwhelming the local context.[34]

These preoccupations revolve around the question of how large-scale tourism can be managed in harmony with local culture, local monuments, and local sentiments. The history of tourism in destinations like Venice goes back centuries, and problems have always existed, but the massive size of the ships, often compared to moving skyscrapers, makes cruise opponents feel belittled, so much so that they play the role of David fighting Goliath, which leads to battle lines but few solutions.

Even as Europe struggles to incorporate cruise line business into its ports, the next generation of opportunity is already taking off in Asia. Sixty percent of young travelers in India are interested in cruise travel. Nearly 30% of the Chinese middle class over 65 are also likely candidates for cruise. Given the demographics in these two economically soaring countries, even if market penetration reaches only 10% of those actually interested, a standard metric for European and North American markets, there is a huge new opportunity for cruise lines in Asia.[35] A study published by CLIA found that 52 cruise ships operated in Asia in 2015, meeting the need for a growing passenger base in the region. Carnival Corporation predicts that Asian passengers will be 20% of their business by 2020, double their 2013 figures.[36]

There is little question this industry is on the rise. And its largest player, Carnival, has become a titan in the world of travel and tourism, with annual revenues in 2012 of $15 billion, with a total of 327,142 cabins available at the end of 2012. It is now the largest travel and leisure company in the world. Compared to the hotel industry, Carnival is surging ahead of even the largest hotel brand names in the world, such as Marriott, which reported nearly $1.8 billion gross revenues in 2012 on 660,000 rooms.[37] While Carnival earns an impressive $45,000 per cabin annually, Marriott earns under $22,700.[38] Net income for Marriott was $198 million and Carnival $1.2 billion, a margin that is lower but reflects a company on the move, expanding and spending a great deal on new ships and ship improvements, at a cost of $40 billion in 2012. Carnival has been extremely successful in handling economic downturns such as the recession of 2009, with much less fluctuation in its profit margin than its competitor Royal Caribbean. As a result, it has maintained net revenues almost 3 times as high as its competitors,[39] quite an accomplishment for a corporation operating in an environment where there are substantial risks.

Cruise business model: business risks

In February 2012, the Carnival Corporation's *Costa Allegra* was aflame in the middle of the Indian Ocean, with over 1,000 passengers that had to be evacuated. The corporation reported a loss of $17 million due to this one incident, and they have since sold the vessel. One month prior, the *Costa Concordia*, another megaship in the Carnival line, ran aground in Italy, resulting in the loss of 32 lives. The megaship hit a reef as a result of cruising too close to shore, foundered, and nearly sank while carrying over 4,000 passengers. Fortunately, an oil spill was averted and no environmental damage was detected, but the cruise ship is filled with paint, solvents, batteries, lubricants, detergents, chlorination products, and flame retardants, which are still being removed as part of the enormous salvage operation.[40] It is the largest ship in history to be abandoned. Media reports were devastating with the story dominating the international news for days. Carnival reported writing off $515 million in losses, while receiving nearly as much in insurance. The total claims against the company by families and injured parties were impossible to estimate, but Carnival advised shareholders that insurance was likely to cover the damages.[41] While the press has repeatedly predicted the cruise industry's demise, there appears to be little long-term damage to the company's reputation. One of 11 brands in the Carnival family, Costa appears to be recovering its reputation nicely based on an accounting test of goodwill, which evaluates how a firm's value is affected by negative publicity.[42]

While risks in corporate annual reports are described conservatively and in the best light for shareholders, Carnival as a brand appears to be relatively unscathed even after these two major disasters. While groundings, fires, and even sinking ships appear to be the most dangerous risks the industry can absorb, the reputation of the cruise brand is a primary concern for the industry. When the *Carnival Triumph* was stranded in the Gulf of Mexico due to an engine room fire, critics and the media were in high gear. With sewage running down the walls and floors, the industry's environmental image was downgraded again.[43]

Advocacy groups such as Oceana and Friends of the Earth fought the cruise industry and brought to light many environmental infractions that raised concerns in the American public in the mid-2000s. Oceana organized rallies in Florida and Seattle and leafleted cruise passengers demanding that the industry treat its waste. In 2000, 53 environmental advocacy groups petitioned the Environmental Protection Agency (EPA) to take action and investigate waste water, oil, and solid waste discharges from cruise ships. In 2007, Friends of the Earth sued the EPA in 2007 for failing to respond to their petition.[44]

The concerns were well-founded. Between 1993 and 1998, cruise lines were found guilty of 87 illegal discharge cases in U.S. waters, 83% of which involved discharges of oil or oil-based products and the balance involving plastics and garbage.[45] The report did conclude that 72% of the discharges were accidental, 15% intentional, and 13% undetermined. The detailed results showed health risks from untreated waste in waters and many inconsistencies in regulatory approaches to pollution from cruise ships. A Clean Cruise Ship Act was proposed in 2009 to regulate waste water dumping beyond the 3-mile limit imposed by international law, recommending a 12-mile limit instead. The Act was reintroduced in 2013 and has not been voted upon by Congress as of May 2016.[46]

Despite losing the battle for a national regulatory framework, the environmental advocacy community claimed a small victory when Royal Caribbean agreed to install advanced waste water treatment technology on all of its ships. With pressure from government and some collaboration with advocacy organizations and environmental NGOs, the cruise line industry started taking firm steps to rectify its environmental image. In Alaska, no ships are operating without AWT, and CLIA members have committed to no discharge of treated sewage within 4 miles of land. Alaska's Department of Environmental Conservation argues in response that there is no current evidence that salmon are not safe and that Alaska is using the EPA's criteria. They acknowledge that the science is evolving, but current studies of copper effects on salmon have been conducted only in fresh water.[47] Alaskans are wary as they are now dealing with the aftereffects of 66,000 gallons of chlorinated pool water dumped into Glacier Bay National Park in January 2013 for which Princess Cruises was fined $20,000.[48] With each new incident, the battle between citizens, advocates, and the cruise industry continues to fester. In 2015, the Alaska Department of Environmental Conservation gave a 5-year permit for cruise ship operations in Alaska, allowing cruise ships to operate if using the most advanced effective AWT available. This ruling was immediately challenged by the Southeast Alaska Conservation Council and other local NGOs.[49]

Prominent environmental advocates, such as James Sweeting, former vice president for Environmental Stewardship at Royal Caribbean, who originally worked on cruise line and other tourism issues for Conservation International, argues that the industry cannot be asked to meet treatment criteria that exceed U.S. national standards for waste water.[50] But the history of cruise infractions, the consistent slip-ups, and the ongoing concerns of residents in pristine waters continue to be one of the greater risks the industry faces.

A middle ground between cruise line business expansion and local needs must be found. The divide between local advocates and the cruise industry has continued to grow wider. This dispute will be ongoing and increasingly fruitless unless more thorough international standards are set for cruise line activity from the International Maritime Organization (IMO). It is time for the IMO to set standards that respond to the special impact of passenger ships, separate from the shipping industry, using science-based mapping and testing of waters for vulnerability and indicators for protection of vital ecosystems such as coral reefs and wildlife, monitoring managed by independent third parties, transparency, and a system for enforcement that is agreed upon by stakeholders and the industry.

Cruise business model labor practices

The cruise industry employs hundreds of thousands of workers who hail from a wide variety of countries. As ship sizes balloon, crew sizes have grown above and below deck. In 2013, Carnival Cruise reported having 9,700 shore-side employees full time and 78,500 crew on their ships. In the same year, Royal Caribbean had 6,200 full-time on shore and 57,000 shipboard workers.[51] Per ship, depending on the size of the vessel, crews run from 300 to 2,100 workers, with an average of 750 to 1,500.[52] Cruise ship work entails long hours, and most of the contract workers on board maintain this schedule for several months at a time. The labor-intensive nature of the industry makes salaries the greatest cost of operating a cruise ship, exceeding fuel or food.[53] The ability to hire inexpensive workers represents an important part of the cruise line business model. A detailed academic study of cruise workers finds that pay, status, and eligibility for higher wages and promotions break down along ethnic lines. This study finds that "the deeper you go in the belly of the ship, the darker the crew."[54] The cruise line industry has effectively taken advantage of global race and ethnic inequalities to manage costs and also dictate what type of worker can attain higher employment status. The loose nature of the maritime labor laws has been notorious for years. In the 1990s, as executive director of The International Ecotourism Society, I sat with a senior legal expert on labor for the cruise lines at an industry luncheon. He was so entirely casual about the lack of legal standards for labor on cruise lines, he was making a joke of it with a braggadocio that was clearly entirely acceptable to the industry. He obviously had little idea whom he was speaking with, nor did he care because the cruise lines he represented were operating within legal labor standards of the sea.

The International Transport Workers' Federation, a global labor association, has fought to change these labor practices for 50 years, and there has been substantial progress in this arena.[55] In 2006, a groundbreaking agreement, called the Geneva Accord, which became law in 2013, consolidated a fragmented set of labor laws into one document under the auspices of the International Labor Organization Convention.[56] The new binding regulation lays down core standards for working and living conditions for the first time. It also breaks through practices of "sub-contracting" laborers which prevented legal regulatory oversight in the past. These laws now apply to all persons who work on board under all flags period.[57] The goal is to create a level playing field for labor standards and costs for ships worldwide. This new accord creates standards for pay, annual leave, termination conditions, and health and social security benefits across the board for all ship employees for the first time. Working hours are regulated to a limit of 14 hours per day, and there are standards for accommodation, food and onboard facilities including heating, ventilation, and noise. Legal analysts and labor activists alike consider this to be a decisive step toward improvement of seafarers' working and living conditions.[58]

The industry has depended on a globalized market, where labor standards are highly variable and wage rates are much lower than in developed country ports. The cruise lines now face rising labor costs to meet the standards of the new

Convention. Carnival Corporation reported in 2012 that full compliance will increase a ship's annual operating costs from $15 to $25 million.[59] Not long after the law was put in force, there were reports of noncompliance. In October 2014, Dutch labor inspectors fined Royal Caribbean €600,000 for violating labor laws when in a Dutch port. Crew members were found to be working in excess of 16 hours per day.[60] Cruise lines that fail to comply will increasingly face expensive fines, litigation, and penalties.

Key legal issues with cruise line environmental management and solutions

The cruise line industry attracts a great deal of scrutiny, certainly much more than the shipping industry to which it officially belongs under the regulation of the International Maritime Organization (IMO). Cruise ships' environmental impacts in U.S. waters have been investigated by the U.S. Congressional Budget Office[61] and the EPA,[62] with mixed results. Even as the industry is brought to task for environmental infractions – and they are caught frequently[63] – they do respond, invest in the necessary environmental management technology, and move forward. The battle lines keep shifting, and data quickly goes out-of-date. One of the other impediments to overseeing the industry's environmental performance is that it is operating in a legal environment that has international law governing most problems without a regulatory body to enforce these laws. An analysis of the legal environment under which cruise lines operate will be given special attention here because of the disquieting gaps that give cruises an almost colonial ability to operate without regulations common on land.

The cruise line industry runs almost entirely under the foreign flags of countries that represent neither where the cruise line industry is headquartered nor where it operates, keeping it safe from many regulations except in their territorial waters and ports of call. The Bahamas is the number one registry state for cruise lines, with Panama and Liberia at second and third respectively.[64] The legal articles that allow registry of ships under foreign legal oversight are very loose, simply requiring that there must be a "genuine link" between the state and the ship.[65] This takes the form of a hefty ship registration fee paid to a flag state, which grants the flag state's nationality to the ships. In theory, these foreign states are responsible for the environmental performance of ships almost never found in their own waters. When a pollution violation occurs, flag states are in charge of managing the problem. Few flag states have any reason to prosecute, nor do they have the desire or power to control the cruise line companies.[66] This economic and sociopolitical reality is a fundamental part of the cruise line business model.

Cruise lines and the entire shipping industry operate in an environment where regulatory procedures largely do not hold water. A detailed report by legal analyst Alexandra Ritucci-Chinni on *International Cruise Ship Pollution* provides an essential primer on the treaties that fail to govern cruise ships and how to correct these problems.[67]

Two international treaties are in place to govern the shipping industry, including cruise lines:

- The *International Convention for the Prevention of Marine Pollution from Ships, 1973*, as modified by the Protocol of 1978 (commonly referred to as MARPOL 73/78). This is the principal treaty that regulates pollution from ships. It entered into force in 1983.
- The *United Nations Convention on the Law of the Sea* (UNCLOS). It was signed in 1982 and entered into force over 10 years later, in 1994.

The International Maritime Organization (IMO) is the body that oversees these treaties, but its role is restricted to recommending regulations. It does not have powers to impose new regulations or enforce existing ones. Many of the provisions of UNCLOS interfere with MARPOL. The UNCLOS article that requires a "genuine link between State and Ship" fails to define this link, as already noted. The majority of cruise ships do not register in their own states. For example, the ships under the Carnival brand are all registered under foreign flags with 70% of their operating ships flagged in Panama, 25% in the Bahamas, and the remaining in Malta. All new Carnival ships in construction will also be flagged outside U.S. territory, in Panama, Liberia, and Portugal.[68] UNCLOS requires that the flag state control administrative, technical, and social matters and the protection of the environment on board.[69] This puts environmental enforcement into the hands of the flag states and not under the sovereignty of either ports of call or homeports. It effectively negates MARPOL's objectives to prevent pollution of the marine environment.[70] An explanation of why this one provision is at the core of the problem and how it can be rectified follows, according to the logic of Ritucci-Chinni's expert analysis.

MARPOL 73/78 has the most specific regulations for environmental protection of the seas. Its Annexes provide the guidelines. There are six Annexes:

- **Annex I** regulates the discharge of oil into water resulting from the operation of ships since 1983. Article 5 of Annex I requires that all parties to this convention accept a ship with valid certificate in ports of call. The Annex does not specify what steps the inspecting port officers can take to stop a ship without valid certification from sailing.
- **Annex II** regulates the discharge of noxious substances that can harm the marine environment since 1987. Conversely, it allows untreated waste to be discharged outside the 12-nautical-mile zone.
- **Annex III** relates to the disposal of items in freight containers since 1992.
- **Annex IV** requires that a ship have proper sewage treatment or a holding tank on board where the ship can disinfect sewage, and a pipeline to discharge the sewage at reception sites since 2003. It allows the discharge of treated sewage over 4 nautical miles from land, but exceptions can be made if a certified sewage treatment plant is on board. It allows sewage discharge closer to shore if the system does not produce "visible floating solids or discoloration of the

surrounding water." Once a ship is 12 nautical miles from land, it can discharge untreated sewage.

- **Annex V** regulates the disposal of garbage since 1988. It allows the disposal of packing materials, and dunnage (materials used to cover and protect freight), and other materials that float 25 nautical miles from land. All plastics are prohibited from being discharged anywhere at sea. Food wastes and all other garbage can be discharged 12 nautical miles offshore. If this waste is ground up, discharge at 3 miles off shore is permitted. Annex V designates Special Areas where waste disposal is not permitted, which include the Baltic, Black, and Red Seas and the Gulfs of Ras al Hadd and Ras al Fasteh. When the ships are in the Special Areas, garbage must be stored on board and delivered to ports where disposal facilities are required to handle the waste.
- **Annex VI** regulates air pollution from ships including sulfur oxide (SO_x), produced by the sulfur content of ship fuels, and nitrogen oxide (NO_x) since 2005. Special Emission Control Areas (ECAs) can be defined. The U.S. and Canadian waters are designated ECAs with protection up to 200 nautical miles off the coast, including Hawaii, effective in 2012. ECAs were adopted for Puerto Rico and the U.S Virgin Islands by the IMO in 2011, made effective in January 2014. Emissions in ECAs cannot exceed 0.1% sulfur by 2015 and must achieve an 80% reduction in NO_x by 2016.[71] In Europe, the Baltic and North Seas have been in ECAs since 2006 and 2007 respectively.[72]

The Catch 22 with MARPOL and its annexes is that UNCLOS provides that only the states where the ships are registered have the power to investigate a ship, and then only for a valid certificate on board. Countries acting as foreign flag states, or offering flags of convenience as it is commonly referred to, gain economically from the tax revenue generated by ships registered under their flags and therefore have no incentive to stop ships from operating, even when they are in clear violation of MARPOL. Any coastal state that seeks to press for enforcement will receive a formal response from the foreign flag state, but most do not have strict enforcement policies. Coastal states can press for an investigation if there has been substantial discharge causing or threatening significant pollution of the marine environment under Article 57. But there is no definition of what substantial discharge means. Even if a coastal state has evidence of substantial discharge, there is no definition of what can trigger the substantial discharge clause. Only the flag state is empowered to investigate in due course. While cruiseforward.org states that flag states offer "robust inspection and enforcement," this is really not the case.[73] There is no time limit on the proceedings (due course is also not defined) and no standardized penalties. It is easy to imagine the frustration of a coastal state with a substantial discharge, no ability to detain a ship, no ability to inspect, and no power to require a timely response. Clearly, this system was designed to be without teeth. In fact, out of 1,000 cases reported to the IMO, foreign flag states have responded to only 534. Of those, 77 of the infractions resulted in fines.[74]

Potential legal solutions

The lack of enforcement capacity of the IMO is the primary problem for establishing any new system of compliance worldwide. Advocates concerned about the growth of cruise lines and lack of oversight might best work on promoting a new international agreement to give the IMO enforcement powers. Without this, cruise lines will face a patchwork of systems within national waters with standards that are always changing. And governments around the world will be faced with having to defend their waters, via legal cases and amendments to MARPOL, rather than having one strong and uniform set of international laws to support them. The Federal Clean Cruise Ship Act in 2004 (FCCSA), which failed in Congress, suggested a "head tax" of $10 for every cruise passenger to finance the enforcement capacity of an agency to oversee cruise lines. An additional fund was to be created to receive all the penalties paid for violations of the Act. The same approach could be implemented by the IMO with participation from all of its signatories.[75]

Revisions to the system of flag state oversight are also urgently needed. UNCLOS provisions that deny coastal states power to protect their own waters by all third-party accounts are not functional. Ships operating under foreign flags need to work within a system where the country they are registered in has the ability and motivation to respond to environmental infractions.

Some of these problems that have now been solved in part via Annex VI, which creates special regulatory areas, but such fixes only aggravate the problem of having variable regulations in different ports worldwide. An international goal should be set to create a uniform set of regulations, which could be achieved via amendment to MARPOL and UNCLOS setting baselines regulations similar to what was achieved by the International Labor Organization (ILO) for labor standards.

Environmental impact management issues and solutions

Cruise ships generate waste streams in specific categories, each of which has its own protocols and systems for treatment.

Sewage or black water: *Water that originates from toilets and medical facilities.*

Issues: MARPOL's regulations are not specific and tend to lack enforcement, as was previously described in detail. In the U.S., the Clean Water Act prohibits the dumping of untreated or inadequately treated sewage into navigable waters. This is enforced jointly by the EPA and the Coast Guard under Section 312. Marine Sanitation Devices (MSDs) are required on board to treat sewage, and specific coliform limits are set for the cruise ships. The ships must use either Type II or Type III MSDs, with limits of no more than 200 fecal coliform per 100 milliliter.[76] However, ship operators are not required to sample, monitor, or report on their discharges, and the Coast Guard lacks bandwidth to enforce these regulations.

Solutions: In Alaska, higher standards remain in place requiring that all cruise ships use Advanced Wastewater Treatment systems that are of higher standards than what is required in the rest of U.S. waters. The Cruise Line Industry Association

(CLIA) membership is committed on a voluntary basis to treat all black water to meet the standards for Type II Marine Sanitation Devices, referred to as Advanced Waste Water Treatment (AWT). CLIA members will not discharge treated black water within 4 nautical miles of shore.[77]

Gray water: *Waste water from sinks, showers, galleys, laundry, and cleaning activities on board.*

Issues: MARPOL has no regulations to date to prevent the discharge of untreated gray water, nor did the EPA in the U.S. until 2013, when it mandated that gray water must be treated prior to discharge under federal law.[78]

Solutions: According to the U.S. EPA, gray water is in especially large quantities on cruise ships and for this reason needs special oversight. Given cruise ships were already using AWT on board, the new gray water regulations were deemed to be economically practicable and achievable. It was estimated that a black water and gray water AWT system costs $7.09 per passenger and crew per season.[79] Gray water has the potential to cause adverse environmental effects because it is high in nutrients and requires oxygen, which can rob ecosystems of their natural balance. How harmful the gray water discharges are requires more research. With 100,000 to 250,000 gallons discharged per day for a 3,000-person ship, AWT installations for gray water on all cruise ships will be the ideal solution.[80]

Oily bilge water: *Water containing solid wastes, lubricants, pollutants, oil, and chemicals.*

Issues: Bilge water may contain solid wastes, chemicals, oils, and other chemicals. A typical, large cruise ship will generate an average of 8 metric tons of oily bilge water for every 24 hours of operation.[81] Bilge water must be cleared out regularly with the oil extracted through a separator to maintain ship stability and remove toxic substances before discharge. Oily bilge water was illegally discharged in U.S. waters by Carnival in 2002 and by Royal Caribbean in 1998 and 1999, resulting in fines of tens of millions of dollars. Falsified records were used to cover up this illegal dumping, which both companies have worked hard to prevent since. Nevertheless, Royal Caribbean's attorneys argued that the U.S. lacked jurisdiction in these cases because the offending ship was flagged in Liberia. Liberia asked that the case be dismissed, but a U.S. judge rejected Liberia's claim.[82] In weaker jurisdictions outside the U.S., for example in the Caribbean, it is little known whether the same scrupulous standards are maintained because there is no monitoring.

Solutions: CLIA's members are said to use holding tanks and have added systems to prevent illegal discharge by crew members, or the oil is separated from the water to a legal level and then discharged.[83]

Ballast water: *Water used to stabilized the vessel and compensate for changes in the ship's weight as fuel and supplies are used.*

Issues: Ballast water is the leading source of invasive species contamination in the world. It typically contains a variety of biological materials, including plants, animals, viruses, and bacteria. No international treaty covers the problem for any part of the shipping industry. In the U.S., the EPA rejected a petition to regulate ballast water as a pollutant. Its stance was contested in court in 2005, and the EPA lost. The EPA chose to appeal, but the district court upheld the original judgment ruling ballast water an official pollutant in U.S. waters in 2008.

Solutions: All boats over 79 feet in the U.S., except commercial fishing boats, must file for permits to discharge ballast water under the National Pollutant Discharge Elimination System (NPDES) as of 2013.[84] It is anticipated by the cruise industry that expensive ballast water treatment systems will be required in the coming years. Systems to remove sediment with any biological life before pumping ballast water are being investigated.[85]

Solid waste: *Glass, paper, cardboard, aluminum, steel cans, and plastics, which can be non-hazardous or hazardous.*

Issues: Cruise ships can dispose of garbage only outside of territorial waters, with plastic banned worldwide. Cruise ships are managing their solid waste via source reduction, minimization, and recycling. About 8 tons of solid waste are generated during a one-week cruise, most of which is treated on board, incinerated, pulped, or ground up for discharge overboard.[86] One study, done by Harvard Extension student Yvette Acevedo in the exclusive Royal Caribbean port of Labadee, Haiti, found no recycling facilities for aluminum and glass, despite the fact that the port was brand-new and designed for Royal Caribbean's megaships. Based on the average metric of 1.15 pounds per passenger offered by Royal Caribbean Cruise Lines, Acevedo estimates that over 800,000 pounds of solid waste was being left behind in Haiti in 2010, without processing or facilities to manage it. Local entrepreneurs were seeking to sell the waste, and there was no evidence of where the rest was being disposed of.[87] Many ports struggle to handle the glass and aluminum that cannot be disposed of at sea or incinerated.

Solutions: Royal Caribbean achieved a 47% decrease in solid waste per passenger per cruise day between 2007 and 2010.[88] According to GIS software used by researchers, about 800 vessels travel through the Lesser Antilles of the Caribbean every year, generating approximately 3.25 million tons of waste which reach local ports. If this waste were to be used to generate electricity, it would be possible to generate 1.5 million kilowatt hours, worth roughly $1 million. While these are rough numbers, the idea of generating electricity from waste in the Caribbean has genuine potential and can be an option for joint consideration of all ports and cruise line companies.[89]

Hazardous waste: *Photo-processing, dry-cleaning, and equipment-cleaning waste. Medical waste, batteries, fluorescent lights, spent paints, and thinners.*

Issues: Hazardous waste is governed under MARPOL Annex II, which oversees 250 hazardous substances and prevents its discharge within 12 miles of land. More stringent restrictions apply to the Baltic and Black Seas. New categories were added to the Annex in 2007, including waters from ballast that may contain hazardous substances that can cause harm to ecosystems.[90] In the U.S., the Resource Conservation and Recovery Act (RCRA) governs hazardous waste management, but its provisions for cruise ships are less than clear.[91]

Solutions: The Convention on the Prevention of Marine Pollution of Wastes and Other Matter prohibits the dumping of any wastes listed in the Convention's

Annex, but only 16 countries have ratified this convention. Members of the International Council of Cruise Lines have agreed not to discharge any hazardous substances.[92] The effectiveness of this agreement is unknown.

Air pollution: *Contaminants generated by the cruise ships' diesel engines that burn high-sulfur-content fuel, known as bunker fuel, which produces sulfur dioxide, nitrogen oxide particulates, carbon monoxide, carbon dioxide, and hydrocarbons.*

Issues: MARPOL's newest annex, Annex VI, has become progressively stringent on NO*x* and SO*x* emissions, and emissions of ozone-depleting substances. In the past, sulfur standards were at a maximum of 4.5% with newer ships required to reduce sulfur content to 3% (of 1,000 ppm). The new Annex VI regulations require sulfur content to be at 1% by 2012 and 0.1% by 2015. Records are required on board ships, and fuel suppliers must ensure that sulfur content meets standards. Emission Control Areas (ECAs) are subject to even more stringent standards, including all of North America up to 200 miles from the U.S. As of 2016, new engines operating in ECAs must achieve the 0.1% standard for sulfur and an 80% reduction in NO*x*. Non-U.S.-flagged ships are subject to examination while operating in U.S. waters, and if there is a violation, the ship can be detained. Individual liability for false information is subject to civil penalties and even criminal liability.[93]

In 2016, the U.S. government gave a temporary exemption to cruise lines when they pledged to install advanced emissions purification on their ships in stages. Carnival Cruise line agreed in 2013 to cut the amount of sulfur from 32 ships out of 102 and to "explore the possibility of adding scrubber technology to more ships."[94] Exemptions are granted by the U.S. EPA based on the level of cooperation achieved with the company.[95] Fuel cost $2 billion for Carnival in 2014, 14% of Carnival's annual operating costs. Carnival estimates that the company would spend $180 million on the scrubber installations for the 32 ships between 2013 and 2016.[96]

Solutions: While MARPOL lacks an international regulatory body, new standards in Annex VI are allowing coastal states to take action. Air pollution regulations now reflect a more up-to-date set of standards and close some legal loopholes both from the regulatory and legal perspectives. The fact that ships can now be inspected in coastal waters, without the participation of the flag state, is a significant advance. Allowing civil and criminal penalties in the coastal state for those ships that do not meet standards or cannot provide logs to prove the use of lower sulfur fuels is new.

The cruise line industry has begun to invest in lowering their SO*x* and NO*x* emissions. The main options are scrubbers, lower-sulfur fuels, or liquid natural gas. Scrubbers are considered to be the best short-term investment to comply with regulations. The cost for oil companies to refine low-sulfur fuel for cruise companies is estimated to cause price increases of 87% on fuel. Liquid natural gas could be the solution of the future, but the infrastructure is not yet available.[97]

In Singapore and Hong Kong, voluntary initiatives were established, with the anticipation that sulfur emission control areas (SECAs) would be established in the region. Hong Kong officially announced its plan to require low-sulfur fuels in 2014.[98]

Environmental management solutions and cruise line reporting

The management of CO_2 emissions for cruise lines remains unregulated but coming under increasing scrutiny and is already part of a larger effort to reduce cruise line impacts. The cruise industry is undertaking more international reporting via the Global Reporting Index (GRI) and the Carbon Disclosure Project (CDP). Carnival Cruise lines was placed on the Carbon Disclosure Leadership Index from 2008 to 2012 by achieving a score within the top 10% of the 35 companies reporting in 2012. Metrics include transparency and absolute reductions in emissions of CO_2.[99] While these reports appear to provide more transparent information on cruise line activities, researchers who have looked carefully beneath the surface are finding many gaps. In-depth analysis shows that environmental management practices are reported according to the discretion of the company without consistency. For example, waste water management is reported without mention of ballast water practices. Or there may be emissions reporting without information on the types of fuels being adopted. There are few reports on labor–management relations.[100]

Of the 19 cruise lines analyzed, each goal was reported on by a different subset of companies without consistency. The goals reported on by the greatest number of cruise lines were waste reduction, reduction of GHG emissions, energy conservation, and air pollution on an aggregate level by the umbrella brand corporations, Carnival, Royal Caribbean, and TUI Travel. The lack of distinction between individual cruise ships, cruise lines, and their parent companies in reporting was cited as one of the most difficult aspects of interpreting the reports. There was a lack of verification for reporting and "a clear need to evaluate the truthfulness of claims."[101] Without more outside verification and consistency of reporting, there will be continuing problems with confirming the progress of the corporate social and environmental responsibility initiatives of cruise lines.

Sociocultural challenges of cruise lines

Cruise lines bring unusual challenges to the shores of their destinations that are also sociocultural. When the cruise industry arrives, beloved villages and towns are transformed. Residents do not immediately perceive the changes because the village's metamorphosis into a cruise industry destination does not take place overnight. But inevitably locals start to see their way of life change. Their place has become a product, which is being marketed and sold to visitors. On the island of St. Thomas in the U.S. Virgin Islands, Charlotte Amalie was once a charming town with local village-style services concentrated in the historic, old city. Since the cruise industry arrived, it has become an industrial port, with huge ships docked right next to the old city. Historic buildings that once housed local services now display luxury items such as watches and jewelry to tourists. No local person would shop in this area anymore. Taxis find a way to go around this area of town to help overnight visitors avoid the unpleasant congestion, now standard fare in the old city.

Foreign investment that excludes local ownership and that builds according to specifications that do not respect local architectural heritage or incorporate local thinking or needs results in places that have no "authenticity." The local people feel drowned out by how fast the growth is, overwhelmed by the number of tourists, and forced out by the cost of living. After their way of life has been drastically altered, the cruise industry can move on because it has little financial incentive to be concerned. A local way of life is not their business, and the globalization of cruise is pushing them to build ships and ports on an unprecedented scale. The market is not insisting on authenticity of place either. Cruise passengers are willing to accept and enjoy port visits that highlight the equivalent of duty-free shopping, with cookie cutter development that has no relationship to the culture of the destination.

Ports around the world that were once beautiful or historic places are transformed by the number of visitors as well. During the high season, St. Thomas absorbs 20,000 additional people a day when six or more cruise ships are docked in Charlotte Amalie. Only 50,000 people live on the entire island.[102] How these places can be preserved is no small challenge. But even when the valuable sociocultural fabric of life is being torn asunder, it can be very difficult for local people to realize and substantiate the problem. Yet if a local way of life is not managed in ways that make certain it is protected, the value of the destination for overnight visitors or tourists not taking a cruise begins to decline. Sociocultural impacts can be severe, and much more must be done in the future not only to protect built heritage but to strategically manage local culture and living heritage.

Sociocultural planning solutions

Efforts to preserve the cultural authenticity of ports have been explored in Falmouth, Jamaica, where extensive consultation took place with local residents on the community's heritage with support from the U.S. Agency for International Development (USAID), the United Nations Educational, Scientific and Cultural Organization (UNESCO), and the Canadian International Development Agency (CIDA). Falmouth's distinctive features were mapped, including a phosphorescent lagoon, its historic district, and its cultural heritage. Critical zones for preservation of culture were mapped to ensure its Georgian architecture and public piazzas were preserved. Local residents were involved at every phase of the discussion, zoning ordinances were passed, and traffic measures were put into place. Required infrastructure was financed, including utilities, sewage, harbor cleanup and government purchase of properties to create a recreational heritage zone.[103]

As the next generation of tourism planning for cruise destinations is discussed, the value of sociocultural assets must be built into the cost of doing business. Already in Europe there are many concerns that the cruise industry will undermine or help destroy the invaluable cultural assets of cities like Venice. Cruise ships, branded "floating skyscrapers" by opponents, drop 35,000 passengers a day in Venice. Opponents are concerned that their wakes may cause damage to the city's fragile underwater foundation, and the air pollution produced by the idling diesel engines not only cast a haze over the city but could also damage its historic monuments. Over 1.8 million passengers per year are already visiting, landing just 300 feet from St. Mark's Square. Local protestors demand a cruise dock offshore, but other community members insist this could jeopardize some 6,000 jobs that the industry has brought to the area.[104]

Destination managers must quickly determine what sociocultural indicators to measure, in order to develop guidelines for arrivals in cruise ports for stakeholder comment and review. Good planning procedures and indicators are well-known by sustainable tourism planners, but this information needs to be deployed. (See Chapter 8.) The ratio of local populations to tourists should be maintained at a level acceptable to the community or city, usually well below 50% in peak season, to ensure that locals can continue to live and work in the place tourists are visiting.[105] Once these limits have passed, protests can only go so far. The damage is done once the beneficiaries to the cruise economy become entrenched. Tourism is not a model that responds well to last-minute fixes. One of the most challenging parts of planning for ports is balancing local commercial interests for these lucrative floating markets full of buyers, with astute analysis of how to manage the number of visitors without permanently compromising local values. It is a very tough set of decisions.

Sociopolitical impacts of cruise line industry growth

The cruise line industry's unique environmental and sociocultural impacts distinguish it from other maritime shipping concerns. This is also true of its sociopolitical impacts. With the thousands of passengers disembarking daily in ports around the world, cruise lines work with local governments on many issues related to port infrastructure and the management of their passengers on shore. Cruise companies work together with local tour operators, some of which are independent and some of which are owned by the cruise companies, to manage all tourism on land. These companies are usually united in purpose and are deeply involved in lobbying for more favorable commercial enabling environments. They can receive significant assistance from the international offices of CLIA and local related cruise line associations when sticky problems arise. Their influence on local governments is significant, as I had the opportunity to observe firsthand in my work in Belize. Their primary tactic is to present statistics on their economic and labor benefits. The industry carefully accounts for and promotes its gross economic impacts, but it does not present net figures, which can more clearly portray how much a destination is benefitting. As one researcher observes, they are highly opportunistic in their use of economic reports.[106] Accordingly, it is essential that government analyze the underlying assumptions of these economic reports to understand their significance and validity. With the help of several studies, it is possible to understand more about the net contribution of the cruise industry to Caribbean nations.

In the Caribbean, the total earnings per destination showed a significant decrease in 2012. Total passenger visits and expenditures were down by 12% between 2000 and 2012. In the Mediterranean, cruise growth was 20% in 2011, and new destinations within the Caribbean, such as Colombia, experienced 9% growth.[107] This leaves the traditional Caribbean ports in a precarious position: do they invest more in maintaining the cruise economy, or do they diversify more now in alternative economic development approaches?

Make no mistake: the economic significance of the cruise industry in the Caribbean is of titanic proportions. In 2012, direct expenditures reached $2 billion.[108] But the trends are not encouraging. While the number of passengers visiting is increasing, the amount they spend is decreasing. The local expenditures of cruise tourists have never been terribly high. Visitors spent just over $130 per person per destination in 2009. As of 2012, that average is now under $130. These expenditures are broken down via types of purchase.

- Watches and jewelry
- Shore excursions
- Clothing
- Food and beverage
- Local crafts and souvenirs

Watches and jewelry make up about 50% of the purchases an average cruise passenger will make, and shore excursions represent approximately 30%. In these two categories, there was a slight increase in purchasing, while the rest of the categories declined.[109] The explanation from the cruise industry publication responsible for these statistics is that customers have become more selective in what they purchase during their destination visits.[110] But there could be other explanations: (1) they are spending more on board than before and/or (2) they receive much more targeted information while on board, in the form of cruise-sponsored shopping lectures about the opportunities to purchase watches and jewelry in ports – ports that are often partly owned by the cruise lines, together with the international jewelry trade.

And this is the bottom line. While direct expenditures tell one story, basic information regarding passenger expenditures made available by cruise lines indicate that roughly half of all gross expenditures leak out again because of the types of purchases cruise customers make. While there is little question that the gross economic benefits of cruise lines have high value, particularly for certain local suppliers and vendors, the overall question of how well these benefits translate into a higher standard of living for local people or a better quality of life, or adequately reach government coffers has been researched far too infrequently, with only a few examples available.[111] These examples point out that there is a low percentage of disembarking revenue, high commissions for tourist packages charged by the cruise lines, low value of taxes paid per passenger, and a concentration of contracts for service providers to passengers in the hands of a very few land-based operators.[112] All of this can add up to a bad deal. Destination governments around the world need to do their accounts and not simply use the statistics the cruise industry supplies.

Destination management solutions

If one region of the world needs immediate and special measures to improve their deal, it is the Caribbean Sea. The Caribbean once attracted over 50% of all cruise visits but in 2015 was 36% of total available bed days. Not unlike the other cases presented here, the Caribbean Sea is suffering an ecological breakdown. Their fisheries are overexploited,[113] and in 2012 its once vibrant coral reefs were declared to be on the verge of collapse by the International Union for the Conservation of Nature (IUCN), one of the most prestigious scientific organizations in the world.[114] A wide variety of environmental NGOs seek to protect the integrity of the Caribbean Sea and coastal regions, but there is no region-wide coordinated effort.[115] The fragmented response has limited the region's ability to work proactively on pressing problems of port management and waste treatment, all of which could benefit from a more proactive, region-wide set of solutions. For example, the Caribbean Sea is not a Special Area under MARPOL: Annex V because not enough of the islands can afford to dispose of cruise waste, a requirement under the Annex. By and large, Caribbean governments are encouraging cruise investment but not developing long-term solutions for an economy that will be to their benefit in the 21st century.[116]

Solutions to this problem may require a whole new way of thinking. How would the cruise industry sell the Caribbean without the white sand beaches of the Caribbean Sea? These beaches are a natural part of the ecosystem that were created and are now protected by coral reefs. Coral reefs have helped to form entire island systems, such as the Bahamas islands which are famous for their pristine waters, white sand beaches, and protected calm waters for boating, swimming, snorkeling, and scuba diving. Yet 33% of the Bahama's reefs are under threat.[117] The value of the reefs to the cruise industry should be calculated. But instead of creating a number that supplies a value of service as ecologists have done in the past, why not look at the business value of the ecosystem as a product to cruise lines? The Bahamas offer an interesting example of this because they are the number one destination for cruise ship passengers in the world.

Bahamas valuation: case example

In 2012, the Bahamas received nearly $400 million in direct cruise expenditures, with over 8,500 employed, earning nearly $150 million in wages.[118] These statistics are impressive, but any business must and does build in a substantial profit margin by working to be certain the majority of what is spent by its consumers is "on board" to fuel its own future.

The cruise industry business model depends on countries, like the Bahamas, to serve as suppliers of beautiful aqua blue waters, white sandy beaches, and bright colorful fisheries to thrill snorkelers and divers. It looks to local governments to supply essential services such as solid waste management, recycling, electricity generation, and sewage treatment on land. Due to the size of cruise operations, small island governments frequently do not have the ability to cover these infrastructure needs for millions of visitors. Passenger taxes, called head taxes, are charged to cover some of these costs. Each supplier nation (city or state) benefits from the cruise industry according to their ability to efficiently and strategically negotiate pricing for the services cruise lines need. The pricing they set is significant and the stakes high. Realistic evaluations of their pricing strategies must be based on a triple bottom line approach, which will allow them to preserve their most vital assets, which is not only port infrastructure – it is beautiful clear water and pristine beaches and coral reefs.

While the nation of the Bahamas established a goal of protecting 20% of its marine ecosystem in Marine Protected Areas, which led to the creation of 10 new marine national parks – and local expertise is there to manage the parks – funding, staff, and equipment are lacking. There is no system to finance the protection of these assets. Adjacent to the Bahamas, the Turks and Caicos Islands have established an important form of monetary support, called the Conservation Fund, which is a 1% share of all tourist and accommodations taxes.[119]

The cruise industry must pay its part of this hypothetical formula. With nearly 4 million passengers a year visiting the Bahamas, each paying roughly $1,000, the industry is earning gross revenues of $4 billion (with an average 10% profit value, worth roughly $400 million annually to the cruise industry). If the Bahamas ecosystem would not exist unless there is an investment, isn't it sensible for the industry to pay 1% of its profits for a destination ecosystems investment of $4 million annually?

The ecosystem of the Caribbean must be viewed differently in the future by the cruise industry. It is not just a "service"; it is a vital product that is being sold at no cost to business. The cruise industry is dependent on beautiful, preserved natural resources to survive. If the Caribbean Sea becomes a tired, overused, and degraded destination, the region's value will plummet for cruise lines and undermine their bottom line directly. It is essential that a formula is created that is offered in simple business terms, with a review of how conservation taxes can tap a legitimate percentage of cruise profits to fully maintain and preserve the beautiful waters, coral, culture, and historic places upon which cruise depends. Is it not entirely possible, before it is too late, to build the needs of these countries into the business plan in a responsible way, so that there is a permanent value to the region, which can be delivered for cruise passengers in the long term?

Belize cruise management: case study

Like many Caribbean nations, Belize was once a strictly agricultural economy dependent on sugar. Investments in agricultural and fisheries helped to diversify the economy in the 1980s and 1990s. But by the mid-2000s, tourism was becoming the number one source of foreign investment.[120] Foreign exchange earnings from tourism rose to 32%, and the industry was growing at 6.8% annually from 1990 to

2006.[121] Its waterfront attractions grew rapidly in the 1990s. At that time, Ambergris Caye, its primary tourist destination on the coast, was a classic, sandy Caribbean village just being discovered by tourists, with a small town center and a few motels. By 2000, the Caye's coastline had been built up with hotels, restaurants, and small boat marinas for several miles. A 2011 study by Cornell students, referenced in Chapter 2, confirmed that development did not preserve the environment or sustain the well-being of local people. Despite these downfalls, Belize's active environmental and ecotourism business communities seek to maintain their homeland as a more environmentally friendly port of call as development continues.

When cruise ships arrived in 2000, it was a clear challenge to the existing model of tourism in Belize. A new "hypergrowth" model that relied on more infrastructure and investment than the country had ever attracted before was introduced and quickly took hold. By 2005, revenue from cruise was already accounting for 24% of all tourism revenue in the country. By 2006, nearly 700,000 passengers were arriving in Belize on cruise ships annually, equivalent to an annual growth rate of over 60% since the cruise lines first arrived in 2000. According to a Belize Central Bank study, a number of steps were taken to maximize the benefits of the cruise industry.[122] This study is summarized here as a model of reviewing the economics of cruise tourism for a country both before and after cruise business is introduced.

The Belize Tourism Board policy crafted in 2005 sought to keep business in local hands as much as possible and prevent economic leakage. It recommended that a nonconflicting strategy be implemented to attract niche/ecological tourism, stayover mass tourism on the coast and cruise, with an annual rate of 1 million cruise passengers per year. The goals were as follows:

- To increase the number of cruise ship calls and arrivals in a sustainable manner based on acceptable visitation limits
- To optimize revenues generated from cruise passengers
- To increase benefits from cruise tourism (and prevent leakage) by linking needed goods and services to cruise lines from Belizean suppliers
- To expand the number of visitor attractions
- To further develop port facilities
- To identify suitable anchorage sites on the coast
- To promote overnight stays in Belize to cruise passengers

The policies also required cruise lines, port agents, and tour operators to:

- Suspend all forms of entertainment on board while the cruise ship is in port,
- Encourage shore visitation by passengers,
- Promote overnight stays and multiple destination visits to encourage and maximize visitor satisfaction,
- Utilize a cross-section of services and avoid the growth of monopolies, and
- Encourage the creation of unique local activities that will enrich the visitor experience.

Despite having taken these steps, the Central Bank study states that there was a widespread sentiment that the explosive growth of cruise tourism was putting the country's niche positioning as a high-end provider of eco-based tourism at risk. The increase from 3,000 to 8,000 cruise passengers daily was one of the primary causes of concern. Belize City has remained largely unchanged because the country did not invest in a major port facility, partially because the country's barrier reef presents logistical challenges. Instead, Belize fostered a locally owned tour business to take visitors out of the shopping area and explore Mayan archeology and enjoy its most popular tour, which is tubing down a pleasant river through limestone caves. A pioneer of channeling visitors onto local tours, Belize has one of the most successful tour participation rates in the Caribbean. Other countries are now following this model. Nonetheless, Belize City lacks the type of port facility that the industry wants, and there is substantial pressure to construct a new port and docking facility to meet industry needs.

The bulk of direct expenditures from cruise in Belize takes place from shopping and tours. Other direct benefits do come from payments to port agents and the small boats, called tenders that ferry passengers from the ship to land. For Belize, the combined benefits of both passenger expenditures and fees paid to the port and small boats in 2006 were just over $90 million. About 60% of all revenues from passengers in Belize comes from tours, some of the most popular in the Caribbean. Of that amount, $13 million went to 4 large operators and $2 million to 51 small operators.

Fort Street Tourism Village (FSTV), found at the dock where cruise tourists arrive, was paid for by an outside investor. The primary shops on the premises are the cruise lines' "preferred shops," including Diamonds International. It was built via a license agreement with the government guaranteeing exclusivity. In 2004, FSTV was purchased by Royal Caribbean and Diamonds International. By 2007, the head tax charged to finance cruise infrastructure was raised to $7.00 from $5.00, with the lion's share (over 55%) going to the FSTV.

Belize incorrectly outsourced the cost of its port, according to the Central Bank, and as a result had to assign too much of the head tax to outside owners of the port. This left the government unable to channel sufficient resources into planning and management of the tourism zone. In 2010, the new Belize Tourism Board leadership faced a serious problem with the cruise port zone outside the FSTV, which had become an unpleasant, overcrowded, and unsafe place for any passenger to visit. At that time, I was hired to write the Action Plan for the Belize Tourism Board[123] and observed the problem directly. All passengers who left the FSTV jumped immediately onto tour buses, which lined up by the dozen to take them away from the port. A noisy, smelly exhaust-filled zone, crowded with buses, was redesigned, not with money from cruise but with funds from a loan for improvement of the port, a negative on the country's balance sheet.[124] A loan from the Inter-American Development bank was used to help clean up the area, create parking for the buses, and set up a local vendor village for selling crafts, but this was only putting Belize further in debt, a required step in order to fix the chaos in port that was not part of the original arrangement with the cruise lines.

Shopping in Belize is not only limited; it is much less economically beneficial than tours. Although the shopping sector is worth $10 million, 85% of consumer dollars are spent in the Fort Street Tourism Village, the port area for cruise passengers, where 90% of the goods sold are imported. This is a very high percentage for any retail environment. Based on a look at the value of the imported goods, the Central Bank determined that the net value of all shopping in Belize was just under $5 million, or half of what was spent. In other words, 50% of all shopping expenditures was leaking out of the country before the new arrangements were made outside of the FSTV to attract more shopping for local goods.

The domination of just a few successful local tour operators has also caused concern, and Belize is still seeking to try to diversify benefits to more tour operators.[125] Tours to local attractions in the country remain local and competitive compared to its Caribbean neighbors. But efforts to diversify the benefits to more local owners were not successful, and 70% of the benefits were flowing to just a few wealthy business owners by 2010.[126] Efforts to correct his problem were high priority for the Belize Tourism Board in 2010, and the Action Plan my firm helped to write called for training 40 tour operators over a period of 2 years to manage a higher percentage of local tours.[127]

Solutions for the cruise port in Belize

Belize secures funds for the protection of its ecosystems via their head tax. This tax, which generated $4.6 million annually in 2006 (10% of the total revenue from the cruise industry) is divided three ways. Fifty-six percent is recycled back to the owners of FSTV (Royal Caribbean). Twenty-two percent goes to the Belize Tourism Board to manage tourism impacts and infrastructure, and 20% goes to the innovative Protected Area Conservation Trust (PACT), a value of nearly $1 million in 2006. PACT provides funds for the support of conservation and environmentally sound management of Belize's natural and cultural resources. It is also supported by a $3.75 fee paid by overnight visitors to the country, a price that needs to be updated.[128] This effort to invest in the maintenance of ecosystems for the nation is a model for the Caribbean.

In the end, the country was seeing business and employment benefits, but with insufficient funds to manage the costs of development, resulting in additional debt to the country. It would be extremely wise for all cruise destination countries to look at not only the current costs, but the costs to the nation in the long term: managing the ports, handling the volume of passengers, and providing the training and marketing support required for small cruise suppliers. However, what Belize policies have achieved is noteworthy. Few nations have successfully charged for the cost of maintaining their ecosystems as part of any cruise package.

Conclusion

Ports around the world need to be ready to negotiate. The cruise line industry brings many direct economic and employment benefits that can lift ailing ports from decay to decadence. But the effort to manage the volume of passengers, the air emissions, waste water, and solid waste systems is costly. The protection of local marine ecosystems cannot be left as an afterthought to be handled with grants for small NGOs. A new conservation tax approach is required to ensure that island nations in particular have the funds to protect their valuable natural assets in the long term.

In Venice, a Green Ports initiative was launched to protect this historic city and ensure that the industry does not undermine its historic value.[129] This multibillion-dollar effort seeks to showcase new technologies and solutions available to port cities in order to revolutionize how ports are managed by focusing on four main areas: air, water, soil, and energy. Environmental sustainability is one of Venice Port Authority's main objectives, and it must conserve the natural balance of the delicate city and its unique lagoon environment to achieve this goal.[130] This effort is an important indicator of costs that other ports may face.

Air particulate measurements were made in port, and the contribution of ships is 14–15% of the total air pollution of Venice when ships are docked. High-sulfur bunker fuel is the issue, and as of 2011 the shipping industry pledged to use lower-sulfur fuels in port. The port has already launched a €15 million effort to develop a portside electrical power system (called Cold Ironing) to enable ships to turn off their engines.[131] This is considered to be the best solution from the point of view of the industry and local regulators.

Polluted discharges have been the most contentious of all matters in the history of cruise management. Venice has chosen to invest €11 million for a treatment plant. It is the largest plant in Europe to use StormFilter technology, which protects waterways from polluted discharges[132] for an additional investment of €1 million.

Soil runoff and groundwater effects have been remediated in the vast port area for €24 million, and all of the banks have been made watertight to avoid contamination with an investment of €166 million. Additional investment was required to dredge the port to make it accessible to megaships. The waste created by the dredging also required proper disposal. This called for the erection of plants to treat and dispose of over 3 million cubic meters of sediment. The total cost is listed as €700 million.[133]

While this may be one of the most expensive port renovations in history, it will not be the last, and it will set the standard for other large ports that seek to host both cruise and shipping simultaneously. It does provide a reasonable indicator of cost to environmentally manage and protect historic ports. It is noteworthy that the effort to avoid contamination and manage the dredging waste was 90% of the cost in Venice.

Renovation of ports in Asia is already under way, and it is happening quickly. Historic ports such as Shanghai or Calcutta face similar issues. Hong Kong has already renovated its port, for $1 billion, at the old airport terminal for the city. One of the busiest ports in the world, Hong Kong has only begun to pick up cruise traffic.[134] Singapore just opened their new port to host megaships in late 2012.

In Alaska, best practice management guidelines for the Port of Juneau may be some of the most extensive efforts yet to create a public private management approach that includes the participation of citizens and small businesses in Juneau.[135] This initiative began in 1997 and is continually updated. A hotline is provided that encourages the community to give input and report issues. The guidelines program encourages participants to notify one another via their website system when guidelines are not followed. Each participating business must sign an agreement that certifies that they have read, understood, and intend to abide by the practices outlined in the document. The best practice management guidelines cover transportation, "flightseeing" trips, tours, cruise ships, docks, harbors, airports, restaurants, hospitality businesses, shoreside tour brokers, and downtown retail.

Air emissions standards for cruise ships are monitored via an Alaska emissions law for all marine vessels operating in Alaska.[136] The standard requires that visible emissions (opacity) from vessel smoke stacks be no greater than 20% opacity for 3 minutes while docked or anchored. The Department of Environmental Conservation (DEC) of Alaska does monitor this. From 2000 to 2009, there were 47 violations.[137] Citizen participation has been a vital part of this initiative as well. The DEC responds to public complaints, with a reported 54 complaints in the same period.[138] Marine vessel operators in port are encouraged to minimize idling to reduce the problem of emissions, but expensive Cold ironing solutions (where ships can plug into the electric grid in port) are difficult to deliver in small ports such as Juneau.

The cost of waste management is monitored, and the cost of management is shifted in part to tour operators who must remove all of their own waste from docking facilities. Shoreside merchants must secure their own waste through private contract. Even tour promotion fliers are discouraged to avoid needless waste.

Juneau represents a good model for smaller ports that are not able raise billions of dollars and offer Cold Ironing to lower emissions or offer waste treatment for the entire port system. The important lessons learned from the Belize case are that the entire cost of port management cannot be shifted to the private sector. There are public costs. Those costs must be built into the agreement with the cruise lines as new ports are built. Cities like Venice or perhaps wealthy cities in Asia may be able to underwrite the entire cost of the environmental management of cruise lines, to the tune of $500 million–$1 billion. Small ports in emerging countries will not have this capacity. Like Juneau, they will need to be ready to negotiate strategically and ensure that head taxes or other fees go directly to the cost of port management first, before allowing it to revert back to the investors, who are in many cases, related to the cruise lines themselves.

Economic leakage from cruise lines hovers in the 50% area. Each nation must review both its costs and benefits much more carefully before investing in cruise infrastructure. Costs must include not only port services but environmental management and ecosystem protection. The cost of preserving local heritage and culture must also be carefully evaluated via participatory planning indicators, with the cost of mitigation of the impacts of mass visitation on fragile monuments factored in. The outcomes of this advance work can help a cruise destination build in the protection of their valuable heritage and environment and prevent the inevitable prospect of devaluation of their home waters and heritage attractions.

If environmental degradation is already a problem, as it is in the Caribbean now, the cost of remediation and protection of coral reefs must be considered as part of the price of offering island attractions. Coral reefs are an ecological system that are essential to the valuable business that cruise lines offer to passengers; they cannot be allowed to collapse. Their maintenance must be a part of doing business. Just as Venice has remediated its historic port to protect it from the next wave of cruise development, the Caribbean must ensure that real funds are invested in the protection of its marine and coral reef assets. Without coral reefs, fisheries are imperiled, and the destruction produced by storm surges becomes much more threatening to the same villages that cruise lines visit.

The protection of heritage continues to be unaddressed in most destinations. The volume of passengers in ports like Venice has forced local people to abandon their city, and this is not the only city to face this problem. Citizen initiatives are needed to create guidelines for visitation to historic sites that limit numbers and densities.

Finally, the legal status of the cruise industry, which operates without oversight in most regions of the world, needs to change. Harmonizing the international legal standards of the IMO and UNCLOS is the highest priority. A new international enforcement program needs to be on the global agenda that will be funded to operate and oversee the new legal standards in cooperation with national governments. While voluntary reporting systems are a temporary fix, they are only scratching the surface and not providing real accountability. Cruise has grown quickly and has created a booming market based on larger-capacity ships that offer more entertainment per dollar than may have ever been achieved in history. But this business model has been developed based on a risky and ineffective international framework that has inadequate enforcement and requires urgent updating. The industry's efforts to meet current environmental standards should not be measured by how well they meet a highly flawed international legal system full of loopholes but by stepping up to the plate to create genuine measurable standards that the global community agrees will sustainably maintain the marine environment and fragile destinations upon which it depends.

Notes

1 Alaska Senate Passes Bill Relaxing Cruise Ship Wastewater Discharge Rules, http://ktoo.org/2013/02/19/alaska-senate-passes-bill-relaxing-cruise-ship-wastewater-discharge-rules/, Accessed October 2, 2016
2 Officials Outline Problems with Alaska Cruise Picture, http://travelagentcentral.com/home-based/conferences/officials-outline-problems-alaska-cruise-picture-20371?page=1, Accessed June 25, 2013
3 Cruise Lines to Drop Suite if Alaska Head Tax Cut, http://archive.boston.com/business/taxes/articles/2010/04/13/cruise_lines_to_drop_suit_if_alaska_head_tax_cut/, Accessed October 2, 2016
4 Alaska Gov Signs Bill Cutting Cruise Ship Head Tax, https://victoriaadvocate.com/news/2010/jun/24/bc-us-alaska-cruise-tax/, Accessed May 3, 2016
5 Cruise Lines International Association, 2010, *CLIA at 35, Steering a Sustainable Course*, Cruise Lines International Association, Washington, DC
6 2015 Cruise Industry Outlook, http://cruising.org/docs/default-source/research/2015-cruise-industry-outlook.pdf, Accessed March 29, 2016

7 Jennings, Hele and Kai Ulrik, 2016, *Cruise Tourism: What's below the Surface? Tourism Concern Research Briefing*, Tourism Concern, Croydon, UK
8 Cruise Forward Regulation: Frequently Asked Questions, http://cruiseforward.org/accountability/regulation/regulation-faqs, Accessed March 30, 2016
9 Klein, Ross A., March 1, 2012, *Prepared Statement, Oversight of the Cruise Ship Industry: Are Current Regulations Sufficient to Protect Passengers and the Environment*, U.S. Senate Committee on Commerce, Science and Transportation, Washington, DC, page 49
10 Ibid.
11 Copeland, Claudia, February 6, 2008, *Cruise Ship Pollution: Background, Laws and Regulations and Key Issues*, Congressional Research Service, Washington, DC, page CRS-25
12 Ibid., page 2
13 2015 Cruise Industry Outlook, CLIA,, http://cruising.org/docs/default-source/research/2015-cruise-industry-outlook.pdf, Accessed October 2, 2016
14 Ibid., Accessed October 2, 2016
15 Record Number of Cruise Passengers Visit Stockholm, http://portsofstockholm.com/about-us/news/2015/record-number-of-cruise-passengers-visit-stockholm/, Accessed October 2, 2016
16 Cruise Baltic Market Review 2016, https://cruisebaltic.com/media/94490/cruise-baltic-market-review-2000-2016.pdf, Accessed March 28, 2015
17 Andersen, Jesper H., Samuli Korpinen, and Maria Laamanen, 2010, *Baltic Marine Environment Protection Commission, Ecosystem Health of the Baltic Sea*, HELCOM Initial Holistic Assessment, Helsinki Commission, Helsinki, Finland
18 Johannessen, D.I., J.S. Macdonald, K.A. Harris and P.S. Ross, 2007, *Marine Environmental Quality in the Pacific North Coast Integrated Management Area (PNCIMA)*, British Columbia, Canada; A summary of contaminant sources, types and risks, Canadian Technical Report of Fisheries and Aquatic Sciences 2716, Fisheries and Ocean Canada, Sidney, BC, Canada
19 Prevention of Pollution by Sewage from Ships, http://imo.org/en/OurWork/Environment/PollutionPrevention/Sewage/Pages/Default.aspx, Accessed March 29, 2016
20 Baltic Marine Environmental Protection Commission, 2013, *HELCOM Ministerial Declaration, Interim Guidance on Technical and Operational Aspects of Sewage Delivery to Port Reception Facilities*, Helsinki, Finland
21 Prevention of Pollution by Sewage from Ships, http://imo.org/en/OurWork/Environment/PollutionPrevention/Sewage/Pages/Default.aspx, Accessed March 29, 2016
22 U.S. Environmental Protection Agency (EPA), 2013, *Final Issuance of National Pollutant Discharge Elimination System (NPDES) Vessel General Permit (VGP) for Discharges Incidental to the Normal Operation of Vessels*, Fact Sheet, EPA, Washington, DC Section 7.1, pages 159–169
23 Carić, Hrvoje and Peter Mackelworth, 2014, Cruise tourism environmental impacts: The perspective from the Adriatic Sea, *Ocean and Coastal Management*, Vol. 102, 350–363, page 352
24 Ibid.
25 Ibid.
26 Ablaey, Ramil, Emily Bonnell, Lingfang Chen, Antonio Modestini, Tadashi Soga, and Sanchit Talwar, April 2011, *Strategy Project: Cruise Line Industry*, HUILT International Business School Strategy Team Paper
27 Royal Caribbean's Splendour of the Seas Will Feature an iPad in Every Cabin, http://appleinsider.com/articles/11/11/28/royal_caribbeans_splendour_of_the_seas_will_feature_an_ipad_in_every_cabin, Accessed October 2, 2016
28 Business Research & Economic Advisors, 2010, *Executive Summary, The Contribution of the North American Cruise Industry to the U.S. Economy in 2010*, Cruise Line Industry Association, Miami, FL
29 Duffy, Christine and Jim Berra, 2013, *2013 Cruise Industry*, Cruise Line Industry Association, Miami, FL

30 Ablaey et al., *Strategy Project: Cruise Line Industry*, page 7
31 About CLIA, http://cruising.org/about-the-industry/about-clia, Accessed October 2, 2016
32 Cruise Market Watch, Growth, http://cruisemarketwatch.com/growth/, Accessed October 2, 2016
33 Regal Princess – Now Arrived!, http://princess.com/learn/ships/gp/, Accessed May 3, 2016
34 Pogioli, Sylvia, In Venice, huge cruise ships bring tourists and complaints, NPR, July 15, http://npr.org/blogs/parallels/2013/07/15/202347080/In-Venice-Huge-Cruise-Ships-Bring-Tourists-And-Complaints, Accessed October 2, 2016
35 Frost & Sullivan, January 2013, *Shaping the Future of Travel in the Asia Pacific: The Big Four Travel Effects*, Amadeus, Madrid, Spain
36 *China Business Newsweekly*, December 1, 2015, China Cruise ship industry expanding. NewsRx LLC
37 Marriott International, 2012, *Annual Report*, Marriott International, Bethesda, VA
38 Carnival Corporation & PLC, 2012, *Annual Report*, Carnival Corporation & PLC, Miami, FL
39 Ablaey et al., *Strategy Project*
40 Costa Concordia: Not a Disaster for the Environment, http://thisisitaly-panorama.com/top-stories/costa-concordia-not-a-disaster-for-the-environment/, Accessed July 19, 2013
41 Carnival Corporation, 2012, *Annual Report*, Note 7, page 20
42 Ibid., non-financial instruments that are measured at fair value, page 27. Note the following calculations are included: the principal assumptions used in the cash flow analysis related to forecasting future operating results including net revenue yields; net cruise costs including fuel prices; capacity changes, including the expected deployment of vessels into, or out of, Costa; WACC for comparable publicly traded companies, adjusted for the risk attributable to the geographic regions in which Costa operates and terminal values, which are all considered level-3 inputs
43 Busted Toilets, Hot Rooms, Headaches, after Fire Strands Cruise Ship in Gulf, http://cnn.com/2013/02/11/travel/cruise-ship-fire, Accessed July 11, 2013
44 Environmental Group Sues EPA over Cruise Ship Pollution, http://seattlepi.com/local/article/Environmental-group-sues-EPA-over-cruise-ship-1236917.php, Accessed May 3, 2016
45 Copeland, *Cruise Ship Pollution*, page 2
46 S 1359 (113th): Clean Cruise Ship Act of 2013, https://govtrack.us/congress/bills/113/s1359, Accessed May 3, 2016
47 Frequently Asked Questions – Cruise Ship Wastewater Discharge Regulation and HB 80, http://dec.state.ak.us/water/cruise_ships/faq.htm, Accessed June 26, 2013
48 Walker, Jim, 2013, Princess Cruises fined for dumping 66,000 gallons of chlorinated water into Glacier Bay in Alaska, *Cruise Law News*, January 29, http://cruiselawnews.com/2013/01/articles/pollution-1/princess-cruises-fined-for-dumping-66000-gallons-of-chlorinated-water-into-glacier-bay-in-alaska/, Accessed October 2, 2016
49 Ruling Expected soon on Wastewater Permit, http://msiak.net/aca/266/03.html, Accessed March 30, 2016
50 Sweeting, James, 2011, *Presentation in Harvard Extension Classroom*, Environmental Management of International Tourism Development, Cambridge, MA
51 Sustainability Accounting Standards Board, 2014, *Cruise Lines Research Brief*, Sustainability Accounting Standards Board, San Francisco, CA, page 22
52 Terry, William C., 2011, Geographic limits to global labor market flexibility: The human resources paradox of the cruise industry, *Geoforum*, Vol. 42, 660–670
53 Ibid.
54 Ibid., page 663

55 Politakis, George P., 2013, Bringing the human element to the forefront: The ILO's Maritime Labour Convention, 2006 ready to sail, *Aegean Institute of the Law of the Sea and Maritime Law*, Vol. 2, 37–51
56 Sustainability Accounting Standards Board, *Cruise Lines Research Brief*, page 3
57 Politakis, Bringing the human element to the forefront, 37–51
58 Ibid.
59 Sustainability Accounting Standards Board, *Cruise Lines Research Brief*, page 23
60 Ibid., page 23
61 Copeland, *Cruise Ship Pollution*
62 U.S. EPA, 2008, *Survey Questionnaire to Determine the Effectiveness, Costs, and Impacts of Sewage and Graywater Treatment Devices for Large Cruise Ships Operating in Alaska*, Washington, DC
63 Op. cit., Copeland, Claudia
64 Cruise Ship Registry, Flag State Control, Flag of Convenience, http://shipcruise.org/cruise-ship-registry-flags-of-convenience-flag-state-control/, Accessed October 2, 2016
65 Ritucci-Chinni, Alexandra, 2009, The solution to international cruise ship pollution: how harmonizing the international legal regime can help save the seas, *Dartmouth Law Journal*, Vol. VII, 1
66 Ibid.
67 Ibid.
68 Carnival Cruise Line, http://en.wikipedia.org/wiki/Carnival_Cruise_Lines, Accessed June 27, 2013
69 It is very important to note that these provisions also keep cruise lines from being subject to labor laws of the countries they visit and even in the countries where they have their corporate headquarters. This has led to excellent investigations into the problems of cruise line labor issues, not covered here but of equal importance, with good coverage in *Overbooked* by Claudia Becker and *The Final Call* by Leo Hickman
70 Ibid., page 45
71 Ocean Vessels and Large Ships, EPA, https://www3.epa.gov/otaq/oceanvessels.htm, Accessed October 2, 2016
72 New Emission Control Area, http://2wglobal.com/news-and-insights/articles/features/New-emission-control-area/#.V_GISOOa3RY, Accessed October 2, 2016
73 Cruise Forward Frequently Asked Questions, http://cruiseforward.org/accountability/regulation/regulation-faqs, Accessed March 30, 2016
74 Ritucci-Chinni, The solution to international cruise ship pollution
75 Ibid.
76 Copeland, *Cruise Ship Pollution*, page 11
77 Cruise Lines International Association, *CLIA at 35*, page 20
78 U.S. EPA, *Final Issuance of National Pollutant Discharge Elimination System*
79 Ibid.
80 The Ocean Conservancy, May 2002, Cruise Control, *A Report on How Cruise Ships Affect the Marine Environment*
81 Sweeting, James E.N. and Scott Wayne, *A Shifting Tide, Environmental Challenges and Cruise Industry Responses*, Center for Environmental Leadership in Business, Conservation International, Washington, DC, page 17
82 The Ocean Conservancy, Cruise Control, page 8
83 Cruise Lines International Association, *CLIA at 35*, page 25
84 http://cfpub.epa.gov/npdes/vessels/background.cfm, Accessed June 27, 2013
85 Cruise Lines International Association, *CLIA at 35*, page 26
86 Copeland, *Cruise Ship Pollution*, page 5
87 Acevedo, Yvette, 2011, *Royal Caribbean's Solid Waste Management Practices in Labadee, Haiti*, Environmental Management of International Tourism Development, Department of Sustainability and Environmental Management, Harvard Extension School, Cambridge MA
88 Royal Caribbean Cruises, Ltd., n.d. *2010 Stewardship Report*, Royal Caribbean Cruises, Ltd., http://viewer.zmags.com/publication/49a26be2#/49a26be2/1, Accessed November 20, 2011

89 Corti, Alberto, 2013, *Differentiating the Methods of Waste Treatment in the Wider Caribbean Region, Introducing a Comprehensive Data-Collecting Model to Promote Waste-to-Energy Practices*, Master's Thesis, Uppsala University, Department of Earth Sciences, Uppsala, Sweden
90 Carriage of Chemicals by Ship, International Maritime, Organization, http://imo.org/en/OurWork/Environment/PollutionPrevention/ChemicalPollution/Pages/Default.aspx, Accessed October 2, 2016
91 Copeland, *Cruise Ship Pollution*, page 13
92 Sweeting and Wayne, *A Shifting Tide*, page 13
93 United States Coast Guard and Environmental Protection Agency, June 2011, *MARPOL Annex VI Pollution Prevention Requirements*
94 Sulfur Content Exemption For Cruise Company, Enviro.BLR.com, http://enviro.blr.com/environmental-news/air/motor-vehicle-fuels-and-fuel-additives/Sulfur-content-exemption-for-cruise-ship-company/, Accessed April 1, 2016
95 Brooks, Philip A., January 15, 2015, *Memorandum, EPA Penalty Policy for Violations by Ships of the Sulfur in Fuel Standard and Related Provisions, Air Enforcement Division, Office of Civil Enforcement*, U.S. EPA, Washington, DC
96 Sustainability Accounting Standards Board, *Cruise Lines Research Brief*, page 10
97 Helfre, Jean-Forent and Pedro Andre Couto Boot, July 2013, *Emission Reduction in the Shipping Industry: Regulations, Exposure and Solutions*, Sustainalytics, Amsterdam, Netherlands
98 Ibid.
99 Carnival Corporation, Leading Responsibly, http://carnivalcorp.com/phoenix.zhtml?c=140690&p=irol-recognition, Accessed October 2, 2016
100 Bonilla-Priego, Jesus, Xavier Font, and M. del Rosario Pacheco-Olivares, 2014, *Tourism Management*, Vol. 44, 149–160
101 Grosbois, Danuta de, 2015, Corporate social responsibility reporting in the cruise tourism industry: a performance evaluation using a new institutional theory based model, *Journal of Sustainable Tourism*, Vol. 24, No. 2, 1–15, DOI:10.1080/09669582.2015.1076827
102 http://cruisecritic.com/ports/newport.cfm?ID=17, Accessed July 17, 2013
103 Sustainable-Heritage-Tourism-Model-For-Falmouth-Jamaica-Towards-Cultural-Heritage-Preservation, http://scribd.com/doc/51277667/Sustainable-Heritage-Tourism-Model-For-Falmouth-Jamaica-Towards-Cultural-Heritage-Preservation, Accessed October 2, 2016
104 Legorano, Giovanni, 2013, Venice looks to calm cruise ship waves, *The Wall Street Journal*, June 21, http://online.wsj.com/article/SB10001424127887323300004578557293160084744.html, Accessed October 2, 2016
105 UNWTO, 2004, *Indicators for Sustainable Development of Tourism Destinations, a Guidebook*, Madrid, Spain, page 65
106 Grosbois, Corporate social responsibility reporting
107 Business Research and Economic Advisors, 2012, *Economic Contribution of Cruise Tourism to the Destination Economies, Volume I Aggregate Analysis*, Florida-Caribbean Cruise Association, Miami, FL
108 Ibid., page 2
109 Ibid., page 53
110 Ibid.
111 Center on Ecotourism and Sustainable Development, November 2006, *Cruise Tourism in Belize, Perceptions of Economic, Social and Environmental Impact and January 2007*, Cruise ship tourism in Honduras and Costa Rica, Center for Responsible Tourism, Washington, DC
112 Center on Ecotourism and Sustainable Development, January 2007, *Cruise Ship Tourism in Honduras and Costa Rica, Center for Responsible Tourism*, Washington, DC, page 18
113 Ibid.

114 Harvey, Fiona, 2012, Caribbean coral reefs face collapse, *The Guardian*, September 9, http://guardian.co.uk/environment/2012/sep/10/caribbean-coral-reefs-collapse-environment, Accessed June 27, 2013
115 McCarthy, Pamela M., 2008, *Masters Project for Environmental Management Degree*, Nicholas School of the Environment and Earth Sciences, Duke University, NC, no page numbers
116 Beach, Ivy, January 2014, *Carnival Cruise Incorporated and the $30 Million Sustainable Procurement Strategy in St. Vincent and the Grenadines*, Harvard Extension course International Development of Sustainable Economies, ENVR-175, Cambridge, MA
117 Ibid., page 34
118 Business Research and Economic Advisors, *Economic Contribution of Cruise Tourism*
119 Burke, Lauretta and Jonathan Maidens, 2004, *Reefs at Risk*, World Resources Institute, Washington, DC, page 43
120 Novelo et al., 2007, *Assessing the Direct Economic Impact of Cruise Tourism on the Belizean Economy*, Central Bank of Belize, Belize City, Belize, page 4
121 Ibid., page 5
122 Ibid., page 8
123 EplerWood International, 2010, *Belize Tourism Board Action Plan, Destination Planning*, Belize Tourism Board, Belize City, Belize, Chapter 1, page 4
124 Ibid.
125 Ibid.
126 Ibid., page 7
127 Ibid., page 50
128 About PACT, http://pactbelize.org/AboutPACT.aspx, Accessed October 2, 2016; Drumm Consulting, no date, Sustainable Finance Strategy and Plan for the Belize Protected Area System, https://facebook.com/DrummSustainableTourismConsulting, Accessed November 30, 2016
129 Green Port, https://port.venice.it/files/page/pdvbrochuregreenport2_5.pdf, Accessed October 2, 2016
130 Green Port, https://port.venice.it/en/node/164, Accessed October 2, 2016
131 Green Port, page 5
132 SW Treatment for Ports, web page, Accessed July 2013
133 Green Port, https://port.venice.it/en/node/164 page 15, Accessed October 2, 2016
134 Wassener, Bettina, 2013, Hong Kong's old airport reopens as cruise ship terminal, *New York Times*, June 11, http://nytimes.com/2013/06/12/business/global/hong-kongs-old-airport-reopens-as-a-cruise-ship-terminal.html?_r=0, Accessed July 11, 2013
135 Tourism Best Management Practices, 2013, *2013 Best Management Practices*, Juneau, Alaska
136 Division of Water, Alaska, Commercial Passenger Vessel Environmental Compliance Program, http://dec.alaska.gov/water/cruise_ships/pdfs/2009_Cruiseship_Air_Summary.pdf, Accessed October 2, 2016
137 Ibid., page 2
138 Ibid.

8

DESTINATIONS

The heart of tourism sustainability

Introduction

Deep in the Ecuadorean rain forest, indigenous people hold the keys to preserving some of the most important biodiverse rain forest reserves on the planet. The Huaorani Ecolodge, part of a tourism program established 20 years ago, is a landmark of cooperation between the ecotourism industry and local people, which helps the Huaorani people to preserve over 135,000 hectares of pristine rain forest (333,000 acres),[1] three times the size of Costa Rica's famed Corcovado National Park. Only 50 years ago, this was an uncharted territory with uncontacted Huaorani people who lived amidst mega biodiversity without outside interference. Now it is the home of one of the most ambitious ecotourism projects in the world, which won the 2015 World Legacy Award from National Geographic for Engaging Communities.[2] The lodge, which is owned by the Huaorani, is managed together with the tour operator Tropic Journeys and their owner Jascivan Carvalho who took over from founder Andy Drumm in 2010. Carvalho has moved the business from a small enterprise to a medium-sized one with good business fundamentals and has helped the Huaorani to take full ownership of their own tourism program, while using it as a means to protect their rain forest livelihood and defend their territory.

This accomplishment is remarkable on a world scale. Within Ecuador, it is of even more significance. This award celebrates the longevity of ecotourism enterprise in one of the most remote communities on earth and the empowerment of local people who are living in the rain forest to preserve their territory on their own terms. There is so much to be celebrated here, but ironically, despite the significance of the award, the threats to this region have dramatically increased, and there is great distress in Huaorani territory.

While ecotourism is a tested tool to protect this invaluable destination, it is nonetheless being opened up by Ecuador's President Rafael Correa to potential oil exploration. The Huaorani are located in a strategic location for the nation. Their territory once included what is now Yasuni National Park, which encompasses over 2.5 million acres,[3] an area larger than Yellowstone National Park. Beneath the vast jungles of the Oriente (East) of Ecuador are hefty oil reserves that have long attracted prospectors. The Oriente has been the target of numerous periods of oil exploration, several of which subsequently became the subjects of international lawsuits for the contamination of ecosystem and endangerment of local health. In 2007, President Correa cagily created an ambitious plan to persuade the world to ante up $3.6 billion in revenue to avoid greenhouse gas emissions from the reserves and compensate the nation for keeping its oil in the ground.[4] Few in Ecuador believe that Correa expected this initiative to succeed, but what was achieved is a cover story for destroying this region in return for oil revenue.

In 2013 the proposal to protect the region was declared dead. The world "stood up" to Ecuador, according to Correa, and "he blamed 'the great hypocrisy' of nations who emit most of the world's greenhouse gases."[5] Foreign exchange is the fuel that stokes the fires of Ecuador's prosperity, and oil and mining attract outside investment on a scale well beyond tourism. It is well-known that Ecuador depends on its petroleum reserves for more than half of the country's reported earnings and 40% of public sector revenues.[6] Oil pays the bills and provides what Correa believes Ecuador needs in order to catapult it from a poor antiquated nation to a modernized, efficient economy.

Meanwhile in the rain forest, Huaorani Federation leader Moi Enomenga, the original cofounder of the Huaorani tourism program, is facing great pressure and tough choices. In my return visit after 20 years to Huaorani territory in 2015, I found him in a very tight spot. He would choose protecting his reserve and investing in more ecotourism long before he would ever permit oil to be explored there, and he has said that he would rather kill those that enter Huaorani territory than allow them to come in without permission.[7] But the oil company's paid emissaries were arriving in Huaorani territory daily with promises to buy trucks for local villagers, and Correa's government is promising free schools. How does Moi make a choice between education and his ancestral way of life?[8] This is a devil's bargain. Oil development brings substantial risks, placing the Huaorani permanently on guard and under threat from oil leaks and the contamination of their ecosystem. Indigenous people, like Moi, struggle to preserve their sense of well-being in connection to the land and their traditional livelihoods as society rapidly modernizes around them. Their commitment to biodiversity conservation and rightful longing for better access to modern communications, schooling and health facilities, make them both the perfect allies of destination preservation and conflicted victims of political schemes to sell off their homelands to the highest bidders in return for some improved social and health services.

In 2015, a surprising ally emerged who grabbed the world's attention and spoke directly to the question of how to sustainably develop our planet, right as I was visiting Moi. Pope Francis released his groundbreaking call for protection of our planet in the encyclical *Laudato Si* (On Care for Our Common Home) on June 18,

2015 in Ecuador. This document states that short-term economic gains are the root cause of the global environmental crisis now taking place.[9] It points out that modernization is causing high impacts on biodiversity and that caring for ecosystems "demands far-sightedness, since no one looking for quick and easy profit is truly interested in preservation."[10] But sadly in May 2016, Tropic was forced to close the Huaorani Ecolodge due to seismic testing in the area by the Chinese oil company Sinopec.[11]

This drama that involves presidents and popes puts Huaorani Territory and Moi Enomenga at the heart of the question of how to create a concrete formula for judging the long-term, financial value of tourism destinations. Could ecotourism preserve livelihoods and protect natural and social capital to the degree that governments would arrest oil development? In the case of Ecuador, the Correa administration has aggressively invested in the opposite strategy without wasting any time. "We will exploit oil, as all countries in the world do," Correa stated in 2013. "We cannot be beggars sitting on a sack of gold."[12]

In my own analysis, after years of working as an ecotourism consultant and NGO leader in the Amazon, Serengeti, and many other of the world's most biodiverse regions, ecotourism provides a palpably more gentle and less brutal form of development, compared to the uphill battles local people face against large industrial exploitation, represented by extractive oil, mining, and logging. Ecotourism also generally offers a diversified means to spread economic development to rural and traditional peoples living far from major economic centers. In Ecuador, it has generated profits at the every level of society ranging from tour operators in Quito, to small hotels and historic inns in the Andes, to local agricultural cooperatives that produce crops and handicrafts for the tourism economy, to community-based ecotourism destinations in the jungles of the Oriente."[13] These benefits must be captured via an alternative valuation system, as discussed in Chapter 3, which places a higher value on local livelihoods and on natural and social capital to help countries like Ecuador to appropriately factor in both the ecosystems and the traditional cultures they have protected so effectively until now. Science-based academic methodologies show that ecotourism is the best candidate to achieve species conservation outcomes if threats from extractive industries are halted,[14] and it is one of the most promising approaches to improving land protection if tourism management strategies are both funded and implemented.[15] This research sets the stage for using tourism revenues as an important tool to help value our planet's common heritage.

Destination management: key issues

In 1990, leaders in the original movement to establish ecotourism as a tool for sustainable development made it clear that tourism must contribute to conserving ecosystems and the well-being of local people.[16] As world populations begin to close in on the earth's most valued resources, these proven principles must be applied on a much larger scale. This chapter asks readers to reassess how nations derive and allocate their income from tourism and how tourism impacts are measured in future to ensure that natural, cultural, and social capital are fully protected.

The threats of overdeveloping vulnerable tourism destinations, once accepted as a natural part of tourism life cycles, must be anticipated by local citizens, local authorities, and national governments, and steps must be taken to protect valuable landscapes, ecosystems, cultures, and local monuments and attractions before the value of the destination is compromised, often for generations. This chapter illustrates the long-term negative impacts of overdevelopment (see the case of Jamaica), why the threats are accelerating, what types of accountability must now be introduced, and the solutions available, and emphasizes the new methodologies required to achieve this. Cases throughout the book demonstrate that in fact the entities with the most responsibility for protecting natural, cultural, and social capital are receiving inadequate financial and technical resources and have few regulatory or legislative mandates for achieving destination protection. At present there are (1) a lack of governmental regulatory power in almost all emerging economy nations to limit and manage the growth of tourism on landscapes outside of protected areas, (2) little relevant revenue generation capacity to finance sustainable infrastructure to service the growth of the tourism economy, and (3) a lack of technical training in both destination accounting and management at the local authority level, where it is most needed. At present, policy makers rely on voluntary action often fostered by nongovernmental and intergovernmental bodies without the budgets or legal mandate to achieve genuine policy change and measurable outcomes.[17]

This chapter suggests that local authorities hold the key to the future of tourism destinations and planning. Local authorities need guidance to ensure that they are managing budgets in ways that account for tourism's costs, and they need legal authority, in the form of land-use planning, to undertake long-term planning of tourism destinations that protects natural, cultural, and social capital. The chapter discusses the many forms of land-use planning as a primary solution for protecting destinations in future.

This chapter also stresses that scholars and destination managers must develop a new method for placing value on the earth's most valued human and biological heritage, which can leverage its sustainable development potential through tourism. World Heritage designations, as defined by UNESCO, have already been placed on many of these most threatened destinations, but this does not guarantee their safety. The Valley of the Kings in Egypt, the Great Barrier Reef in Australia, the Galapagos Islands in Ecuador, and the Taj Mahal of India all face threats of inappropriate development and unsustainable tourism development despite their renown and World Heritage status. This chapter seeks to articulate a new vision, which asks governments to prevent exploitation and to identify mechanisms to finance protection as a high priority to preserve the value of tourism destinations in the long term, using lessons learned from the protected area community, which is increasingly improving mechanisms for tourists to pay for protecting ecosystems.

According to the most recent global analysis, ecosystems are rapidly declining globally, and there is an immediate need for more accurate tools to "comprehensively measure the contributions of ecosystems to human well-being."[18] This chapter proposes that key cultural and natural tourism assets can be valued by governments using "balance sheets" that base the value of tourism destinations not on their gross revenue-generating capacity or service to human well-being but rather on the net revenues accrued from tourism development to governments calculated in terms of revenues minus liabilities.[19] Such net balance sheets will allow local authorities, working together with the private sector, to appropriately value the precise revenues they require and assess the revenue generation mechanisms they will need to meet those needs.

New policies are urgently needed to transform voluntary frameworks into legislated, financed systems of national development. These policy frameworks must be supported by a global, measurable system of indicators that capture priority data for local destinations to measure the level of protection they are achieving. Voluntary destination planning mechanisms without measurable indicators need to be replaced by GIS-powered systems that capture data on vital ecosystem services, such as the availability of water in watersheds and the health of coral reefs, over the long term. Researchers testing a GIS data-gathering approach on a well-known tourism case study area, Kangaroo Island in Australia, found that GIS systems effectively bring tourism planning into the 21st century and plug into real decision-making systems.[20]

Critical new policies and measurement systems for protecting global assets may seem a tall order, but without them citizens will have no basis to measure and predict the changes on their landscape and no way to call their public policy leaders to account. This chapter seeks to lay out precisely what systems need to be in place.

Destination management fundamentals

The concept of a destination is very broad and must be defined according to the needs of the locale that is seeking to manage it. The UNWTO defines destinations as "physical spaces in which a visitor spends a least one overnight."[21] Destinations are largely managed by entities that seek to promote them, often known as Destination Management Organizations (DMOs) – which can be multinational, national, state/provincial, and local. These organizations are generally funded through a combination of public and public–private sources and use tourism revenues to market destinations and maintain their competitive marketing advantage in an increasingly crowded tourism marketplace. Although there is no global study to document exactly what percentage of tax revenues goes toward marketing tourism, what is known is based on a survey of 51 countries in 2009, which revealed that total funding for marketing, for those participating in the study, reached $2.7 billion. An estimate of the total global spend by Oxford Economics for destination marketing was $4.2 billion in 2009. The average national budget was $53 million. The range of budgets was from $0.7 million for Madagascar to $173 million for Mexico.[22] Such revenues for marketing, without significant allocation for product "maintenance," are contributing to the overheated growth of tourism worldwide. The reallocation of such funds is one of the most important dialogs that should transpire to ensure that global destinations are not focused only on marketing but rather on protecting their destinations from overuse and a lack of planning.

There is a clear justification for the idea of reinvesting a portion of tourism revenues into local destination management. A destination is more than a place that attracts overnight visitors. It also is a place that has local residents, and only some of these residents are part of the tourism industry. To help residents and local governments to prepare for tourism growth and pay for the costs of managing this growth, every destination needs to prepare and develop participatory plans, as will be demonstrated in the following research. This chapter recommends that all local authorities undertake a strategic plan that defines a mission, strategic objectives, and desired results for the community, region, and nation, as well as a master plan that is regulated to manage the use of land under the authority of their government. But the process cannot stop there.

All destination management requires the establishment of indicators: measures of change over time that can be tracked using consistent systems that record how the destination is being transformed. Monitoring must be long-term and involve significant public participation. Measuring how tourism is transforming landscapes must be done with citizen agreement and participation. In addition, governmental managers in charge of sustainable destinations must review their indicator measurements on a regular basis with civil society and local citizens in order to discuss the changes and new requirements for protecting natural and cultural assets.

Destinations are subject to a wide variety of economic, social, and environmental changes with many options for indicators. The best volume on indicators to track destination sustainability is a large reference book published by the UNWTO, *Indicators of Sustainable Development for Tourism Destinations, A Guidebook*,[23] which provides an extensive number of scenarios for managing tourism destinations change over time. This encyclopedic volume serves as the best reference for all destination managers considering establishing a program of monitoring with indicators. It also offers standard systems for public consultation and participation, along with suggestions of how to establish a set of indicators that are suitable to a specific destination. An indicator program is crucial, but it cannot be useful without a plan for managing sustainable tourism that has a long-term scope. As local citizens see their small towns become the subject of tourist consumption, much needs to be done to put planning systems in place.

Destination life cycles

Destinations around the world are found in highly different stages of development. Humans naturally love to discover new destinations, especially when they are in the bohemian stage, and voice crushed psychological disappointment as they change. Oddly, we travelers see no relationship between our arrival and the inevitable changes that come over time. We love being there first. This is a very deep part of the human psyche, likely related to our early migratory behavior that has allowed the human species to spread relatively quickly across the planet, as documented in the Pulitzer Prize–winning book, *The Sixth Extinction*.[24]

The classic article by Stanley Plog, "Why Destination Areas Rise and Fall,"[25] brilliantly captures the human desire to explore places and then leave them behind once they lose their distinctive character. Communicating what we discover, sharing it with others, allows destinations to grow, sprawl, and ultimately become overcrowded. Most of us then seek to move on. The fact that TripAdvisor is the largest travel website in the world speaks volumes about the human need to discover places and share our opinions with others.[26] In 2014, this website received 225 million online reviews annually. The deep human instinct to roam and tell about it is without question a massive and growing social phenomenon, and it may be a human instinct, given our migratory nature.

Every year the best destinations are selected, and every year we see travelers rushing to explore them. Travel and Leisure named Chengdu, China – the panda capital – with a new "72 hour no-visa policy and a packed lineup of hotel openings" as one of the top new destinations for 2014." Having discussed China tourism policies at Harvard and in the first and third chapter in this book, we know that the potential for bringing sustainability to China is great and that China itself has stated that more planning is required, but the process has barely begun as of 2016. With global tourism on the rise and the rise of new destinations spreading more rapidly than ever due to social media, there are legitimate concerns about the future of a destination when it is growing quickly. It is very useful to understand the phases of a destination life cycle, based on a free market approach that assumes little government intervention, planning, or effort to manage growth.

In the 1980s, R.W. Butler sought to characterize these main stages of destination life cycles in his well recognized tourism modeling research.[27] To make the process simpler and easier to understand, the life cycle information will be narrowed in this book to four phases using my own version and experience of this phenomenon. Figure 8.1 presents a standard graphic of this cycle.

- **Emerging Destinations** are being discovered, and they do not have large investment in local tourism infrastructure. There is a high level of informality in tourism offerings and a low number of visitors who are discovering the area compared with the number of residents. Local residents are generally welcoming to the visitors and still living a life that is similar to before tourism arrived.
- **Partially Developed Destinations** have been discovered. There is an increasing amount of private investment in local tourism infrastructure, a medium number of visitors, and land speculation by outsiders is beginning to take off. There is generally a lack of tourism regulations, and residents are heavily influenced by the free market atmosphere, all seeking to get a piece of the pie if possible. Some residents are beginning to have their doubts about the process of tourism development.
- **Highly Developed Destinations** are becoming crowded with travelers, particularly in high seasons. There is a heavy amount of private tourism

infrastructure, sprawl, and traffic, with increasing competition for land, and water may even become less accessible to local residents. Energy demands are skyrocketing, often causing blackouts, and there is an inflation of prices on goods, services, and land, which can push local residents out of the areas. The inner tourism core begins to be unattractive to higher-paying visitors, forcing developers to create luxury hotel developments outside of the crowded areas. Innumerable small shops, often with the same goods, can sprawl for blocks in the crowded areas. Essential services such as waste management are ineffective or absent, and residents are now living a life that is highly influenced by tourism development with growing disenchantment.

- **Overdeveloped Destinations** often have little open space, and the main tourism areas are characterized by large-scale sprawl and poorly planned use of the waterfront landscape. Beach erosion often becomes a serious problem. Land and room prices are declining in the original core areas of the tourism offer, and the luxury sites move farther and farther away from the core destination to outlying areas. Environmental contamination is common, and former local residents have moved away or are increasingly resentful of tourists. Local cultures are trivialized or disappear. Prostitution and drug use become more common, and immigrants live in poorly served slum areas, without public services and with poor education prospects and few opportunities to improve their lives.

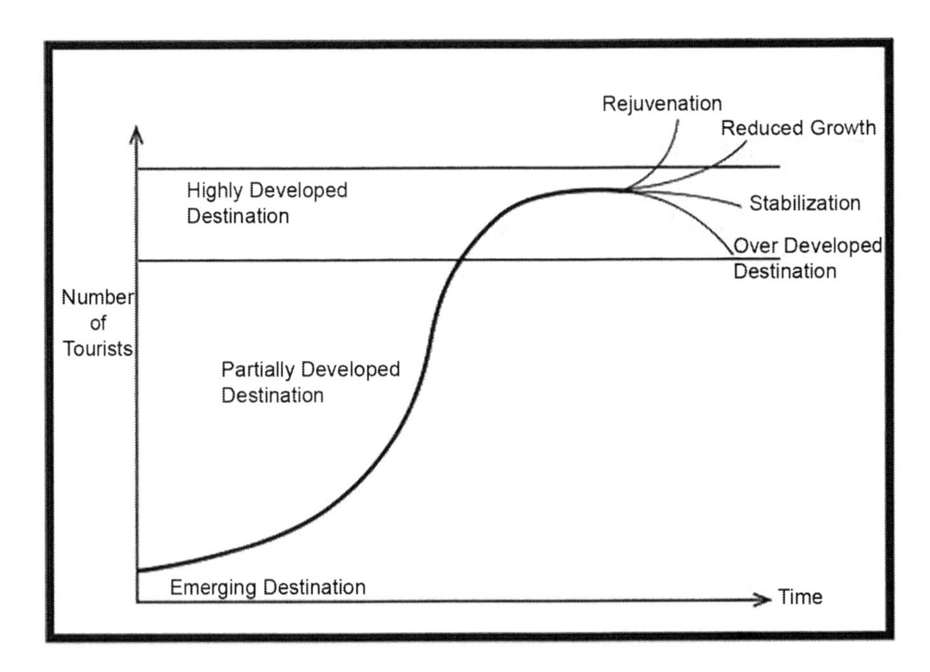

FIGURE 8.1 The tourism life cycle adapted from R.W. Butler

The tourism life cycle review is a tool to help experts to develop strategies for a destination to avoid its falling into decline. As depicted in Figure 8.1, there is a "critical range" when destinations are either stabilizing or following a pathway of decline or rejuvenation. It has been tested, with some success, as a predictive model that can guide innovation and the diversification of tourism product development to ensure that destinations do not decline.[28]

Excellent case study work has also been offered on how island destinations can anticipate the problems of growth, stagnation, and decline.[29] The model has also been explored by researchers concerned about the growth of tourism in the 21st century and its impacts on the environment with a discussion of why there has been a lack of policy responses to the potential of destination decline worldwide.[30] All of these studies indicate that interventions are needed to prevent destination decline. Yet there are few investigations of what precise policies can prevent decline, except for adjustments in marketing and product selection. Regulatory and financing solutions are barely discussed, such as how land-use planning and investment in new green infrastructure may help to arrest these problems.

A well planned destination can hold value much longer if it is maintained. This is a basic principle in business but seems to have eluded the tourism destination planning community. Without investment in products and maintenance, it is certain that a business will fail. Destinations are no different and require investment to be maintained. How these investments are made is of significant importance, and the formula to achieve this must be considered in depth.

In the future, new systems of valuation of tourism's assets will be required and can provide the basis for investment in green destinations. This will help to attract financing through green bonds, which are frequently offered by municipalities, bilateral trade and development agencies, private banks, and multilateral banks to support investment in renewable energy, energy efficiency, sustainable waste management, sustainable land use, biodiversity conservation, clean transportation, and sustainable water management.[31] Trust funds might be also established, such as the Protected Area Conservation Trust in Belize, which places public direct tax funds into a "lockbox" for reinvestment in ecological and community well-being, as explored in Chapter 7. If reliable, funded and audited public financing instruments are established to protect destinations, this will establish that nations are placing a value on their tourism assets. This type of investment will be essential to the future of destinations seeking to maintain value in the long term, avoid decline, and generate net benefits for local people and ecosystems.

Residents and their local authorities must recognize that their destination is subject to cycles that can lead to the destruction of their most beautiful places. Very frequently, local stakeholders become so disillusioned with the overcrowding of their paradise that they can become angry and even jaundiced about prospects of arresting unsustainable growth. It is difficult to have our dreams sullied when destinations begin to decline. Understanding the growth of destinations and the exact demands this growth will have on local resources and residents is the best antidote. Developing systems to anticipate and calculate this growth and pay for the services and preservation of valuable assets that will be required is the only way to avoid the worst aspects of insistent demand.

Different steps are required to manage sustainability, depending on which phase of development the destination is in. The Emerging Destinations experience rapid climbs in visitor numbers, which can seem exciting and often bring on a delirious boom of development instead of a wise response to the potential of a bust. After the first flush of growth, medium-scale numbers with continuing growth can be the best time to capture the steady-state revenue potential of tourism. But this requires policies to protect land, avoid speculation, and preserve culture and the environment. Unless this is done, there comes an inevitable point when the destination becomes saturated and reaches overdevelopment. Saturation can lead to a decline of demand. If the site is overcrowded and not protected, it can also cause a decline in the pricing of the rooms and attractions, causing a death spiral and full devaluation. All stakeholders are and must be responsible for managing destination cycles and arresting decline. Each sector must be accountable in new ways, as is now discussed.

Private sector accountability

While destinations and their policy makers have yet to place a value on their essential natural and cultural attractions, socially responsible corporations have moved dynamically to demonstrate the value of tourism enterprises in local economies.

A pilot study in Cyprus by PricewaterhouseCoopers (PWC) was performed with support from the TUI Group for the Travel Foundation, to create a Total Impact Measurement and Management (TIMM) system that seeks to estimate the value of economic, environmental, social, *and* tax impacts to help decision makers and stakeholders to review their policy choices. In Cyprus, tourism is a major source of economic development, with 6.8% of its gross domestic product directly contributed by tourism in 2013.[32] The study looked at the direct, indirect, and induced impacts of the TUI Group's operation in Cyprus, using eight hotels and their supply chains. Not all impacts were considered, such as biodiversity conservation, but the study offers a unique view of how governments, together with business, can look at the impacts of tourism's business activity in their country to understand what is presently covered in terms of the cost of doing business in Cyprus and what is not.[33]

This research demonstrates that TUI customers are covering the costs of doing business in Cyprus, with a total tax impact of TUI customers in Cyprus of €13.7 million annually, or €25 per customer per night, one-third of which is directly taxed. The research indicates that the cost of managing tourism was €4.20 per tourist – well below the tax revenue generated per customer, making tourism a net positive form of development.[34] *But* the TIMM model does not explore how direct tax receipts are allocated, which leaves open the question of how well the destination is being maintained with these funds. Given that airport tax represented 50%

of all direct tax revenues,[35] it is likely that the costs for managing the airport are covered, but there is no further exploration of how the additional tax receipts are spent by the government.

Accor Hotels undertook a comprehensive socioeconomic footprint as part of their Planet 21 research program[36] on their $13.75 billion brand published in 2016, which was a similar exercise in understanding the economic impacts of their brand. This ambitious study breaks new ground as it makes very good estimates of the full global impact of the corporation, including hotels under management subcontracts and those with franchise contracts *and* their supply chains on local economies in 86 countries within 25 sectors of the economy. A total of €22 billion in direct and indirect economic impacts was generated by the corporation in 2015 with a 3.1% multiplier effect caused by their total supply chain. Significant tax support to local economies, estimated to be €800 million annually in direct tax and indirect tax revenue generation, was documented, a useful set of statistics that also point to the need for more research on how those funds are used to protect the natural and cultural products upon which this corporation depends.[37]

These studies give a very strong indication that responsible multinational companies are in fact paying adequate tax to cover the cost of doing business within nations. But how countries could allocate those taxes in future to secure their destinations needs much more research.

Public sector accountability

There is very little information worldwide, from any nation, on the uses of tourism taxes. Only the work of Dr. Linda Ambrosie, who diligently reviewed the use of taxes in Cancun, Mexico, in her PhD dissertation and subsequent publications, reveals that approximately 80% of tourism taxes are being used for marketing and promotion and how little even the remaining tax funds are being used to manage tourism impacts.[38] Without an understanding of the use of tourism revenues, it is truly difficult to recommend improvements on their allocation, but a general statement from The International Council for Local Environmental Initiatives (ICLEI) offers good guidance:

> National governments and tourism businesses, which receive disproportional economic benefits from tourism, share an interest in maintaining tourism-related hard currency inflows, and should either directly provide infrastructure or ensure local authorities receive funds for this purpose. Without such support, powers should be provided to local authorities to increase local revenues from tourism, through local airport taxes, hotel and service taxes and development fees, to finance infrastructure construction and maintenance. Only through such economic instruments can the public costs of tourism activity be internalized in the local tourist economy.[39]

An international study is needed that builds on the work of Accor and TUI to discover how tourism taxes are utilized. Such an effort would point the way to the best public and private solutions to protect tourism destinations. A literature review of tourism tax policies reveals that there are extensive studies on how tourism tax dollars influence the success of a destinations in the U.S. and whether they impact hotel occupancy. But there are no research papers on optimal ways to invest tourism tax dollars in tourism-supporting infrastructure.[40] In future, the policy community needs to move quickly to research how tourism taxes are allocated and what fair proportion of these taxes might be invested in local and regional sustainability and triple bottom line goals. It is not safe to assume that the socioeconomic impacts so vividly illustrated by Accor and TUI will be used by policy makers to support sustainable growth. In fact, there is every reason to believe that is unlikely to be the case.

NGO accountability

The rapid growth of tourism puts global experts under the gun to effectively lay out the most intelligent steps for governments to achieve tourism sustainability. And there has been no lack of effort by NGOs to foster and guide sustainable tourism decision making at the local level. But real, measurable guidance on the sustainable management of tourism is still not in evidence. In the past, aspirational documents from national governments seemed sufficient progress, but as global renewable and nonrenewable resources are diminishing, and waste and waste water are becoming an unmanageable burden for destinations across the globe, aspirational goals led by voluntary organizations working on short-term grants are not an adequate solution, and they seem to be leaving governments without the mandate for policy reform and investment that is genuinely required. Only a few NGOs have had the capacity to raise these important issues on a consistent basis using validated research projects, such as Tourism Concern and the Center for Responsible Travel, but their efforts, while beneficial, are generally sorely underfunded.

The tool most touted by NGOs for over a decade has been certification for destinations. The Global Sustainable Tourism Council, an NGO formed with funding from a wide range of foundations and international donors, released criteria for destinations in November 2013.[41] These criteria and their suggested performance indicators largely measure national-level policies and guidelines that foster sustainability. Sustainable Travel International worked with the Global Sustainable Tourism Council to select 16 destinations that were recognized as Early Adopters to pilot the criteria. According to their literature, destination leaders "helped to validate the range, applicability, and clarity of the criteria and indicators."[42] The published reports look at the results of stakeholder meetings and the aspirations of the different destinations but did not establish indicators to measure performance once the project funding was over.[43] As of 2016, web updates suggest there will be a blog, GSTC Early Adopters Revisited, that will profile these destinations,[44] a nice marketing plug but with little evidence of further accountability. There is no mention of moving toward regulatory processes or financing the cost of managing more tourists in any of the GSTC literature available. And the destinations that made an original commitment to assessing themselves according to GSTC criteria are unlikely to be subject to any further accountability.[45] This raises the question of how NGOs manage the question of the impacts of their own programs.

In discussions on sustainability policy that took place at Harvard University in 2014 with a half dozen tourism ministers in a room behind closed doors, not one was moving toward legislated regulatory systems for implementing sustainability for tourism destinations. Nor were they considering the potential value of regulating the tourism industry or taxing it in new and different ways to support sustainability policies.[46] All were familiar with the GSTC criteria, and several were pursuing them. All acknowledged that regulatory mechanisms are the most important tools to achieve long-term policy, but few felt there was much chance to achieve this. Because the GSTC certification process does not require that they progress toward regulatory systems with legislation to protect vital resources, and there is no independent auditing program that reviews what destinations are accomplishing over time, there appears to be a process here that effectively leaves tourism ministries free to tout sustainability as a goal for their governments without clear assessments or clear measures to pursue policies.

A Cornell research team investigated the different voluntary destination certification programs to determine whether they functionally assist government decision making not only at the national level but importantly at the municipal level, where the real issues reside. They found that Earthcheck, a GSTC-approved certification body, has successfully supported decision making in Huatulco, Mexico, and Kaikora, New Zealand, the only communities with publishable, audited figures as of 2014 that have sought to use certification to manage their destinations.[47] These limited examples are laudatory and could lead to further success. For example, both Huatulco and Kaikora monitor the costs of various utilities to deliver energy and water per person – and project the needs of their town based on the forecasted growth of populations and tourists.[48] These simple spreadsheet exercises have helped town members to actively monitor their homes with the help of municipal officials, a solution that could easily be employed by other small towns that have transparent and willing officials to work with. But no one could suggest this is adequate progress after millions of dollars have been spent to date on the destination sustainability certification program fostered by GSTC and its allies.

An alternative is to further train and empower local municipalities to do the work. Ideas for managing destination sustainability must be kept practical and doable, especially in support of municipal decision makers. The fact is that most destination management costs for municipalities are seen as "externalities" by the private sector that they are not responsible for, even if it means they will receive better public services. Such concepts are changing, and enlightened companies like TUI and Accor Hotels are exploring how to generate the revenue their destinations require. Global industry leaders are cautiously acknowledging that investment in city planning and infrastructure will be more useful than cries for reduced taxation.[49] But the cost to preserve environmental and social capital in destinations is "invisible in the vast majority of corporate decisions, accounts and economic models," according to the Chartered Institute of Management Accounting.[50] This makes it extremely difficult for local authorities to make the case that they need more revenue to manage their destination.

NGOs and civil society need their own checks and balances with third-party auditing of their results, particularly when millions of dollars of public and foundation funds are put in their hands to achieve tourism sustainability goals. While aspirational statements of policy support for sustainable tourism were the model of the past, NGOs in sustainable tourism must hold their own systems accountable using long-term systems with measurable results. They could follow the example of the agriculture community via the Committee on Sustainable Assessment (COSA), which evaluated the impact of certification labels and made many recommendations to improve results that have been adopted by a wide array of certifying bodies working in agriculture.[51] NGOs in sustainable tourism could seek to work together with a third-party agency in just such an approach to measure their many tools to determine a more impactful approach that will protect destinations in the long term.

Solutions for local authorities: indicators and accounting systems

Destinations of the future must set goals to ensure that they can measure the evolution of their place and not be caught by surprise with its degradation. This needs to be done by local authorities in consultation with the private sector and civil society. Each destination requires a strategic plan that can be frequently updated, which is written based on participatory processes. For example, many towns experiencing tourism booms quickly learn they need to stress affordability – as prices of land and goods start to inflate. Locals also find a much larger demand on their energy services, which can spike energy costs, and see plastic refuse mount on the streets in places that once had only organic waste, as was discussed in the cases of Malaysia and Macchu Picchu town in Chapter 3. Local municipalities are often unfamiliar with how to cope with these new demands and lack experience to prevent these problems. A new paradigm in destination accounting is required to allow local authorities to review costs for offering important public services as a starting point.

At a very simple level, spreadsheets can be created that review the new costs of offering public services to outside visitors and then review those costs against the new revenues that are being generated. A simple ledger sheet will be a simple and effective way to manage the information. Most municipalities need to track the following costs:

- Water consumption and price per unit, per day per resident and per day per tourist

- Energy consumption and price per unit, per day per resident and per day per tourist
- Transport types, patterns, costs, and emissions, per day per resident, per day per tourist
- Urban waste and price per unit treatment, per day per resident, per day per tourist
- Waste water management and price per unit treatment, per day per resident, per day per tourist

For many communities, little has been done to evaluate how to maintain quality of life, provide affordability for local citizens, and ensure that waste does not overstress existing basic systems. A new approach is needed that proves the value of sustainability services to both citizens and tourism corporations alike. There is a justifiable concern that graft and corruption will siphon off the major revenue benefits that tourism generates, and without systems to ensure funds generated are the funds spent, there will be little hope of a reliable system to manage tourism sustainability in future. If municipalities were audited by outside experts and given higher bond ratings if they achieve essential efficiencies that sustainability infrastructure can provide – this could be an essential step to raising investment to maintain destinations in future, ideas that can mirror the type of reporting and financing discussed in Chapter 5 for the Dominican Republic. Auditing municipalities may sound like an uninspiring task, but it may well be how the next wave of investment in sustainable tourism is justified.

Keeping indicators simple and fundamental to maintaining quality-of-life goals is critical to achieving the desired goals. Establishing smart indicators, some of which are designed to trigger specific action, is a way of monitoring without too much elaborate work, for example:

- **Land use:** Affordable land must be maintained in a percentage of the municipality.
- **Air emissions:** Carbon monoxide must not exceed levels that endanger public health.
- **Carbon emissions:** Carbon emissions should be measured and capped, with excess reduced over a set time period.
- **Water quality:** *E. coli* should not exceed a level that exceeds health standards for swimming at beaches.
- **Toxic runoff:** This should not reach residential communities from waste dumps and must be diverted and treated.
- **Historical city or town architecture that maintains valuable heritage:** This should be preserved in a certain percentage of town centers.
- **Community gathering places:** These should be maintained, such as plazas for residents or beachfront that is accessible to the public, in a certain percentage of the core areas where tourists also visit.

- **Tourism workers' children:** The children should receive education in adequately sized classrooms with a recommended number of teachers per classroom size.
- **Tourism workers health care:** This care should be provided according to the basic standards available, with health clinics within reach of tourism centers.

Calvià, Spain: case example

With these goals or others that are equally suitable, there should be a civic structure and budget in place to measure each of the indicators on an annual basis. In Calvià, a city on the island of Mallorca in the Balearic Islands of Spain that became overdeveloped in the 1980s, a 1998 law was passed to enforce limits on accommodation growth and protect 40% of the natural areas on the region in the island, while 20 strategic lines of action were set.

Growth management goals: Calvià

- Contain human pressure in order to limit growth and favor comprehensive restoration of the coastal areas
- Favor quality of life of local residents
- Maintain land and sea natural heritage and promote the creation of a tourist and regional eco-tax
- Recover historical, cultural, and natural heritage
- Promote comprehensive restoration of the residential and tourist population centers
- Improve Calvià as a tourist destination, substitute quality for growth
- Improve public transport and favor cycling
- Introduce sustainable management in key environmental sectors: water, energy, and waste products
- Invest in human knowledge and resources
- Innovate government policies and develop opportunities for public–private investment.[52]

Calvià undertook the process because their destination was in crisis due to overdevelopment and achieved quite a few notable outcomes as a result of this process They put the following planning procedures in place.

Calvià Planning Procedures

- Ecological urban planning
- Waste management and recycling
- Training for workers in the tourism industry

- Public transit initiatives
- Creation of pedestrian zones
- Regulations for moorings in their harbor to limit density
- Prevention of dredging for beach restoration
- Recycling and urban waste program to ensure 70% is separated at origin[53]

Calvià succeeded due to well-crafted communications on their sustainability strategy, a flexible and clear set of goals, and a focus on quality of life for local residents.[54] For destinations around the world, the planning process must stress outcomes for residents that improve their well-being. If local residents do not see the results from planning tourism, they will not support it. Equally, businesses, which may resist any additional costs, may bend their views and help to support destinations if waste is recycled, beaches are kept clean, and sewage is treated. If costs for vital municipal services are clearly outlined and justified, a stronger case for public private cooperation can be made.

Solutions for local authorities: land-use planning

The discipline of land-use planning is an essential tool for tourism planners of the present and future. It is the means for anticipating the stresses that tourism growth will place on local resources and thus for containing negative impacts. Most tourism activity transpires outside protected areas and must be managed by municipal authorities. They are responsible, often without adequate training or authority, for planning tourism activity on the land in all areas outside of federal jurisdiction.

Land-use planning can put restrictions on use and help to improve the quality of life for residents, while also maintaining the value of tourism real estate. This section of the chapter will rely heavily on the *Environmental Land-Use Planning and Management* text by John Randolph,[55] who provides an encyclopedic reference to all tools that are applicable to the management of development of any kind on the landscape. Randolph points out that the conversion of natural and productive lands to human use, sprawling patterns, and inappropriate location of development, road, and building construction have broad implications on human environmental health and natural area protection.

> Growth management is achieved through policies, plans, investments, incentives and regulations to guide the type, amount, location, timing and cost of development to achieve a responsible balance between the protection of the natural environment and the development to support growth, a responsible fit between development and necessary infrastructure and quality of life.[56]

Open space and green corridors are increasingly seen as essential to urban planning and should be seen as fundamental to all tourism destination planning. As our global population increases above 7 billion and demands on land continue to accelerate, not all valuable habitat and open spaces can be expected to be put in federally or state protected area status. There are many reasons to develop land and increasingly few to preserve it. Tourism should be viewed as a primary economic reason to protect not only federally or state protected land but also urban, semi-urban, and rural green corridors. These green corridors help tourism destinations to maintain a higher level of value to the consumer.

More than 100 years ago, Frederick Law Olmsted, the founder of the field of landscape planning and the designer of New York's Central Park, conducted a study of how parks help property values. To justify the expense of the $13 million spent by New York City on developing Central Park, Olmsted provided Manhattan with a 17-year study that demonstrated there was a $209 million increase in the value of property impacted by the park. This landmark study has been confirmed hundreds of times. Municipal revenues are increased if green corridors and green spaces are both created and maintained. According to the American Planning Association, property tax and sales tax benefits for urban areas that maintain parks are nationally understood to generate value-added tax dollars to destinations, such as the U.S. Riverwalk Park in San Antonio Texas, which has overtaken the Alamo (Texas's most famous historic attraction) as the most popular tourism destination in Texas, helping to continually boost the city's $3.5 billion tourism industry.[57]

Tourists are not the only beneficiaries of green spaces. Local residents, whose quality of life must be preserved as part of tourism planning, prefer to live near green corridors. In the U.S., the National Association of Realtors found that the majority of home buyers prefer to purchase near parks and open space and are willing to pay 10% more for property that has access to parks and protected open space.[58]

Acquiring land and development rights

There are numerous means to protect land for green corridors and open space in order to control overbuilding and sprawl. Randolph details these as follows:

- Government land-use zoning and districting
- Use-value tax assessment
- Green infrastructure planning
- Land trusts
- Purchase of development rights
- Conservation easements
- Development impact fees

Government land-use zoning and districting

Governments must have the authority to constrain the use of property according to different types of uses in order to implement land-use planning laws. These laws must include regulatory teeth to contain tourism growth. Legislatures must agree that land-use planning can advance legitimate state interests, prevent public harm, and provide public benefit. A wide variety of land use regulations in the U.S. are used for growth management and environmental protection. One of the most successful current efforts to protect urban and semi-urban green spaces can be attributed to Overlay Zoning. Zoning simply defines the size of buildings, lot size, and maximum percent of lots that can be covered. It is an effective tool that can be used to preserve corridors of special value to the community that are also key to preserving the value of tourism destinations. This allows government land-use planners to place special restrictions on the development of vulnerable lands, green spaces, wetlands, or wildlife habitat.

Use-value tax assessment

As tourism grows and sprawls on the landscape, it can devour useful agricultural land and open spaces that were previously of low property value but of high importance to local quality of life. Agricultural land and open space can be taxed at different rates to encourage preservation. Reduced tax assessments encourage the preservation of valuable agricultural land and foster the preservation of real estate for agricultural, horticultural, forest, and open space use in the public interest.[59]

Green infrastructure planning

The protection of green corridors is now known as green infrastructure (GI) planning among land and natural area planning circles in the U.S. This movement aims to work in concert with land developers in a proactive system that helps to create a holistic view of the properties of a growing destination. GI planning seeks to identify areas for preservation in high-growth regions before they are developed, tag their value, and help the development community to better preserve the value of their places. The U.S. State of Maryland was a pioneer in this methodology. They created a Greenprint program with the goal to "preserve an extensive, intertwined network of land vital to long term survival of plants and wildlife and industries dependent on clean environment and abundant resources."[60]

Maryland's Greenprint program provides many interesting benefits for their GI planning program that could offer guidance to destination planners.

- Protection of natural land is a vital investment
- Preserving open space stimulates spending by local residents, increases property values, increases tourism, attracts businesses, and reduces public costs
- Developers, private landowners and others benefit from having a clear understanding of where the most ecologically valuable lands are located and where targeted conservation activities will be directed
- Using green infrastructure maps and data, local governments can enhance their efforts to provide open space, recreation lands and natural areas that retain the unique character of their communities.[61]

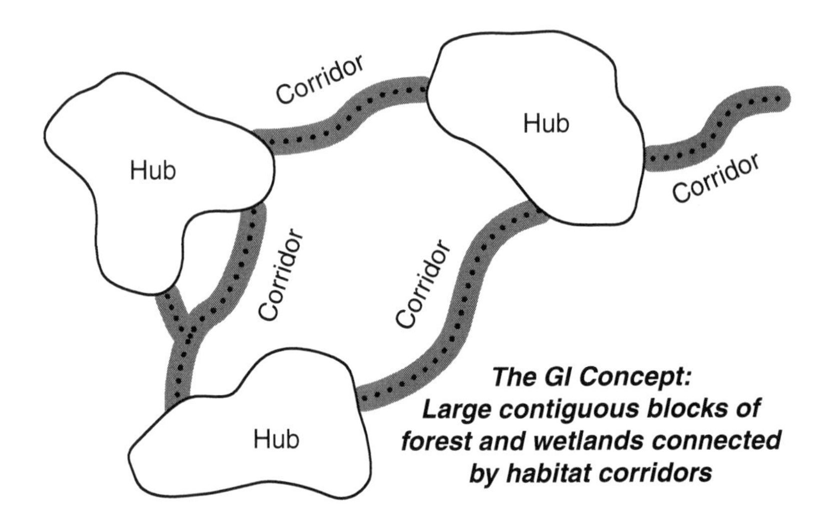

FIGURE 8.2 Green Infrastructure concept planning[63]

Maryland's Green Infrastructure Assessment tool uses geographic information systems (GIS) to identify key areas of ecological and natural resource value. Once the assessment process is completed, lands can be acquired by the state or national government according to a prioritization system that values areas of greatest natural or ecological or cultural value the most highly, especially if there is a connection to existing corridors.[62] (See Figure 8.2.) The hubs referenced in Figure 8.2 could easily be not only green zones but also special historical/cultural protection zones protected by law to achieve a mix of connections to both biodiversity and sociocultural/historical areas.

Land trusts and conservation easements

Land trusts are nonprofit conservation programs that set land aside via purchase, donations, or conservation easements. Easements are negotiated with landowners, developers, or local governments and require a legally binding plan on land use, restricting development and resale rights in perpetuity.[64] Conservation easements protect established conservation value at much lower cost than the purchase of land. And they benefit the landowner as well, in the form of payment or tax benefits.[65]

Purchase of development rights (PDR)

Development rights can be purchased from landowners by governments to protect open space and agricultural land. PDRs have successfully protected 1 million acres of farmland in the U.S., with Pennsylvania, Maryland, and Vermont as the three leading states.

Development impact fees

Municipalities can provide incentives or disincentives to steer developers toward clustered development patterns to prevent sprawl. Undeveloped land is often the target of speculation in tourism destinations, and coastal areas are the first to be snapped up. Development impact fees could be used to raise the cost of the most valuable locations for tourism development. Even if governments lack the authority to zone coastal areas, they may be able to assess development impact fees to anticipate and protect land that has the highest long-term value for development. Development impact fees could even be assessed to protect tourism land based on ecological and cultural conservation value. Much lower fees or zero fees could be charged for low-priority inland sites with lower ecological and social value to the state and local residents.[66]

Solutions for local authorities: master planning

Tourism requires planning, but what type of document is best to guide the planning process? Master plans are excellent as guiding documentation for governments to lay out their policies or strategies for tourism development. But there is a double meaning to the term "master plan." In tourism development, the term is often used as a synonym for a strategic plan. The leading sustainable tourism planning firm Solimar, which offers master planning, defines it as a "road map which offers a 10–20 year strategic vision and action plans for developing tourism in a sustainable manner for the benefit of local residents, investors and tourism operators."[67] In the field of land-use planning, master plans are the government entity's plan for the utilization of particular areas for residential, commercial, or tourism uses and the corresponding environmental impacts.[68] In any discussion of master planning, there must be a clear delineation between a strategy document, which sets out aspirational goals, and a master plan, which seeks to regulate change.

Master planning, when applied to regulate land use, could be at the center of the next generation of planning for tourism. Governments, donors, and NGOs need to recognize the value of using growth limits, green spaces, zoning of uses, and other reforms in policy to retain the value of preserved landscapes and heritage sites. Master plans that regulate land use are the vehicle for achieving this. Nonetheless, master plans are not always the ideal approach to managing impacts on landscapes if they are legislated with inadequate public input. Top-down legal instruments of any kind can become a means to undermine the public will. But a new generation of master planning techniques is emerging that are carried out with interactive procedures and therefore can be reflective of the public mind. The book *Master Planning Futures*, by Lucy Bullivant,[69] gives numerous examples of how master plans can advance the vision of society while tying it to mandates to prevent nonrenewable resource overuse, overcrowding, and poor environmental planning.

Bullivant writes that Google Earth and Green Map systems are easily used by active citizens to document green spaces, biking path corridors, and endangered habitats. It won't be long until citizens who are experiencing the negative impacts of tourism growth can prepare their own master plans using Internet-based mapping tools and indicators that can be tracked easily with citizen-driven data. Visionaries in the field of geodesign are now laying out a dynamic process for planning that depends not just on data but on simulations to help citizens and designers to "see the real impacts of design decisions" and to use data from citizens to directly respond to growing impacts.[70]

Dynamic new uses of master planning can be an inspiration for tourism planners. For example in South Africa, low-income townships were isolated and excluded from capital growth and sidelined in a once brutal government-led system called apartheid, legislated to segregate black communities and limit their legal land rights. Master planning has been used to address this entrenched, inequitable land use to bring together mixed-race communities. New housing developments were designed for integrated living for people of all incomes. Modern homes, financed together with schools, clinics, parks and recreational sites, commercial, and retail were laid out, skirted by and interwoven with environmental conservation areas, all with good infrastructure. More than 30 open spaces were designed and planned in the first phase, including traditional parks and sports fields.[71]

The urban, semi-urban, and rural landscapes of the world will become increasingly dependent on good planning in order to ensure that the burgeoning middle-class needs are met. A new society can be fostered that allows mixed-income individuals to live and work in the very same "destination" where tourists also recreate. Together, travelers and local people can experience local life, culture, and natural beauty in mixed-use communities. It will be increasingly wise to intersperse tourism with agricultural uses to help blend societies and develop linkages to healthy food economies. The prevalent model of creating the compounds around a pool or along a beach isolates tourists from experiencing a living geography of people living within their places. The demand for "authenticity" is a steadily growing market trend that may help to break the mold of enclave tourism and allow residents and visitors to create market demand for shared spaces that are designed to foster genuine interchange within corridors designed to protect heathy ecosystems, farms, and living culture and history.

Lowering the exclusionary aspects of tourism will also help prevent income inequality by allowing travelers to naturally mix with locals and purchase local goods, not just in fair trade stores but in local markets. The integration of different members of society can be achieved with the new innovative concept of developing a "sociopolis" where "shared habitats" trigger social interaction between inhabitants. The concept of the sociopolis is being watched avidly by a new group of planners. They are being built in Valencia, Spain, to both enhance citizens' well-being and bring rural spaces back to urbanized environments. If a more rural sense of place can grow within urban areas, a new, more balanced society can be rooted there.[72]

According to Bullivant, master plans should be living documents, born of sensing places giving people a sense of mutual objectives, not documents that create all-encompassing blueprints. She also notes that our global community is moving toward mixed-use neighborhoods with efficient transportation that will allow for the less wasteful use of energy and natural resources. And instead of creating top-down gated dynamics, she argues that we need a new DNA of alliance based on partnerships that allow social entrepreneurs to build social equity and sustainable outcomes.[73]

As long as tourism heavily relies on creating artificially designed, gated environments, there will be challenges to creating places that are based on mutual needs of the host and guest. A new revolution of perception is required, both from tourism planners and from travelers, that demands the shared enjoyment of the same beautiful spaces and that does not focus on isolating tourists in privileged settings without any contact with local people except for service personnel.

Jamaica Master Plan: case study

The island of Jamaica in the Caribbean is home to some of the earliest all-inclusive resort complexes in the world. These resorts have left social and environmental scars that make residents doubtful and even resentful about tourism development. Jamaica is an island that reached the Overdevelopment phase of a destination life cycle more than 20 years ago, and the buildup left many islanders without access to their own beaches. The harassment of tourists has been a long standing problem for Jamaica. In the 1990s, a visitor satisfaction survey revealed that 56% of respondents reported harassment. By mid-2001, three cruise lines pulled out of Jamaica to ports in Mexico and Puerto Rico. Heightened security was brought in, and guests were warned to stay away from restaurants, taxis, or other vendors who were not hired by the all-inclusive hotels.[74]

In response to overdevelopment, in 2002 Jamaica published a hefty *Master Plan for Sustainable Tourism Development*. This comprehensive plan offered a smart and well thought-out set of goals for improving the sector, financed by considerable efforts to promote more tourism for the island.[75] In short, it sought to reduce the antagonism toward enclave tourism, diversify the tourism product environment, involve communities, protect parks, and create better environmental management for its existing, aging all-inclusive hotel clusters.

One of the primary objectives of the plan was to foster more spending outside the all-inclusive enclave sector by developing the country's heritage assets and using tourism as the basis for urban renewal. The plan focused on financing these goals and created a Heritage Fund that could invest in selected sites. Each town was to receive a significant grant contribution of up to $2 million toward its approved plans to develop cultural and heritage sites, if they could provide 20%. Community-based ventures were to receive seed capital of up to $50,000 and loan financing averaging $100,000 per loan if projects met criteria for funding.[76] Additionally, the plan did not shy away from assessing the cost of improving Jamaica's aging resort centers, Montego Bay, Negril, and Ocho Rios. It found that these three resort complexes were far past "carrying capacity" in terms of environmental services and called for a Resort Partnership with a mandate of developing essential civic infrastructure using the financial resources of the private sector together with the regulatory revenue-raising powers of the public sector.[77]

A clear path was set to arrest the degradation of Jamaica's most densely populated tourism sites and clean up the areas. The plan called for extending and upgrading the provision of sewerage and the extent of hookup to the system by hotels, businesses, and residents. It also targeted solid waste collection and disposal in squatter communities to protect Jamaica's marine parks. It noted that because of the ribbon development of large hotels along the coast (otherwise known as sprawl), the environment *would continue to degrade even if the hotels themselves were certified and environmentally friendly*. The plan identified the practice of awarding beach licenses to all-inclusive hotels that prevent local use as one of the prime factors that antagonized the Jamaican people and made them angry and unpleasant with visitors.[78]

This piece of master planning offers a rare, transparent view of how tourism can create an environment of distrust and exploitation and foster chronic misuse of valuable natural resources. The end result is a form of sociocultural and environmental damage that can take generations to fix. The planners not only identified the problems, they identified funds for cultural preservation and for small, local business to help give the island a better reputation and local people a better chance. The plan presented a legislated process for long-term change, not just aspirational goals, and set out on a path to create a more palatable environment for locals and travelers alike. Unfortunately, it was not fully successful, partly due to economic factors.

According to reports from the Inter-American Development Bank (IDB), Jamaica was not meeting the visitor growth and foreign exchange generation goals in the Master Plan. In response, the government offered tax incentives and a 15-year tax holiday to attract foreign hotel investment for hotels of 350 rooms or more. IDB experts suggested this was not wise and noted it was undermining the Master Plan goal of managing and paying for the impacts of growth. Nonetheless, a Tourism Enhancement Fund (TEF) does collect $10 per tourist who arrive by air and $2 per traveler arriving on cruise lines. These funds are invested in heritage, health, resort enhancement, community tourism, and environmental management – a major achievement for Jamaica. The plan prompted investment in sewage systems in resort areas and a highway linking the north and south coasts. While reinvestment in Jamaica's tourism environment is transpiring, no study on the cost versus benefits to Jamaica was completed.[79] Such a study is sorely needed.

The idea of sociopoli planning could easily be adopted in Jamaica. A new vision of bringing once excluded societies into better contact with tourists could help to avoid the negative behaviors caused by the social exclusion. Just as the design process in South Africa sought to break down the barriers of the townships and create mixed neighborhoods with green space, so can sociopoli planning achieve this in the future in Overdeveloped tourism destinations. If tourism has built a cultural divide, it must be recognized, and integrated planning can help to repair the damage.

Koh Lanta Yai, Thailand Master Plan: case study

Thailand has a growing tourism economy that has been centered for decades on the island of Phuket in southern Thailand. The country's 1976 National Tourism Development plan identified Phuket as a major tourism resort, and in the span of one decade, tourism in Phuket became the dominant social, economic, and political reality of the region, reaching 50% of gross provincial product by 1991. Unchecked growth resulted in significant environmental damage. As an island with limited amounts of fresh water, Phuket became subject to frequent water shortages, magnified in peak tourism seasons when rainfall is at its lowest. The clean, translucent water of Phuket's coastline became turbid due to soil erosion and a lack of sewage treatment. The Sino-Thai elite on the island gained substantially from the rapid rise of tourism services on the island, while others were marginalized. A proliferation of brothels, massage parlors, and girlie bars in Phuket led to public health challenges, including AIDS. While Thailand announced awards for sustainable tourism in 1996, and ecotourism became the catchword, research shows that in practice Thailand continued to develop tourism in a highly unsustainable manner.[80]

The many island destinations within Southern Thailand became an escape valve for tourists who sought to explore more pristine beaches in the Andaman Sea and enjoy outstanding scuba diving. Kho Phi Phi became known as a backpacker destination in the late 1980s/early 1990s and gained immense popularity as the island featured in *The Beach* starring Leonardo Di Caprio. According to Dodds and Graci, in their book *Sustainable Tourism in Island Destinations*, Kho Phi Phi had many poorly constructed hotels when the December 2004 tsunami hit. Tragically 70% of the buildings on the island were destroyed, and all the hotels were wiped out,[81] and the government shut down the island after the disaster. For a brief time all development stopped, and many stakeholder discussions transpired with the goal of improving Kho Phi Phi's development process. Locals sought to raise key issues of protecting cultural and natural capital, including the fact that Kho Phi Phi is located within a marine national park, which justifies a higher level of planning consideration and protection. But little was agreed upon, and the island redeveloped as before.[82]

As of 2014, Kho Phi Phi was receiving more than 1,000 tourists a day, and according to local reports, its ecosystem is under threat and fast disappearing. Visitors produce about 25 tons of solid waste a day and up to 40 tons in high season. Policy makers now charge a $0.57 USD fee per visitor to cover the cost of waste management, which reporters say is inadequate.[83] As the trash continues to increase, authorities are considering collecting fees based on the weight of the trash that must be removed. On Kho Phi Phi, there is no waste water management. Individual septic devices are suspected to be overwhelmed, and brown water is being seen in the main bay. The national government has not made additional assistance available to the island for waste water treatment because tax rates are based on the resident population, not the visitor population. Visitors have written the newspaper saying that year after year, Phi Phi Island is losing its beauty.[84] Just as Phuket was left to decline, history is repeating itself with Kho Phi Phi.

In February 2005, the president of the Thai Public Policy Foundation contacted the chancellor of the University of California at Berkeley to explore collaboration after the devastating 2004 tsunami. They agreed that a joint research program with graduate students from Chulalonkorn University and University of California Berkeley would develop a plan for sustainable tourism development in Krabi Province, where Kho Phi Phi was located. Because of the condition of Kho Phi Phi after the tsunami, they chose the island of Koh Lanta Yai. Ominously, it had been designated by the Thai government as the "next hub of the Andaman."[85]

The Master Plan written for Koh Lanta Yai represents a model of environmental planning that can be applied to any case study destination in the world in the Partially Developed phase of a destination life cycle. The research teams were tasked to look at three development scenarios in light of three global economic prosperity scenarios. Local people were given full access to information on what tourism development can bring, before it actually transpires. The goal was to create an equitable and self-sufficient economy by integrating the process of regional planning into the tourism development program. Working groups were formed in ecology, infrastructure, and economic-social-cultural-political areas to gather the data necessary to inform the process.[86]

What the model does not include are standard reviews of the tourism industry's demand, flow, and dynamics, its evolution, its education system for tourism, or its characteristics of local enterprises. It also does not look at competition or positioning of tourism product. It also does not create a map of present tourist attractions with a review of how each type of attraction can bring different types of demand. While all of these items are ultimately part of a tourism planning process, they are what is normally included. In contrast, this plan provides what is almost never included and should be.

Ecological planning

This working group assessed the current ecosystem health of the island to determine best management practices to sustain local ecology as part of a tourism development strategy. They approached it from a watershed perspective, viewing the watersheds as "veins" of the island that needed to remain healthy to protect the ecosystem. Geographic information science (GIS) analysis was used, and watersheds were categorized according to the level of development and the sources of contamination, including tourism.[87]

Infrastructure

This working group assessed the roads and transportation system, water resources, sewage treatment, and solid waste management. It also investigated existing examples of sustainable infrastructure on Koh Lanta Yai.[88]

Economic-social-cultural-political (ESCP)

This working group interviewed island residents both in formal town hall settings and in homes. They examined how local cultures were affected by tourism development and the level of political capital each ethnic group had to influence the future development planning process.[89]

Master Plan vision

The Koh Lanta Yai Master Plan was presented as a process for all members of the community to take part in, not a top-down development exercise being carried out by government. The research teams made a holistic effort to present the process as a means for local people to hold onto the value of their place not only as a tourist destination but as a home and source of well-being for local people. They aimed to "unite and balance the needs and ambitions of the local residents, the tourism industry, and the government agencies into a single sustainable tourism development strategy."[90] The master plan did not focus on maximizing tourism development or growth. Instead, it is a "study of better management practices to promote a sustainable tourism strategy for the island."[91]

Documenting development trends on Koh Lanta Yai

The teams found that the hotels and bungalows were largely not located in the villages, occupied large parcels of land with direct access to the beach, and generally had long paved roads to access the resorts, many of which were guarded or gated. The variety of hotel design styles gave no sense of local aesthetics, and resorts were characterized by quick development methods using concrete, poor site planning, and the apparent need to achieve short-term profits.[92] Comparative maps allowed researchers and local residents to observe how the island was moving in the direction of the problems similar to Phuket, due to unconstrained development and rapid destruction of natural ecosystems.[93]

Watershed threats

The primary attractions on Koh Lanta Yai are the white sand beaches with turquoise water. Any disturbance to the landscape surface leads to erosion and sediment runoff into the perfect waters of the island. Sedimentation is also a threat to coral reefs, which are an important local attraction. The team assessed the watershed looking at all land alterations including plantations, shrimp farms, urban areas, and tourism. The percentage of cover of each was used to place the alteration impacts into different categories and review which watersheds were the most impaired at the time of the research. Projections were performed, using so-called build-out scenarios that included tourism, shrimp farms, and plantations to uncover how all of the watersheds would be affected in future, allowing researchers to present to residents the potential impacts of growth on the quality of their waters in the next 25 years.

Ecological services to residents and tourists

Rain forests, riparian corridors (the vegetated zones adjacent to water bodies), beach forests, beach and intertidal zones, mangroves, and coral reefs were all assessed for baseline information on ecological integrity. Each was rated for their contribution to the overall sustainability of Koh Lanta Yai. The most highly rated ecosystems for ecosystem services provided were riparian corridors in the zone where rivers meet the sea, as these areas serve as barriers to sedimentation caused by erosion and keep other ecosystems alive. Rain forests, mangroves, and coral reefs were rated as the second highest contributors to island well-being. Rain forests stabilize soil and recharge aquifers, and mangroves maintain fisheries and protect the sea from sedimentation. Coral reefs protect the coastline, maintain fisheries, and serve as tourism attractions for scuba diving and snorkeling. Beaches were given the next level of priority because they serve as primary tourism destinations but are largely dependent on the ecosystem services of riparian corridors, forested areas, mangroves, and coral reefs, making them a secondary target for protection.

Infrastructure assessment

The working groups concluded that the tourism industry places significant demands on the island's finite resources and found that the island lacked any system for valuing the cost of those demands on ecosystems and local residents. They assessed roads requirements, a proposed bridge to the mainland, energy resource and water resource use, waste water and sewage practices, and solid waste management.

In the case of *water use*, they measured rainfall, reviewed water levels of local aquifers, assessed the water quality in the aquifers, the amount of water used by residents (23 liters per day), and the amount of water consumed by tourists per day (260 liters per day). They also reviewed the type of access to water available to residents and tourists, via either private wells, springs, or purchase of bottled water. The teams reviewed locations where saltwater intrusion either was occurring or would be likely to occur due to overextraction of aquifers.

In the case of *waste water and sewage* practices, they reviewed what types were available, such as latrines, septic tanks, or direct discharge into the sea. The team reviewed regulations for the treatment and the type of compliance that was observable. They reviewed how many resort owners managed to avoid regulations by building just under 90 rooms, which exempted them from treating sewage and permitted direct discharge into streams or the ocean. They checked water quality according to the drinking water standards and mapped the presence of fecal coliform where it was discovered. Other nutrient pollution problems were also documented, such as ammonia and phosphorous coming from both fertilizer and animal waste.

Solid waste measurements were taken per person for locals and tourism facilities. Researchers found that Thais on average produce 0.65 kilograms of waste per day, while tourists produce between 3.23 and 10.10 kilograms of waste per day (5–15 times that of the local islanders). They reviewed solid waste management practices at the dump on the island and waste picker practices, which were the only form of separating waste at the dump. The team evaluated concepts of appropriate-scale modern sanitary landfill practices, their applicability, cost, siting criteria, design, and operating standards if corrective action to current waste management at the dump was to be taken.

Economic, social, cultural, and political assessment

The team analyzed the level of cultural and educational capacity to benefit from tourism, which coincided strongly with religious affiliation. They found that Thai Buddhists were much more likely to benefit from tourism than Thai Muslims. Thai-Mai, who were landless, nomadic peoples known as sea gypsies who follow traditional customs and have little to do with tourism, were the least likely to benefit from tourism development.

Researchers found that basic resources such as water and land were unequally distributed among these ethnic/religious groups, creating inequitable capacity to gain from tourism development, and predicted that tourism growth would exacerbate these differences. The team concluded that the stresses tourism brings would cause tensions in the local cultural fabric because the Muslim population's needs were likely to be overlooked even though they represented the vast majority of islanders. Much of these results mirrored the issues presented by scholars who had already studied Phuket's development in the 1970s to 1990s, indicating that little had changed and that Koh Lanta Yai would likely be subject to the same fate as Phuket unless a planning intervention had more regulatory power to manage growth.

Tiered payments for fair share of infrastructure costs

The Koh Lanta Yai Master Plan presented a strong, calculated assessment of the cost to preserve the destination, with the fairest possible fee schedule based on calculated use of local resources and infrastructure. (See Figure 8.3.) This is real sustainable development, couched in the reality of local costs, political systems, and stakeholder needs with well thought-out suggestions to achieving political consensus. The researchers recommended that "a new, tiered fee schedule to charge for resource consumption, waste production and transportation would help the tourism industry to pay its fair share of infrastructure costs on the island."[94]

To carry forward the planning process, they recommended that a Lanta Ban Rao Council is formed to provide a centralized system for measuring and monitoring the needs of the island and island businesses.[96] Similar to the idea in Huatulco,

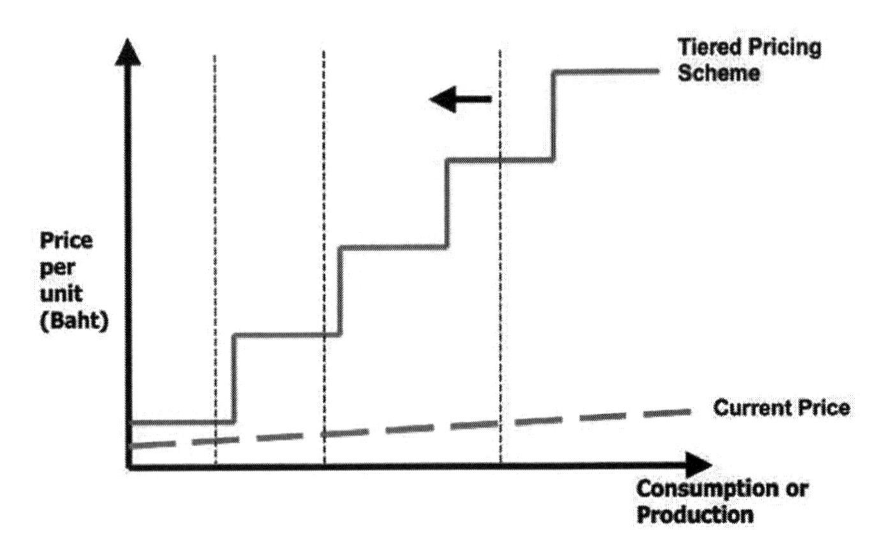

FIGURE 8.3 Koh Lanta Yai, Thailand, proposed fee schedule for tourism-based consumption[95]

Mexico, where a coordinated centralized green team brought together the vision and the data for the community, the Lanta Ban Rao Council would be central to the future of planning to preserve the island, as would be a tiered fee schedule for resource consumption, waste production, and transportation to be sure the costs of tourism development were equitably covered by the industry and local society based on use.

This cost-effective plan offers an ideal template for managing tourism destination planning by local authorities in future. It depends on local civil society cooperation, government will, and technical support from local and international universities. The outcomes of such procedures would ensure the proper management of waste products from tourism, protect ecosystems and culture, and ensure that land use is planned in a way that creates a mixed-use development process that discourages luxury gated communities, encourages authentic mixing between travelers and locals, and allows a mutual dedication to covering the costs of protecting island resources in the long term. Such master plans would not be expensive and would build educational capital and capacity to deliver on tourism sustainability worldwide.

Solutions from the protected area community

A dynamic plan to garner investment to secure local tourism assets must be a central component of all sustainable tourism plans for destinations. Much can be learned from the Protected Area (PA) community, which has sought to use tourism to pay for the preservation of planetary resources for 50 years. Placing a value on ecosystem services generated by tourism use should be expanded worldwide as part of the strategic effort to finance global tourism assets.

In Kenya, a plan to finance 70% of all costs of operating the nation's Kenya Wildlife Service (KWS) was put in place in 1989 but failed due to corruption and illegal use of the entry fees paid in cash. KWS moved to replace the cash collection gate fee system, with the use of a new Safari Card, which puts all entry fee payments (based on a sophisticated willingness to pay system for parks of varying value) through a secure digital set of transactions paid on the card. This has raised the amount of funds reaching parks substantially. The Belize Protected Area Conservation Trust, explored in Chapter 7, is another successful mechanism in this category. The island of Palau established a Green Fee of $15 per visitor, generating millions of dollars that are invested in regional capacity to manage ecosystems.[97] Such strategies demonstrate that tourism has great capacity to support the management of ecosystems and local communities, and, though still largely underutilized, this capacity has the potential to be scaled up considerably.

Regions outside of parks that offer invaluable tourism assets, such as the coral reefs of the Bahamas, discussed in Chapter 7, need new conservation fee support based on strategies to pay for the cost of managing tourism ecosystem services. The new Safari Cards in Kenya are just one small example of how revenue collection for ecosystems could become part a system that is paid for by the swipe of a card, based on algorithms that effectively place a value on the use of resources that travelers are willing to pay, and the cost of the infrastructure services required locally. Ideas from the next generation of destination managers should be flowing and quickly, based on the successes documented in Kenya, Belize, and Palau alone.

New standardized methodologies to place a value on public resources must be part of the global sustainable tourism management frameworks. Consumer awareness is also fundamental to success. The question travelers should be asking is not, "How do I get there before it is too late?" but rather, "How can I pay the price required to protect the places I love?" A mainstream trend of taking responsibility for planetary resources has to spread and soon. Tourism policy makers will need to see that travelers mean business and are willing to pay to protect the resources and beauty spots they love. Such a movement, based on our digital sharing economy, cannot start soon enough and will be discussed further in the Conclusion (Chapter 10).

Implications of climate change on destination planning

Academic research on tourism planning for climate change is in its infancy.[98] While it is fully understood that climate change planning will be essential for destinations, very few tools have been developed specifically for this purpose.[99] For this reason, this section will be brief. Two studies are highlighted that outline the significance of climate change to destinations and the deeply rooted incapacity to respond to increasing climate risks. Solutions to these issues will be offered based on the spatial planning tools already being deployed by planners in other sectors of the economy.

Coastal tourism is an economically dominant segment of the tourism industry, which is the most likely to be affected by climate change with vast consequences for investors, the hotel industry, and all of its many suppliers worldwide. Surprisingly, no regional or global scale studies on projected Sea Level Rise (SLR) impacts have specifically examined the potential damage to the tourism sector. This may be due to the lack of geospatial data sets on coastal tourism assets, including resorts, beaches and local infrastructure.[100] Even with this lack of data, scholars and lay observers have noted that tourism destinations, their governments, and the private sector remain remarkably unconcerned about rising sea levels, despite well-known projections by the Intergovernmental Panel on Climate Change (IPCC).[101]

Research on the Caribbean coast has revealed that it is highly susceptible to coastal erosion that will accelerate with SLR. Of the 906 major coastal resort properties inventoried in an analysis of 19 Caribbean island nations, 29% were found to be at risk for partial or full inundation with a 1-meter SLR, with possible inundation losses of more than 50% of coastal properties in 5 of the 19 countries (Anguilla, Belize, British Virgin Islands, St. Kitts and Nevis, and Turks and Caicos Islands).[102] "The greatest vulnerability was found in Anguilla, British Virgin Island and Turks and Caicos – where the relative economic importance of tourism to the national economy is very high."[103] Climate change has major importance for Small Island Developing States in general, which may suffer from detrimental implications on their property values, loss of tourism revenues, growth of insurance costs, and pressures to adapt with retrofits and new investment, which could be particularly difficult to obtain for smaller properties.[104]

In a different part of the globe, southern Africa has also been shown to be vulnerable to severe projected physical climate impacts, with mean annual temperature increases 1.5 times the global average as of 2014 and increasing rainfall variability that may result in both floods and droughts, intense hot spells, and wild fires. The government of South Africa accepts that climate change is a major challenge for sustainable development and has publically stated it is a critical concern for their tourism economy. Destinations across the nation face degradation of environmental resources, including wildlife, beaches and heritage sites, shifts in water availability, biodiversity loss, and reduced landscape aesthetics, among many other risks.[105] While the national government of South Africa seeks to be proactive, researchers have found that local (municipal) governments face innumerable barriers to action on adaptation or planning, largely because they are overwhelmed with the basic needs of their communities, including housing, water, sanitation, and infrastructure. The mandate to respond to climate adaptation is referred to as an "unfunded mandate" by local governments given the meager support they receive to respond.[106] The conclusion of this case study is that small rural towns have almost no capacity to respond to climate change. Tourism falls low on the list of priorities, and there is minimal funding for municipalities to respond.[107]

Solutions to these issues are closely related to the solutions already proposed in this chapter for destination management in general. There must be a greater understanding of the consequences of losing valuable tourism assets at higher levels in government, policies to mandate assessments of risks, participatory programs to develop adaptation strategies, and budgets to make this possible at the municipal level. Spatial planning tools are the preferred tool to enable this process.

1 The risks of climate change affects must be analyzed and validated. Analysis is the quantification of impacts on the tourism economy, such as the study on Caribbean nations just referenced.
2 Development of adaptation strategies must be explored using participatory planning procedures, reviewing both the problems and opportunities caused by climate change illustrated with spatial mapping tools.
3 Design can then proceed in response to the problems and opportunities, based on actual review of the land-use configurations that will have to be adjusted using a variety of visualized scenarios, based on projections for SLR and other climate risks.
4 The final step is the evaluation of visualized scenarios which proceeds via negotiation between stakeholders, with a series of inputs on each scenario and a ranking of the various options.[108]

Researchers who have tested these tools have found success with helping governments and citizens to discuss and review the genuine challenges of climate change on local economies. Performing such analysis for tourism will be highly revealing and informative and will allow stakeholders, especially in tourism-dependent economies, to confront coming challenges in an informed manner for the first time. There are no certainties in this process, and the trade-offs are unlikely to be easy. Anticipating the loss of valuable assets will be painful, but the process will promote a more informed approach that could help to save millions or billions of dollars in tourism assets, both natural and human-made.

Conclusion

Destinations are born, grow, and die within a consumer society that perpetually seeks out the best new places to visit. In this cyclical industry, the inevitable demise of one location after another is not only expected, it is part of tourism culture. Every traveler knows that destination life cycles exist, and they seek to stay ahead of the curve. Countries like Jamaica or Thailand may perpetually struggle to emerge from a cycle of destruction unless regulations and planning become obligatory in the tourism development process, and new financing mechanisms are put in place that support the protection of destinations.

Destination planning is a discipline that needs a complete rethink and will require upgraded tools. It is frequently sold as a set of aspirational goals couched in improving markets to attain more revenue both at the national and the local levels. Accountability is very low. Maintaining strategic tourism assets should not be positioned as part of selling destinations. Such strategies fail to emphasize the commitment it takes from national and municipal governments to cover the cost of destination management.

Governments must recognize that tourism depends on the protection of parks, clean water, coral reefs, cultural monuments, local crafts, biodiversity, and genuine cultural interchange. If these assets were valued in the long term and projected for their tourism value, it is likely that more financing could be obtained to protect what travelers value the most – authenticity and pristine natural resources. Municipal systems to manage reliable data on the cost of managing tourism services will be fundamental to that process, and they must be based on indicators that are agreed upon through participatory planning processes. Such metrics can be recorded and updated using geodesign tools that can be installed locally and managed transparently, even internationally, in cooperation with civil society and universities in future.

Governments in charge of tourism across the world must seriously consider policies that can protect land and create corridors for natural vegetation and biodiversity protection. Such policies are proven to protect the value of real estate and prevent the damaging effects of uncontrolled sprawl. Local authorities must be prepared to arrest the overdevelopment of tourism and prevent the overuse of local resources for private gain. A new age of planning must dawn, which arrests this cycle and focuses on community well-being, equitable uses of land and resources, and full involvement of communities in planning, as illustrated in the case example of Koh Lanta Yai in Thailand.

Lessons learned from the Protected Area community indicate that tourism fees are an important and successful means to establishing the value of ecosystem services. Such mechanisms could easily be expanded to support the preservation of tourism destinations. The Balearic Islands, where the case study island of Calvià is located, have installed an eco-tax, which help the islands to improve their facilities, according to leaders there.[109] This type of tax can be assessed either directly or indirectly via reallocation of existing tax structures for tourism, which this chapter strongly advocates needs full study and review. There is increasing agreement that new sources of revenue for destination sustainability must be discussed, and this dialog must include questions of achieving better planning, long-term monitoring with indicators, participatory procedures, transparency and auditing of the new funds, and, importantly, attracting private sector and green bond funding to address the substantial needs for new clean development infrastructure.

Tourism planners of the future must realize that product development and marketing are not the sum total of their job. Rather the preservation of natural and social capital will need to become a priority for their work. If tourism planners can divorce themselves from antiquated concepts of palm trees and pools along isolated beaches and create dynamic new master plans that instead foster sociopoli, a renaissance of social interchange could blossom. With the creation of mixed-use development and green corridors, social exclusion can be avoided and poverty alleviation and local healthy employment advanced.

The threat of climate change trumps nearly all discussions of how to manage tourism destinations in future, and the data already indicates that billions of dollars of tourism assets are at risk. Surprisingly the tourism development and real estate communities are not fully responding to this challenge, and, as has been presented in this chapter, there are few governmental organizations presently assessing the risks to their national tourism infrastructure. With the signing of the Paris COP 21 accords, there is every hope that a new generation of planning will now begin that accounts for carbon impacts of tourism and assesses the global risks to tourism development. There is little time to waste. Local citizens, especially in small island nations, are likely to suffer the brunt of economic recessions and unemployment if megastorms and sea level rise destroy their source of economic well-being. Solutions to this enormous issue are discussed fully in the Conclusion of this book.

For destinations and for those who are investing in them and those who are their caretakers, what is most essential is a change in the policy environment. There needs to be a genuine, bankable, net value placed on preserved tourism assets, which can be financed via tourism fees and taxes, bonds, and international investment from the private sector and international donor funds. In future, municipal fee systems for the fundamental costs of tourism development cannot be placed solely on local citizens. This chapter presents rational calculations that would allow local authorities to assess costs for tourism development according to resource use, be it fresh water, recycling services, or waste water treatment. With a fair and equitable division of costs, there can be a mutually beneficial investment in the protection of valuable resources and the financing of clean development infrastructure. With such measures, there is no reason that sustainable destination development cannot proceed immediately.

Notes

1 2015 World Legacy Awards, http://travel.nationalgeographic.com/travel/world-legacy-awards/#/tropic-journeys-ecuador_88889_600x450.jpg, Accessed October 3, 2016
2 Ibid.
3 The new age of exploration, *National Geographic*, http://ngm.nationalgeographic.com/2013/01/125-yasuni-national-park/wallace-text, Accessed October 3, 2016
4 Yasuni: Ecuador abandons plan to stave off Amazon drilling, 2013, *The Guardian*, August 15, http://theguardian.com/world/2013/aug/16/ecuador-abandons-yasuni-amazon-drilling, Accessed October 3, 2016
5 Ibid.
6 Ecuador – Economy, http://theodora.com/wfbcurrent/ecuador/ecuador_economy.html, Accessed October 3, 2016
7 Enomenga, Moi, personal communication to Megan Epler Wood on June 28, 2015 at Huaorani Ecolodge, Ecuador
8 Enomenga, Moi, 2015, *Audio Interview on Ecotourism and Oil Exploration*, EplerWood International, Burlington, VT
9 *Laudato si*, https://en.wikipedia.org/wiki/Laudato_si%27, Accessed July 10, 2015
10 The Holy See, June 18, 2015, *Encyclical Letter* Laudato Si *of the Holy Father Francis on Care of Our Common Home*, Vatican, Rome, Chapter 1, *What Is Happening to Our Common Home*, pages 20–21, 37 and 38

11 Chinese oil search shuts Ecolodge at Amazon headwaters, 2016, *Environment News Service*, May 25, http://ens-newswire.com/2016/05/25/chinese-oil-search-shuts-ecolodge-at-amazon-headwaters/, Accessed October 3, 2016
12 New Age of Exploration, http://ngm.nationalgeographic.com/2013/01/125-yasuni-national-park/wallace-text, Accessed October 3, 2016
13 Epler Wood, Megan r, 2007, The Role of Sustainable Tourism in International Development: Prospects for Economic Growth, Alleviation of Poverty and Environmental Conservation, in *Critical Issues in Ecotourism*, ed. James Higham, Butterworth-Heinemann, Elsevier, Oxford and Burlington, MA, pages 158–184
14 Buckley, Ralf C. et al., 2016, Net effects of ecotourism on threatened special survival, *PLoS One*, Vol. 11, No. 2, e0147988, DOI:10.1371/journal.pone.0147988
15 Boley, B. Bynum and Gary T. Green, 2015, Ecotourism and natural resource conservation: the "potential" for sustainable symbiotic relationship, *Journal of Ecotourism*, Volume 15, 2016, Issue 1, pages 36–50, DOI:10.1080/14724049.2015.1094080
16 Epler Wood, Megan, 1990, Original notes from the board of directors of the International Ecotourism Society when the definition for ecotourism was discussed and finalized for use by the organization upon its founding
17 Gössling, Stefan et al., July 2012, Transition management: a tool for implementing sustainable tourism scenarios, *Journal of Sustainable Tourism*, Vol. 20, No. 6, July 2012, 899–916
18 World Resources Institute, April 2015, *Revaluing Ecosystems: Pathways for Scaling up the Inclusion of Ecosystem Value in Decision Making*, eds., Luretta Burke et al., Executive summary, page 3
19 Ambrosie, Linda, 2012, *Tourism: Sacred Cow or Silver Bullet?, Dissertation for Haskayne School of Business*, University of Calgary, Calgary, Alberta, and Myths of tourism institutionalization and Cancun, 2015, *Annals of Tourism Research*, Vol. 54, 65–83
20 Brown, Gregory and Delene Weber, 2013, Using public participation GIS (PPGIS) on the Geoweb to monitor tourism development preferences, *Journal of Sustainable Tourism*, Vol. 21, No. 2, 192–211
21 United Nations World Tourism Organization, 2004, *Indicators of Sustainable Development for Tourism Destinations: A Guidebook*, UN World Tourism Organization, Madrid, Spain, page 8
22 Oxford Economics, The Comparative Economic Impact of Travel and Tourism, National DMO Spending, page 25–26, http://wttc.org/-/media/files/reports/benchmark%20 reports/the_comparative_economic_impact_of_travel__tourism.pdf, Accessed October 3, 2016
23 United Nations World Tourism Organization, *Indicators of Sustainable Development*
24 Kolbert, Elizabeth, 2014, *The Sixth Extinction: An Unnatural History*, Henry Holt and Company, New York
25 Plog, Stanley C., 1974, Why destination areas rise and fall in popularity, *Cornell Hotel and Restaurant Administration Quarterly*, Vol. 14, No. 4, 55–58, DOI:10.1177/001088047401400409
26 TripAdvisor Fact Sheet, http://tripadvisor.com/PressCenter-c4-Fact_Sheet.html, Accessed October 3, 2016
27 Butler, R.W., 1980, The concept of a tourist area cycle of evolution: implications for management of resource, *Canadian Geographer*, Vol. 24, No. 1, 15–12
28 Romao, Joao, Joao Guerreiro, and Paulo Rodrigues, 2013, Regional tourism development: culture, nature, life cycle and attractiveness, *Current Issues in Tourism*, Vol. 16, No. 6, 517–534
29 Graci, Sonya and Rachel Dodds, 2010, *Sustainable Tourism in Island Destinations*, Earthscan from Routledge, Oxon and New York
30 Hall, Michael C., 2010, Policy learning and policy failure in sustainable tourism governance; from first-order and second-order to third-order change? *Journal of Sustainable Tourism*, Vol. 19, Nos. 4–5, 649–671
31 World Bank Group, 2015, *What Are Green Bonds? International Bank for Reconstruction and Development*, Washington, DC

32 Travel Foundation in Association with PWC, July 2015, *Measuring Tourism's Impact, a Pilot Study in Cyprus*, Bristol
33 Ibid., page 8
34 Environmental costs are €4 Euros per customer per night and the overall cost of providing public services to TUI customers is approximately €0.20 per night
35 Ibid., page 16
36 AccorHotels, Our Footprint, http://accorhotels-group.com/en/sustainable-development/planet-21-research/our-footprint.html, Accessed October 3, 2016
37 Planet 21 Research, January 2016, *AccorHotels Socio-Economic Footprint, First Study into the Worldwide Socio-Economic Impacts of a Hotel Group*, Accor Hotels, Paris, http://accorhotels-group.com/en/sustainable-development/planet-21-research/our-footprint.html, Accessed October 3, 2016
38 Ambrosie, *Tourism: Sacred Cow or Silver Bullet?*
39 International Council on Local Environmental Initiatives, 1999, *Tourism and Sustainable Development, Sustainable Tourism: A Local Authority Perspective*, Background Paper # 3, Department of Economic and Social Affairs, Commission on Sustainable Development, Seventh Session, April 19–30, New York
40 Knipe, Tom, 2011, *Bed Taxes and Local Tourism Development: An Outline and Annotated Bibliography*, Cornell University Department of City and Regional Planning, for Mildred Warner's Course: Privatization, Devolution and the New Public Management
41 Global Sustainable Tourism Council Criteria for Destinations, https://gstcouncil.org/en/gstc-criteria/criteria-for-destinations.html, Accessed October 3, 2016
42 Global Sustainable Tourism Council Announces Second Group of Early Adopters to Pilot New Destination Criteria, http://responsibletravelreport.com/index.php?option=com_acymailing&ctrl=archive&task=view&mailid=170, Accessed October 3, 2016
43 Twining-Ward, Louise, July 20, 2015, personal communication by email to the author
44 GSTC Early Adopters, Revisited, http://gstcouncil.org/blog/1234/gstc-early-adopters-revisited/, Accessed May 4, 2016
45 Twining-Ward, Louise, July 20, 2015, personal communication by email to the author
46 *Private Session for Tourism Ministers on Leadership*, November 2014, Center for Health and the Global Environment, Harvard T.H. Chan School of Public Health, Cambridge, MA
47 Crowe, Stephen, Robert Vicencio, and Kimberley Mark, May 19, 2014, *Study on Sustainable Tourism Destination-Management Decision Making*, Center for Sustainable Global Enterprise (CSGE) and the Samuel Curtis Johnson Graduate School of Management at Cornell University, Ithaca, NY, for Sustainable Travel International, page 41
48 Ibid.
49 World Tourism Forum Lucerne, 3rd Think Tank, April 22, 2015, http://static1.1.sqspcdn.com/static/f/180610/26917043/1458135826473/Summary+Think+Tank+2015.pdf?token=djRTZktyPLQP3NgM10J3MsDW1Zw%3D, Accessed October 3, 2016
50 Maxwell, Dorothy, 2015, Future-proofing against natural capital debt, *Eco-Business*, July 2, http://eco-business.com/opinion/future-proofing-against-natural-capital-debt/, Accessed October 3, 2016
51 COSA, 2013, *The COSA Measuring Sustainability Report*
52 Graci, Sonya and Rachel Dodds, 2010, *Sustainable Tourism in Island Destinations*, Earthscan from Routledge, Oxon and New York, Overcoming Challenges
53 Ibid.
54 Ibid.
55 Randolph, John, 2004, *Environmental Land Use Planning and Management*, Island Press, Washington, DC
56 Ibid., page 39
57 American Planning Association, *How Cities Use Parks for Economic Development*, City Parks Forum Briefing Papers, #3
58 Ibid., Key Point #5

59 A Citizen's Guide to the Use Value Taxation Program in Virginia, https://pubs.ext.vt.edu/448/448-037/448-037.html, Accessed October 3, 2016
60 Randolph, *Environmental Land Use Planning and Management*, page 100
61 Maryland Department of Natural Resources, Land Acquisition and Planning, Maryland's Green Infrastructure Assessment, http://dnr.maryland.gov/land/Pages/Green-Infrastructure-Mapping.aspx. Also see Conservation Fund, 2004, Maryland's Green Infrastructure Assessment and GreenPrint Program, Case Study, http://conservationfund.org/images/programs/files/Marylands_Green_Infrastructure_Assessment_and_Greenprint_Program.pdf, Accessed October 3, 2016
62 Ibid.
63 Maryland Department of Natural Resources, Land Acquisition and Planning, Maryland's Green Infrastructure Assessment, http://dnr2.maryland.gov/land/Pages/Green-Infrastructure.aspx, Accessed October 2, 2016
64 Randolph, *Environmental Land Use Planning and Management*, page 83
65 Ibid., page 90
66 Ibid., page 167
67 Tourism Master Plans, http://solimarinternational.com/strategic-planning/tourism-master-plans, Accessed October 3, 2016
68 Master Plan, legal definition, http://yourdictionary.com/master-plan, Accessed October 3, 2016
69 Bullivant, Lucy, 2012, *Master Planning Futures*, Routledge, Oxon and New York
70 Ahead of his time, Dr Stephen Ervin celebrates the coming of age of geodesign, 2012, *ArcNews*, Fall, http://esri.com/news/arcnews/fall12articles/ahead-of-his-time.html, Accessed October 3, 2016
71 Bullivant, *Master Planning Futures*, Lion Park, *Urban Design Framework*, extension to Cosmo City, Johannesburg, South Africa section
72 Scholtus, Petz, 2010, Sociopolis, the urban housing project brings the campo to the city, *Treehugger*, October 28, http://treehugger.com/sustainable-product-design/sociopolis-the-rurban-housing-project-brings-the-campo-to-the-city-photos.html, Accessed October 3, 2016
73 Bullivant, *Master Planning Futures*, Epilogue
74 McElroy, Jerome L., Peter Tarlowe, and Karin Carlisle, 2007, Tourist harassment: review of the literature and destination responses, *International Journal of Culture, Tourism and Hospitality Research*, Vol. 1, No. 4, 305–314
75 Commonwealth Secretariat, 2002, *Master Plan for Sustainable Tourism Development*, Jamaica, London
76 Ibid., page 8
77 Ibid., page 9
78 Ibid.
79 Panadeiros Monica and Warren Benfield, March 2010, *Productive Development Policies in Jamaica, Inter-American Development Bank, Department of Research and Chief Economist*, IDB Working Paper Series, 128, Washington, DC
80 Kontogeorgopoulos, Nick, 1998, Tourism in Thailand: patterns, trends and limitations, *Pacific Tourism Review*, Vol. 2, 225–238
81 Pomonis, Antonios, Tiziana Rossetto, Navin Peiris, Sean Wilkinson, Domenico Del Re, Raymond Koo, Raul Manlapig, and Steward Gallocher, 2006, *Indian Ocean Tsunami of 26 December 2004*, Mission Findings in Sri Lanka and Thailand, Earthquake Engineering Field Investigation Team, Institution of Structural Engineers, London
82 Graci, Sonya and Rachel Dodds, 2010, *Sustainable Tourism in Island Destinations*, Earthscan from Routledge, Oxon and New York, Chapter 7, Unsustainable Development in Kho Phi Phi, Thailand
83 Special Report: Phi Phi Cries for Help, http://phuketgazette.net/phuket-news/Special-Report-Phi-Phi-cries-help/38250#ad-image-0, Accessed July 27, 2015
84 Ibid.

85 Chulalonkorn University, University of California Berkeley, and Thai Public Policy Foundation, 2007, *A Master Plan for Sustainable Tourism Development*, Koh Lanta Yai, Krabi Province, WP-2007–06, Berkeley, CA
86 Ibid., page 2
87 Ibid., page 3
88 Ibid.
89 Ibid.
90 Ibid., page 4
91 Ibid.
92 Ibid., page 8
93 Ibid., page 10
94 Ibid., page 41
95 Chulalonkorn University, University of California Berkeley, and Thai Public Policy Foundation, *A Master Plan for Sustainable Tourism Development*
96 Ibid., page 44
97 UNDP, August 2012, *International Guidebook of Environmental Finance Tools: A Sectoral Approach, Protected Areas, Sustainable Forests, Sustainable Agriculture and Pro-poor Energy, Chapter 4: Protected Areas*, United Nations Development Programme, New York
98 Hernandez, Ana Beatriz and Gerard Ryan, 2011, Coping with climate change in the tourism industry: a review and agenda for future research, *Tourism and Hospitality Management*, Vol. 17, No. 1, 79–90
99 Ibid.
100 Scott, Daniel, Murray Charles Simpson, and Ryan Sim, July 2012, The vulnerability of Caribbean coastal tourism to scenarios of climate change related sea level rise, *Journal of Sustainable Tourism*, Vol. 20, No. 6, 883–898, http://dx.doi.org/10.1080/09669582.2012.699063
101 Intergovernmental Panel on Climate Change [IPCC], 2012, Summary for Policymakers, in *Managing the Risks of Extreme Events and Disasters to Advance Climate Change Adaptation*, Special Report of Working Groups I and II of the Intergovernmental Panel on Climate Change, eds. C.B. Field, V. Barros, T.F. Stocker, D. Qin, D.J. Dokken, K.L. Ebi, M.D. Mastrandrea, K.J. Mach G.-K. Plattner, S.K. Allen, M. Tignor, and P.M. Midgley, Cambridge University Press, Cambridge and New York, pages 1–19
102 Scott, Daniel et al., The Vulnerability of Caribbean coastal tourism
103 Ibid., page 891
104 Gössling, Stefan and Daniel Scott, 2012, Scenario planning for sustainable tourism: an introduction, *Journal of Sustainable Tourism*, Vol. 20, No. 6, 773–778
105 Rogerson, Christian M., 2016, Climate change, tourism and local economic development in South Africa, *Local Economy*, Vol. 3, Nos. 1–2, 322–331
106 Ibid., page 327
107 Ibid., page 328
108 Eikelboom, T. and R. Janssen, 2012, Interactive spatial tools for the design of regional adaptation strategies, *Journal of Environmental Management*, Vol. 127, S6–S14
109 Harley, Nicola, 2016, British families face tourist tax of up to 75 British pounds to visit Balearic Islands this summer, *The Telegraph*, April 23, http://telegraph.co.uk/news/2016/04/23/british-families-face-tourist-tax-of-up-to-75-to-visit-balearic/, Accessed October 3, 2016

9

CONCLUSIONS

The future of sustainable tourism

Humans are continuously moving across our vast planet, settling, adapting, and dominating every landscape. The industry of travel facilitates our movement, and air travel is at the nexus of the astounding number of peregrinations transpiring annually. Over 1 billion international travelers and 5–6 billion domestic travelers crisscrossed the planet in 2014.[1] International tourist arrivals are increasing by 4% on average and could charge ahead to 5% annually through 2025 due to the growing number of travelers from China.[2] The volume of individuals who are flying makes the airline industry a top ten carbon emitter, with emissions forecasted to escalate by 100–300% by 2050 depending on the actions taken. In this time frame, travel and tourism impacts on the ground will have many consequences for our natural, social, and cultural worlds and will require resources well beyond what local populations use. Tourism's energy consumption is conservatively expected to increase by 154%, water consumption by 152%, and solid waste disposal by 251%, according to both the UN World Tourism Organization (UNWTO) and United Nations Environment Programme (UNEP),[3] and academic research suggest these figures are quite conservative.[4]

The drive to travel is without doubt one of the most important human economic, social, and environmental industrial trends of the 21st century. What is more, the core outbound and inbound markets for travel are rapidly changing. The new middle class in Asia is flying at unprecedented rates, changing airline routes, and fueling the rapid growth of airports and a boom in hotel development throughout the region. South-to-South travel is 40% of all travel flight paths as of 2015, and it will not be long until Europe and North America are no longer the dominant destinations in the world.[5] And the digital economy has developed tools to book and manage travel that are so simple that buyers can instantaneously make decisions that once took months.

Each distinct sector of the complex spiderweb of tourism supply chains is now easily accessed, both directly online or through a variety of outlets all working to deliver increasingly customized product. Each corporate sector is working to gain a larger piece of digital terrain, and their business models are more often than not at arm's-length, with very little liability for any other part of the total supply chain. Hotels, airlines, tour operators, and cruise lines are all explored in this volume, reviewing their distinct business models and the specific environmental impacts and management techniques presently available to their teams. The newest paradigms and alternative approaches to achieving greater planetary neutrality are also explored, not only within corporate boundaries but on landscapes, marine environments, and our global atmosphere. Destinations, the heart of the tourism product, are revealed as being increasingly vulnerable, with very few systems in place to protect them.

Academic research presented here confirms that the tourism economy is creating impacts on the ground that remain largely unaccounted for. Policy makers and local authorities must not only react, they will need to innovate based on clear evidence that destinations are overwhelmed, underfinanced, and unable to curtail tourism's impacts without wholly new forms of governance. No one book can offer all of the solutions, but the Conclusion of this text presents an urgent message to the global community that action is required from both policy makers and industry to rescue the natural and cultural products tourism depends on.

This chapter proposes that a framework of tourism governance that relies on standard systems of checks and balances must be developed at the international, national, and local levels, separating economic development duties from the science-based tasks of environmental management. Industry is part and parcel of this new process and can be an essential player in developing global metrics, benchmarking tools, and investment and financing solutions. If industry and government were to agree on a global plan to investigate what financial resources are available for tourism sustainability and discuss what specific financing mechanisms can be brought to bear, this would be an enormously valuable statement of progress.

The economic development potential of tourism is also ripe for another round of investigation based on the many lessons learned. The pro-poor work done in the 1990s and 2000s has laid the groundwork to amplify the UNWTO system for managing tourism statistics and to form the basis for economic development metrics that can guide donor investment and leverage the economic impacts of tourism to reach those who need it most.

This volume is calling for reform. Such reform will depend on agreement that new forms of governance and more investment is required to measure and manage tourism's impacts on the planet. This Conclusion will review how both industry and government might reconceptualize the tasks ahead.

Holding the tourism economy to account

As the world enters the post-Paris COP 21 era, there is an extraordinary opportunity to create new intelligent systems to pay for and manage tourism's global assets. The vision to tap these resources needs to be martialed, and the benefits need to be fully understood. Tourism can help to protect resources that are not only valuable for tourists but also valuable for every level of society. In 2015, not only were the Paris COP 21 Agreement signed, but there was also an historic commitment to the new Sustainable Development Goals and a growing focus on lowering and offsetting aviation emissions. Each of these global efforts deserves introduction to place the new landscape of policies for sustainable tourism into proper focus.

The new *Sustainable Development Goals (SDGs)* replaced the Millennium Development Goals (MDGs) in 2015. The SDGs set out 169 targets for ending poverty, protecting the planet, fostering prosperity for all humans, creating a peaceful world, and strengthening global partnerships in order to achieve these goals.[6] The SDGs specifically mention sustainable tourism as an important mechanism for meeting SDG Goal 8, which sets out the goal of creating sustained, inclusive, and sustainable economic growth, full and productive employment, and decent work for all. A sustainable economic development agenda from tourism should be considered as part of a larger economic development program for tourism worldwide and the UNWTO is laying out plans for measuring sustainable tourism as an inclusive and environmentally sound economic development tool.[7] While the 169 SDG targets offer a wide variety of aspirational goals, only 29% are well developed enough to provide measurable targets. The overarching goals of the SDGs may well help the global community to adopt metrics that move beyond economic growth and GDP as a stand-alone metric, but they lack critical efficiency targets, such as lowering the use of water or energy, which likely make them unsuited for the measurement of carbon intensity as well.[8]

The *Paris COP 21 Agreement* to combat climate change was signed by 195 nations in December 2015 to keep global temperature rise in the 21st century well below 2 degrees Celsius and to strengthen the ability to deal with the impacts of climate change. It is a new type of agreement, which by all accounts unites developed and developing countries in a common framework. It creates a process for science-based assessments every five years that will inform the implementation of countries' climate plans. Assessment will start in 2023, but countries have agreed to return in 2018 to review the implementation of mitigation measures to inform their 2020 mitigation contributions. Ten thousand new climate initiatives are expected to be launched, with 127 million hectares of degraded land to be restored, $1 trillion in solar investment, and 114 companies to set emissions-cutting goals. Finance will be key to achieving these goals, with $100 billion to be mobilized by 2025.[9] Transparency is a fundamental characteristic of this agreement, and a uniform transparent framework for emissions reporting is required from all countries.

Historically, tourism has not been involved in carbon accounting largely because the Kyoto accords did not include either aviation or maritime transportation in its global mandate. As a result, the travel industry was not brought into larger multi-sectoral dialogs on managing carbon emissions, except through the European Union Emissions Trading System (EU ETS), which unfortunately failed to create an aviation accounting agreement that was accepted internationally. (See Chapter 5.) This has left the travel industry and tourism ministries without global technical input or cooperation. But the post–COP 21 world will set out new requirements that will include the travel industry for the first time.

The COP 21 Agreement has created a planetary resource management system on an unprecedented scale. The framework established by COP 21 will include many industrial sectors, including tourism. Every nation is now budgeting their carbon emissions based on a benchmarking system in order to make a commitment to lowering greenhouse gas emissions, in reports called *Nationally Determined Contributions* (INDCs). While in the past, only rich countries and some multinational corporations were doing carbon reporting, now every country will report using the most up-to-date technical analysis, as part of the Paris treaty. This effort is 100% international, requires systematic cooperation and science-based accounting, and cannot depend on consultants flying in and creating reports for governments that do not correspond with other accounting programs.[10]

The NDC system will include the tourism sector in integrated reporting systems on carbon, requiring systematic budgets for lowering the use of energy, creating waste, and increasing waste water treatment.[11] The process has already begun, and many countries delivered Intended Nationally Determined Contributions (INDCs) before Paris using participatory methodologies that evaluate pathways for lowering carbon emissions.[12] Governments, industry, academic organizations, and civil society all weighed in. Many developing countries will be seeking technical transfer in carbon accounting to meet their reporting requirements. The establishment of a global system of consistent measurements is already spurring innovation and investment. Countries such as Colombia have set out Green Growth strategies that include tourism and seek to make the process positive for all sectors of society. Colombia is aiming for the rational use of ecosystem services and increased resilience in the face of growing environmental issues caused by climate change.[13]

The *International Civil Aviation Organization (ICAO)* decided in October 2016 on a market-based mechanism (MBM) to manage the sector's carbon emissions that exceed the cap to be set in 2020. States can choose to opt into a pilot phase as of 2021, and an additional voluntary phase from 2024-2026. Russia, India and Brazil chose to opt out of all voluntary commitments, leaving some of the most important regions of the world to grow without aviation emissions accounting. But as of 2027, states with carbon emissions higher than 0.5% of the total sector's emissions will be required to enter into the agreement. While many details remain as of this writing, and the quality of the offsets need oversight, this new mechanism represents an important new source of revenue for clean low-carbon development worldwide.[14,15]

These agreements set the stage for both new forms of governance for sustainable tourism and new forms of accounting. A responsible tourism framework must be organized to take advantage of the skills, existing mandates, and research methodologies that are best used for each type of accounting role.

Achieving a multinational integrated system of sustainable tourism accounting

Many international actors will need to fully cooperate in the establishment of a new multinational system of sustainable tourism accounting. Experts in the field of environmental governance suggest that environmental accounting requires specialized agencies that are not also tasked with economic growth.[16] National governments consistently find that environmental protection must be managed through environmental ministries. An environmental accounting system requires third-party data and analysis, with systems that can be nationally and internationally vetted. It is therefore logical that the new system of carbon reporting of tourism sectoral impacts should be monitored through agencies or commissions that have experience with and personnel who can manage science-based mechanisms. The future of all climate change accounting will depend on strengthening the connection between science-based indicators and public policy decision making.[17]

The UNWTO represents the tourism ministries of the world, which have a mandate of economic development and growth. This book has outlined that tourism ministries presently lack the mandate to allocate adequate long-term funding resources for sustainability and generally focus on tourism marketing. While the responsibility of accounting for the environmental costs of tourism development lies with tourism ministries, there is little evidence that they are suited to this task. The true costs for sustainability are generally falling upon local municipalities, which in most instances do not have the capacity to charge tourists for the additional costs accrued. This vacuum of responsibility is one of the most important governance issues that sustainable tourism faces. How the tourism sector accounts for its environmental costs could become part of the global agenda for the first time through the NDC process. Carbon impacts of the tourism industry will also be considered through the SDG process, although the global response to climate change is considered to be best managed through the mechanisms established via COP 21 by many experts.[18]

This book suggests that the travel and tourism field must quickly adopt a dual-channel program. One program, to be led by the UNWTO and tourism ministries, must ensure the success of the SDG process focusing on the economic development goals that are laid out there. And the second should be led by UNEP and ministries with the expertise to foster a science-based program that can accurately monitor the sector's impacts on the national level as part of the NDC process, without pressure to achieve economic growth goals as a primary mandate.

Solutions for sustainable tourism governance

The dual program proposed here separates the duty of enhancing and fostering more inclusive economic growth from the role of managing environmental sustainability. This maintains a healthy checks-and-balances system and takes advantage of the skills of existing players in each domain.

1 Sustainable tourism economic development goals

The enormous potential of sustainable tourism as an economic development tool is laid out in Chapter 3. A full reading of that chapter makes it clear that neither sustainability nor inclusive economic development transpires with simple promotional and marketing tactics or by providing industry a free rein. Competitiveness indicators, according to the research presented here, have also failed to date to properly value the preservation of natural and cultural assets, which in the field of tourism seems especially essential. With the new SDG framework, there could be new indicators developed for measuring human development goals as part of an enhanced Tourism Satellite Accounting (TSA) program. The new TSA program could draw from the valuable results of the Pro-Poor Initiative and lessons learned from the UNWTO's own STEP program.[19] Indicators would include how tourism is improving human development indices, providing opportunity to women and minorities, and bringing new economic opportunity to rural communities. SDG goals strongly recommend reducing inequality within and among countries. But new metrics will be required that better recognize the sources of human well-being, including natural, social, and cultural capital.[20] Studies done by academics, discussed in Chapter 3, conclusively show that there are inadequate statistics to prove that sustainable tourism improves these metrics. With a new TSA–SDG "pro-poor" program, countries throughout the world could gather data, share information, and inform grant- and loan-based funding sources on the most far-reaching and inclusive approaches to improving the well-being of local people with globally accepted indicators.

The potential for this type of global pro-poor statistical system is exciting and could genuinely improve sustainable tourism donor projects, which have floundered despite the enormous sums spent. But the financing for the goals set by the proposed TSA–SDG program should not depend entirely on donor investments. Other options should be carefully considered.

a Ministries and Destination Management Organizations need to study the level of tax revenues they require strictly for promotion. The private sector in all likelihood no longer fully requires destination marketing on the scale that is presently supported by government tax funds. Many questions need to be raised about the high subsidies that are being provided to this industry through tourism taxes, which might be better allocated toward tourism as a form of economic development. Full stakeholder review should transpire, which allows the participation of local authorities and local people and discusses options for tax fund management. The new TSA–SDG metrics could institutionalize this process and give much greater confidence in the program.

b New programs that allow local authorities to determine their fiscal requirements need to be high on the agenda for consideration and built into the review of what is the cost basis for achieving sustainability. Civil society and the private sector should be fully involved in this national process to discover the fiscal requirements to manage tourism for economic benefit and sustainability in the long term. The municipalities of Huatulco and Kaikora, which have managed Earthcheck destination certification programs, as described in Chapter 8, offer one example of how stakeholders could manage the process. These systems could be scaled up and enhanced using the statistical analysis and projection tools originally conceived of for the country of Belize by Cornell business school students in the case described in Chapter 1.

The future of sustainable tourism depends on proper statistical accounting and projection systems to pay for the development of tourism to benefit the poor. If the UNWTO and tourism ministries created and implemented a TSA system that accounts for human development needs, this could guide donor funds for a generation and fully meet the challenge laid out by the SDGs to account for progress in this arena.

2 Sustainable tourism environmental policy goals

The COP 21 Agreement sets the stage for a new approach to managing the environmental impacts of tourism. This process must be managed by scientists, who can manage the measurement of specific environmental impacts on the ground caused by the tourism economy. The management of environmental indicators should be overseen by ministries with environmental science capacity in collaboration with UNEP. These experts, who should not be responsible for improving economic growth, would be tasked with researching the specific environmental impacts of tourism as part of national NDC reporting programs. The tourism industry's relative contribution to carbon emissions will be calculated via the sector's use of different energy sources, land use, solid waste generation, and waste water treatment approaches.

To carry out such a program, financing should be sourced from the new ICAO market development mechanisms funds or from the Green Climate Fund, which has been made available for investment in low-emission climate-resilient development as of April 2016, after the formal signing of COP 21.[21] For countries with a high level of tourism impacts, new programs could be established immediately to move local investment strategies toward clean development mechanisms for sustainable tourism that will qualify for this new source of revenue. Some programmatic design work to make this comprehensible to policy makers would be in order, such as:

a A *Carbon Neutral Destination Program* planning unit within Environment Ministries or ministries with the capacity to manage carbon accounting for tourism. Such a program could set the stage for launching research on the main hotspots of tourism impacts on a national level, establishing baselines of carbon impacts, setting goals for reduction and mitigation, establishing caps for tourism emissions, and seeking investment for mitigation. These investment strategies could seek certification either as part of offset credits and/or emissions, allowance financing strategies through the new ICAO market-based mechanism program or the Green Climate Fund.

b *National Geodesign Tourism Land Preservation Programs* that will track tourism land use with international, cloud-based indicators that are monitored at the national and international levels and establish goals for lowering tourism impacts on key natural resources, particularly forests but also wetlands and coral reef ecosystems. Caps on impacts are set, and contributions to conservation are sought through impact fees, tourism direct taxes in lieu of payments for ecosystem services, and the REDD programs and the Green Climate Fund, which will invest in lowering impacts on forest ecosystems.

c *National Sustainable Tourism Trust Funds* could manage payments toward clean development and conservation goals and oversee the allocation of investments in the conservation of ecosystems and the development of green infrastructure. These funds could be matched by dynamic private and bank-led financing instruments, such as green bonds and impact investment, which would facilitate public–private financing on a scale never seen before for sustainable tourism. All of these investments could be captured by the tourism economy, lower overall operational costs, and help to underwrite much needed sustainable infrastructure. Green economy investors will likely be interested in working with tourism developers especially if trust funds are available from national sources that are overseen by private boards dedicated to delivering fully vetted annual financials that are audited by outside third-party organizations. Together developers and financiers could funnel tourism revenues toward both planetary conservation and local economic development.

d A *Global Destination Preservation Trust Fund* might be one innovation to consider for encouraging more industry support for conservation and sustainability. A small portion of the tourism wholesale and retail price tag could be allocated to a new trust fund that reinvests in global tourism sustainability. This trust fund could invest in the specific costs of maintaining and preserving internationally valued tourism assets, such as biodiversity zones, world heritage areas, and other sites visited by a high percentage of international travelers. Such a trust fund could complement national sustainable tourism trust funds and manage international investments for national use.

Capturing emerging funds to make the tourism economy more sustainable can be achieved only via public private cooperation. While much more discussion and further design work are needed to create such a governance and financing system, there are some important guidelines to consider.

- Economic development strategies must be separated from environmental benchmarking tasks.
- Economic development strategies require inclusive pro-poor/sustainable livelihood indicators to meet SDG goals.

- The new carbon accounting systems must measure physical impacts on the ground, preferably as part of multi-sectoral units that transparently report to the INDC process.
- The costs of sustainable tourism infrastructure must be factored in from the ground up, working with local authorities.

Training and capacity building

Tourism management on this vast scale with such substantial resources at play should no longer rely on the ad hoc voluntary system that is presently used to manage sustainable tourism. In the future, a fully financed, well-staffed infrastructure operating at national, local, and international levels should be the status quo, considering the growing impacts of tourism development. A comprehensive plan to develop new graduates who can meet these demands should be evaluated by a global educational commission to identify educational needs and requirements for this next generation of workers. The amount of work that could be achieved with a disciplined set of quantitative studies coming from universities alone – to address the question of monitoring tourism, allocating resources to managing tourism, creating systems of land-use planning, and fostering investment in green accounting and technology – would be without precedent.

Employees with the skills to undertake carbon accounting, life cycle analysis, hydrological monitoring, GIS land planning, and much more are the professionals needed for the future of managing sustainable tourism. Some of these roles are desk jobs, but many should be field oriented to ensure that the data is coming from real, tangible sources that can be tagged and captured for ongoing analysis in future. The new systems of data should move the sustainable tourism world away from voluntary mechanisms for reporting, such as GRI and CDP, which have been shown in this text to inconsistently measure corporate reductions due to a lack of consistent boundaries. Planetary environmental impacts are best monitored using new science-based frameworks that are consistent, cloud based, and integrated into global research systems, while depending on local capacity to use updated, cost-effective technologies to manage indicators, such as the new geodesign tools.

Improving economic benefits

Tourism has a unique role in economic development, especially in rural regions where the poorest residents of our planet now reside and the earth's most valuable ecosystems are the most intact.[22] The SDG goal to end poverty in all its forms requires social protection and concrete social policies.[23] To ensure that tourism development leads to sociocultural protection, laws that secure local rights to resources are essential. Political and legal reforms solved many of the problems associated with unregulated tourism in Botswana and Namibia and created a system for the fair distribution of equity in the process of developing tourism in a highly desirable wildlife tourism area.

Lessons learned

Where tourism has devalued social equity, as transpired in Jamaica for instance, the local value chain needs to be carefully repaired to allow for local business to return to healthy, legal forms of commerce. The tourism value chain incorporates many sectors of the economy, including telecommunications, retail, and health services. These linkages are underexploited by local business, and some of the most lucrative sectors are owned by foreign investors, tour operators, and airlines.

New approaches

Given the cash flow available from tourism, new business opportunities abound in tourism supply chains worldwide, and donors seeking to generate economic development opportunity need only look at how to improve local business capacity to invest in such outstanding opportunities as renewable energy and the sustainable food economy, alternative waste treatment, and a wide range of energy, water, and waste treatment alternative technologies.

One-third of all tourism expenditures are for food, highlighting how important it is for hotels to research and invest in local food value chains as a likely route to improving economic development in many regions. Mapping the value chain will help hotels to assess where there are missing links that can benefit both local people and enhance product lines for tourism business. Seafood or fair trade cooperatives are already expanding and should be the source of synergy between tourism and other sustainable economic development programs. And investment in such stellar enterprises as Wilderness Safaris, which places a high percentage of its profits into local economic, social, and conservation benefits, is a tested way to magnify the best of sustainable tourism and keep it local.

Tourism often offers a better chance than other industries for marginalized citizens and women to move ahead. Women are more likely to be the "boss" in food and beverage operations than in other industries, and restaurants are often the first step they take to gain access to equity value in the tourism marketplace. Large companies can also achieve lower costs and higher value if they take advantage of local talent, especially in tourism. The more that such benefits extend to local people, the better the service culture will be and the more that visitors will be happy and satisfied.

Tourism projects that succeed at preserving both nature and culture result in more balanced, happier societies. Travelers around the world seek out nature and authentic culture that help them to mentally and physically renew themselves. Economic results from tourism may increasingly come not from exclusive complexes dedicated to foreign travelers or from corporate-owned shopping malls in airports and on piers pumping out uniform product sales environments worldwide, but rather from "sociopoli," which encourage travelers and local society to mix in public space. Such globalized commons, where locals meet visitors in their daily lives, benefit local vendors, reduce crime, improve the opportunity for interchange between cultures, and create an environment of authenticity that also fosters the arts and the creative economy.

Tour operator supply chains

Lessons learned

The changing dynamics of the tourism economy and the globalization of markets are driving down profits in tourism supply chains. Online and mobile booking are driving a growing portion of sales, and new solutions are needed in order to preserve the assets the industry depends on. Traditional tour operators do a great deal of business around the world, which drives significant indirect economic benefits through their supply chains. They depend on transport providers, restaurants, hotels, attractions, museums, protected areas, and monuments, all of which drive economic benefits locally. But as tour operator supply chains are pressed by digital purchasing models to lower their costs, local suppliers are increasingly operating without adequate resources. Under such circumstances, sustainability becomes a low priority. New systems to drive value back into the system and to protect destinations and local businesses must be on the agenda for the industry in future.

New approaches

As the industry continues to create arm's-length global tactics to distance itself from liability or responsibility, there is a need to review precisely how much of the supply chain's assets even reach local shores. More studies of tax offshoring will help nations to tighten their revenue-generation strategies. The use of value chain analysis to locate where gaps in the chain are found is another critical tool that can inform stakeholders and foster cooperation. Life cycle analysis of the impacts of tourism corporations, such as the research achieved in the Planet 21 project of the Accor Hotels company, can also help multinational companies to recognize that their primary impacts may go well beyond their perimeter. Finally, supply chain and procurement policies must be managed with the total impact of tourism's local economic and social footprint in mind through input output analysis. And tour operators can now plug in excellent procurement tools to lower their global impacts and manage their carbon accounts, with Travelife and CARMACAL available to be plugged in to existing tour operator booking engines.

Hotels

A gold rush environment in the developing world is creating a massive real estate boom for hotels, particularly in Asia where properties are built with inadequate permitting and no resource protection practices in place, even in locales that have very scarce resources. Water use is of particular concern, with consumption not uncommonly 10–18 times higher per traveler than local residential use. International hotel brands have only recently begun to require global environmental management systems for their properties to help them to better benchmark their complex families, which are likely to be franchised and operated by companies not directly owned by the main brands.

Lessons learned

Hotels are major energy users, ranking in the top 5 in the commercial building sector. They vary markedly in efficiency, even within the same branded chains, because of operational and philosophical differences between owners. Surprisingly, even with the growing interest in Leadership in Energy and Environmental Design (LEED) certification and energy management worldwide, hotels are using significantly more energy per room per night because of the increasing energy demands of the average client in hotels worldwide. Energy savings of 10–20% can be expected if a hotel undertakes a technical review of equipment and operations and implements conservation measures. Governments can support energy efficiency in this crucial sector by offering training, financing tools, and low-interest loans and tax incentives to help subsidize renewable energy and net metering policies to enable private business to earn funds from renewable energy.

But investment in sustainable energy or any other system requires awareness and a willingness to study the footprint of a hotel, something small property owners frequently eschew. Even as the branded global hotel system becomes more environmentally accountable, global independent and small hotels rarely manage their facilities with environmental efficiencies in mind, even if there is considerable payback for doing so. This lack of interest in lowering impacts may characterize 70–80% of the properties in tourism, whose owners prefer ad hoc approaches over rigorous efforts to measure their impacts.

Conversely, the rise of the sharing economy has brought a new type of hotel room manager to prominence via Airbnb and other home-sharing booking engines. Unlike small hotel property owners, these new "hoteliers" are often covering costs for their existing homes with the additional income from home-sharing booking engines, which may give them a greater incentive to strategically lower their environmental costs to gain the benefit of the extra cash they are receiving. Global research will be required; with Airbnb alone selling 2 million rooms worldwide in 2015, these booking engines may well be absorbing some of the demand for new hotels, a significant environmental benefit.

Many hotels have begun to reduce solid waste and food, and global brands are showing impressive reductions. But most destinations are dominated by small properties and independent hotels, which are not measuring or lowering waste and are often endangering fragile destinations that host major monuments such as Angkor Wat or Macchu Picchu. Even large hotels may not hook up to sewage systems because around the world, only a very small percentage of sewage is even treated. Honolulu is now being forced to pay $14.3 billion for a complete revamp of its municipal sewage treatment system, forcing local citizens to shoulder 175% increases in their water bills. In most regions of the emerging economy, citizens could ill afford such tax increases to pay for the demand tourism is placing on local infrastructure, as described in the case example of Mombasa in Chapter 1.

New approaches

The cost burden of managing water and waste water can be significantly reduced via the introduction of efficiency systems. Net Zero water projects aim to ensure that 100% of storm water and building water is managed on site. To achieve this, gray water, which is 50–80% of all waste water, needs to be separated in new building constructions and piped to distribution in areas that need water for irrigation purposes. Hotels have higher percentages of gray water than most buildings because of commercial kitchens and laundry operations and could greatly benefit from separate gray water systems. Advanced alternative technologies for waste water treatment are becoming increasingly affordable, especially for small and medium-sized resorts. Constructed wetlands are an effective means to meet sanitation standards with proven results and cost 50% less than conventional systems.

As the global bulge in middle-class travel continues its relentless march, especially in Asia, hotels will soon be in the position of competing for water and energy with local users. The case of ITC in India offers one example of how a major company is investing in greening their own supply chain and developing efficiency standards on a very profound level. ITC has already set policies to recycle gray water and use it for toilets and drip irrigation, saving 40% on their water consumption. Shri Y.C. Deveshwar, the chairman of ITC, seeks to use ITC hotels to help correct social imbalances in India and use the travel industry economic and social value for change. To this end, ITC is supplying 30% of its food and beverage operations with providers within 100 miles. This conglomerate has unusual resources for investing in local markets because of its history in agriculture (tobacco), but the fact that it is realigning its goals using tourism shows the potential for the hotel and agricultural communities to cooperate and create more efficiency and equity in local supply chains. Hotels can purchase a significant amount of local produce and potentially deliver on becoming a much greater source of revenue for a wide variety of local suppliers, especially in agriculture but also in energy, water, waste, and waste treatment.

Airlines and airports

Air travel is the source of 2.5% of all carbon emissions from human activity worldwide. If aviation were a country, it would be the 7th largest in terms of greenhouse emissions, and if no action is taken, climate change pollution from aviation is forecasted to triple by 2050.[24] To make matters worse, aircraft emissions are forced into a layer of the atmosphere where it is more vulnerable, making the industry a dangerous and important climate actor that has yet to resolve how to meet greenhouse gas reduction goals.

Airlines: lessons learned

New aircraft designs from Boeing and Airbus will lower the per-passenger impacts of flight by approximately 6%, but the growth of the industry and emissions projected far exceeds this figure. Alternative fuels have extraordinary potential to provide low-carbon gains, but they face daunting market limitations due to considerable extra cost for production, the challenge of production at scales that defy imagination, and the requirement to have enormous storage capacity near airports. Aircraft are now the most efficient in history. But even with the best scenarios imaginable, carbon neutrality, once projected by the International Civil Aviation Organization (ICAO) as an achievable target by 2020, is now understood to be unrealistic even by 2050.

New approaches

Aviation is a field that brings out the best in global thinkers, and there are often surprising breakthroughs in the technology of flight. Air traffic control systems are a humble but important example of how global flight could be made much more efficient. The U.S. Federal Aviation Administration has a road map for installing NextGen air traffic control technology in the U.S., which will reduce air traffic delays by 41% by 2020. It is projected that improved navigation equipment on airplanes will reduce total emissions and cut costs for fuel, but the implementation has been slower than hoped. While GPS-guided navigation seems fundamental in this era, and nearly flawless GPS systems are in every new iPhone, installing these systems in aircraft and in air traffic control facilities is taking time. Even with the many delays, airlines are starting to report fuel-savings benefits.

Airports: lessons learned

Airports are the global hubs of trade and leisure that could be modernized palaces of technology breakthroughs and efficiency or the choke points in which all of aviation's dreamscapes are converted to nightmares. With 6 billion air passages expected by 2020, the industry will need much more support from the ground to make travel a pleasure and not an experience consistently plagued with delays and clogged arteries. The challenge of managing airports is as monumental as these facilities have become. They need special attention to keep valuable landscapes from being destroyed, noise to a minimum, toxic runoff from polluting water tables, and CO_2 and NO*x* emissions from dirtying regional airscapes. Airports are a fundamental part of the rapid development of Asia and the global South. These countries should project their growth patterns and future environmental impacts carefully to make certain their regional footprint does not destroy local well-being and provide more input to local citizens in the process.

New approaches

Airports can begin to achieve better management of their footprint through more sophisticated measurements of noise, air pollution, and emissions using modeling with the intent to capture variability in weather, congestion, and runway patterns of the airport and to avoid averaging out daily results as a substitute for more thorough reviews of emissions and noise peaks. Smoother operations of aircraft on the ground and the reduction of aircraft idling on runways will be one of the most important achievements in airport design in future to avoid emissions and noise largely due to congestion.

Noise is the single most important environmental issue in the management of airports, and the mitigation of the problem is technically hampered by flawed noise measurement systems. The U.S. Federal Aviation Administration (FAA) and other oversight authorities worldwide need to discontinue the practice of averaging noise over a 24-hour period and implement new measurements systems that account for peak noise impacts. Solid waste in airports is also measured in ways that can be misleading or questionably accurate. Most airlines are not recycling because of the difficulty of implementing policies that are dependent on local municipal airport systems, which require each tenant and airline to contract its own waste haulers. Centralized recycling has allowed San Francisco to set a goal of 90% recycling or more by 2020.

Airports face increasing pressures to limit their impacts on local residents in Europe and America. More holistic management and measurements of their impacts will greatly improve results. It cannot be forgotten that vulnerable citizens in their air- and soundscape need special protection from the noise, waste, waste water, and air pollution they produce. Regulatory policies in Asia, Africa, and Latin America need to be established based on the lessons learned in Europe and North America.

Cruise lines

The cruise industry has an average annual growth rate of 9.5%, well beyond the average growth of tourism internationally, which is 4%. A $30 billion market that relies on the U.S., Canada, and the Caribbean Sea for most of its destinations, it is quickly diversifying into European and Asian waters. The cruise lines have created a business model that increasingly relies on very large ships, now known as megaships. Between 2000 and 2013, 167 new ships were launched with enough cabins to jump total passenger capacity by 10% in 2010 alone. While more developed nations, such as the Scandinavian nations found on the Baltic Sea, have successfully developed new regulations to protect their waters, cruise line expansion ambitions will outpace any efforts to protect waters in most regions. International law is flawed and does not require the protection of coastal waters over 12 miles offshore, a policy that needs an overhaul based on the passenger loads of the new megaships now bringing thousands of passengers per day into vulnerable waters, such as the Baltic and Adriatic Seas.

Lessons learned

The current standoff between local governments and the cruise business leaves few options for local stakeholders to more vigorously protect their environment, unless they are willing to put tens of thousands of jobs at stake – a risk few governments can afford to run. The atmosphere is not made any more productive by the environmental community, which offers criticism and not negotiation. The industry receives more scrutiny than most, but it must be recognized that legal standards of the sea are still primitive by comparison to those on land, and it is this flaw that creates so much conflict.

The industry's labor practices are also unique. Cruise employs tens of thousands of crew members from around the world, without national labor law oversight on board. New labor standards for cruise lines were made law in 2013 by the International Labor Organization, which will increase each ship's annual operating costs by $15 to 25 million according to one report but will introduce genuine labor standards to the industry for the first time to the benefit of hundreds of thousands of workers.

Air pollution from ships, including sulfur oxide, has now come under new, stricter regulations, which allow ports to also use Special Emission Control Areas to monitor and enforce strict limits to emissions in their own jurisdictions. Cruise ships depend on high-sulfur fuels to offer competitive prices to their passengers, but the new regulations now require that they lower their emissions in ports.

New approaches

A transition to low-sulfur fuels will be one of the most important environmental actions cruise lines can take in the next several decades. The industry's rapid growth plan needs to build in more overhead for the cost of environmentally friendly fuel. Replacing high-sulfur fuels with an alternative fuel, such as liquid natural gas, is an important opportunity to consider and deserves much more research. Researchers reviewed options for liquid natural gas on cruise ships and found it to be feasible with a possible payback time of 4 to 8 years.[25] For now, scrubbers are considered to be the best short-term options for cruise, as the cost of refining low-sulfur fuel could increase fuel prices by 87%.[26]

Maritime carbon emissions were excluded from the Paris Cop 21 Agreement, leaving carbon reporting to voluntary metrics. Internal documentation from the cruise line industry indicates that the cost of transition to low-sulfur fuels is already anticipated to be a motivation for much higher ship efficiency.[27] While cruise executives have not gone on record since Paris, other parts of the maritime industry would prefer to have transparent carbon accounting with limits and monitoring to achieve a level playing field.[28] But for now the maritime industry and cruise lines can sail with only voluntary carbon accounting.

Sustainable tourism on a finite planet

Lessons learned

The tourism industry has a globally dispersed product distribution system that is becoming increasingly digital with few impediments to becoming one of the most successful industries in the 21st century. It has outstanding revenue streams that reach even the poorest on the planet, making it a very good candidate for financing from donors to create enabling conditions to foster economic development. But it lacks the proper metrics or systems to manage the measurement of progress, and the ability to ensure development is genuinely inclusive and suffers from a chronic overuse of local nonrenewable resources, which is growing by 100–200% or more by 2050. The aviation industry's growing carbon footprint hangs like a black cloud over the travel industry's rosy projections of financial health and environmental respectability. Tour operators, hotels, and cruise lines have a record of overlooking the management of the destinations they sell and avoiding responsibility for the impacts they do not ostensibly control.

New approaches

New systems to finance the cost of managing tourism sustainably and preserving its most valuable assets are proposed in this chapter, combined with governance systems that protect investment and ensure that audited sustainability investments are made. These funds can be financed by growing global market–based carbon allowance and offset systems, the Green Climate Fund, Clean Development Mechanisms, and private impact investment. Qualified agencies, which use science-based accounting and indicators, must guide the process of ensuring that these investments result in lowering carbon emissions and impacts on nonrenewable resources. Lowering the carbon impacts of the tourism industry is a complex challenge that requires the coordinated monitoring of tourism as part of larger global and national monitoring teams, best overseen via the COP 21 treaty process and facilitated via university cooperation worldwide. A ground-up set of procedures that measures locally and responds to urgent national needs for financing sustainable infrastructure will be essential to lower national carbon impacts. The case of the Dominican Republic, outlined in Chapter 4, demonstrates the potential for large-scale tourism businesses to join with government to move entire economies toward national grids based on renewable energy.

Tourism's economic development potential is of clear importance, and the economic impact metrics already used by tourism ministries could be adapted to ensure that sustainable tourism is benefitting the poor. Regional well-being and local livelihoods must be protected, and enabling policies are required to ensure there is adequate local health, energy, road, and educational systems, which are not built by tourism but in fact enable tourism in order for economic development to transpire. This will require a review of how tax funds from tourism are allocated. Municipal balance sheets that account for both liabilities and revenues will help local authorities to both project the needs and to account for the costs to budget adequately for the required local sustainable infrastructure. Audited accounting systems will be required based on the actual additional incremental costs tourism places on struggling local communities, as demonstrated in the case of Koh Lanta Yai in Chapter 8.

More educational programs are needed to build a professional workforce to achieve these goals. It would be a mistake to seek to reassign tourism marketing professionals to the task of managing environmental impacts or overseeing the complex process of auditing local municipal accounts. Instead, new types of professional management of the earth's resources and its peoples will be required for both industry and government. Opening the doors to a new generation of management will bring fresh air to these projects and offer the kind of innovative thinking the tourism field requires. This is no longer an industry that can place sustainability in small departments with low-level staff who have no access to core business decisions. Nor can it divorce itself from its most essential assets. Instead, the industry must place a new value on professional sustainable management of its supply chains and its natural and cultural products. If there are no efforts to invest in the future, local destinations will become devalued, as beach fronts become polluted and overcrowded, local people disenchanted, sprawl the norm, and polluted air typical, with congested airports reaching a standstill and carbon emissions skyrocketing. The most obvious path is to bring more professional management to every part of the industry and government in order to begin the process of conserving the future of both this industry and its destinations.

Notes

1 UNWTO, 2015, *UNWTO Tourism Highlights*, 2015 Edition, World Tourism Organization (UNWTO), Madrid, Spain
2 Global Travel to Rise 5.4% Annually as Chinese Demand Ramps up, http://bigonlinenews.com/global-travel-to-rise-5-4-annually-as-chinese-demand-ramps-up/, Accessed October 3, 2016
3 Steiner, Achim and Taleb Rifai, *Tourism and Sustainable Development: Pathway to Low-Carbon Development*, UNEP Sustainable Tourism Programme, Division of Technology, Industry and Economics, Paris, France
4 Gössling, Stefan and Paul Peeters, 2015, Assessing tourism's global environmental impact 1900–2005, *Journal of Sustainable Tourism*, Vol. 23, Issue 5, pages 639–659 DOI:10.1080/09669582.2015.1008500
5 UNWTO, *UNWTO Tourism Highlights*
6 Transforming Our World: the 2030 Agenda for Sustainable Development, https://sustainabledevelopment.un.org/post2015/transformingourworld, Accessed October 3, 2016

7 UNWTO, June 2016, *A Statistical Project to Support Mainstreaming Tourism in Sustainable Development*, Project Proposal, Provisional Version to be presented to UNCEEA meeting, UNWTO, Madrid, Spain
8 ICSU, ISSC, 2015, *Review of the Sustainable Development Goals: The Science Perspective*, International Council for Science (ICSU), Paris, France
9 Outcomes of the U.N. Climate Change Conference in Paris, http://c2es.org/international/negotiations/cop21-paris/summary, Accessed October 3, 2016
10 NDC Registry, United Nations Framework on Climate Change, http://unfccc.int/focus/ndc_registry/items/9433.php, Accessed November 4, 2016
11 Intended Nationally Determined Contributions (INDCs), http://unfccc.int/focus/indc_portal/items/8766.php, Accessed October 3, 2016
12 INDCs as Communicated by Parties, http://www4.unfccc.int/submissions/indc/Submission%20Pages/submissions.aspx, Accessed October 3, 2016
13 Government of Colombia, 2015, *Intended Nationally Determined Contribution*, United Nations Framework on Climate Change (UNFCC), http://www4.unfccc.int/submissions/indc/Submission%20Pages/submissions.aspx, Accessed April 11, 2016
14 UN aviation brokers first ever global climate deal, October 6, 2016 http://nationalobserver.com/2016/10/06/news/un-aviation-agency-brokers-first-ever-global-climate-deal Accessed October 8, 2016
15 Flightpath, 1.5 degrees, Frequently Asked Questions, http://flightpath1point5.org Accesed October 8, 2016
16 Biermann, Frank, 2014, *Earth System Governance, World Politics in the Anthropocene*, MIT Press, Cambridge, MA
17 Ibid.
18 ICSU and ISSC, 2015, *Review of the Sustainable Development Goals: The Science Perspective, The International Council for Science (ICSU)*, Paris, France, page 63
19 The ST-EP Initiative, http://step.unwto.org/content/st-ep-initiative-1, Accessed October 3, 2016
20 ICSU and ISSC, *Review of the Sustainable Development Goals*, page 54
21 GCF Readiness Week to Focus on Accelerating Countries' Direct Access to the Fund's Resources, http://greenclimate.fund/-/gcf-readiness-week-to-focus-on-accelerating-countries-direct-access-to-the-fund-s-resources?inheritRedirect=true&redirect=%2Fhome-, Accessed October 3, 2016. This fund specializes in transforming energy generation and access, creating climate-compatible cities, encouraging low-emission and climate-resilient agriculture, scaling up finance for forests and climate change, and enhancing resilience in Small Island Developing States, which indicates at the time of writing that sustainable tourism will be a lower priority unless directly tied to these top priorities,
22 Millennium Ecosystem Assessment, 2005, *Ecosystems and Human Well-Being Synthesis*, Island Press, Washington, DC
23 ICSU and ISSC, *Review of the Sustainable Development Goals*, page 15
24 Flightpath 1.5 Degrees, http://flightpath1point5.org/, Accessed April 13, 2016
25 Sulfur Content Exemption for Cruise Ship Company, http://enviro.blr.com/environmental-news/air/motor-vehicle-fuels-and-fuel-additives/Sulfur-content-exemption-for-cruise-ship-company/, Accessed October 3, 2017
26 Helfre, Jean-Florent and Pedro Andre Couto Boot, July 2013, *Emission Reduction in the Shipping Industry: Regulations, Exposure and Solutions*, Sustainalytics, Amsterdam, The Netherlands
27 International Maritime Organization (IMO), August 2014, *Further Technical and Operational Measures for Enhancing Energy Efficiency of International Shipping*, Submitted by BIMCO, CLIA, ICS, Intercargo, Intertanko, IPTA, and WSC, Marine Environment Committee, 67th session Agenda item 5
28 Lakshmi, Aiswarya, 2015, Maritime Sector Split on Paris Climate Deal, December 15, Marine Link, http://marinelink.com/news/maritime-climate-sector402235.aspx, Accessed October 3, 2016

INDEX

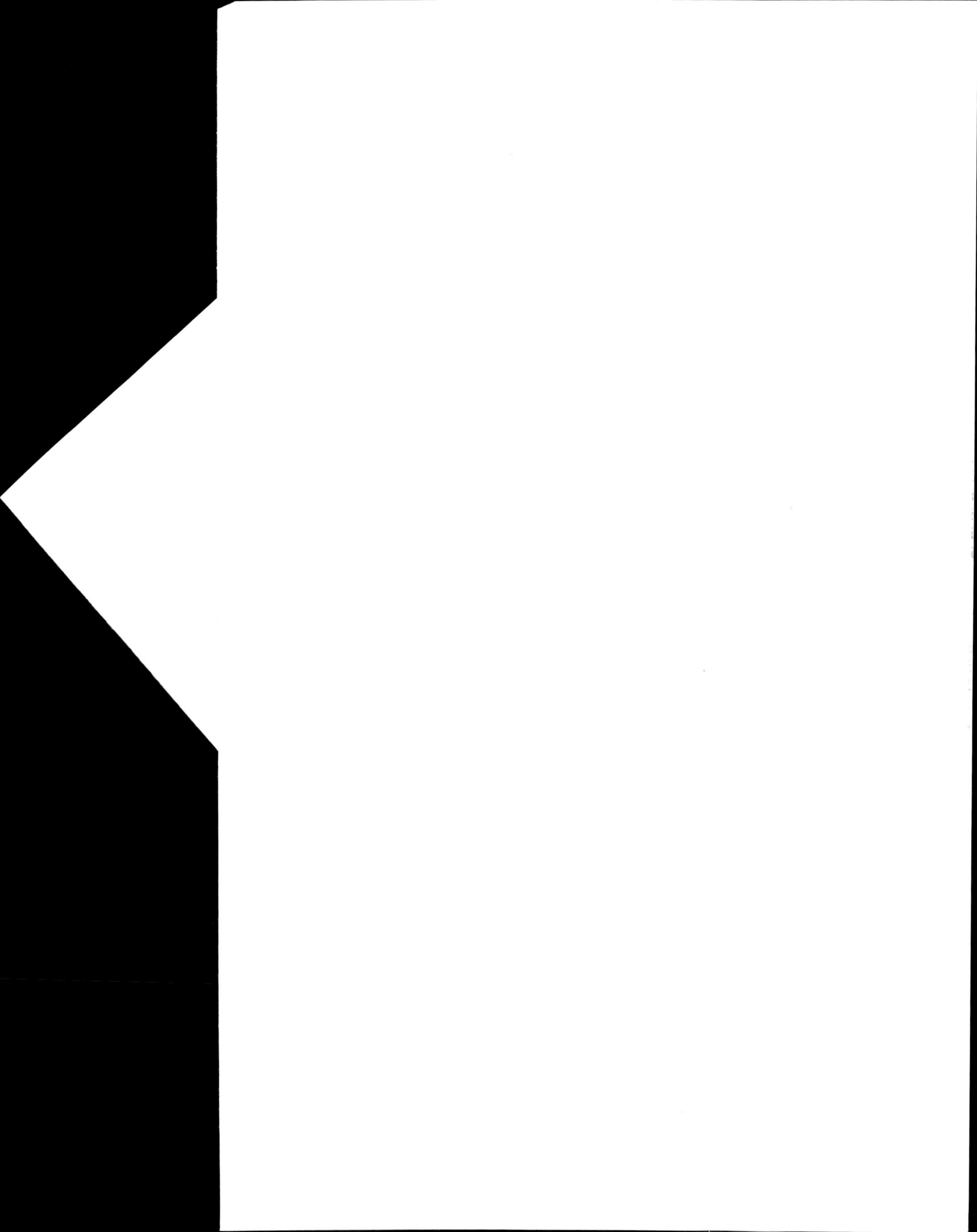